Stability and Controls Analysis for Delay Systems

Stability and Controls Analysis for Delay Systems

JinRong Wang

Michal Fečkan

Mengmeng Li

ELSEVIER

ACADEMIC PRESS

An imprint of Elsevier

Academic Press is an imprint of Elsevier
125 London Wall, London EC2Y 5AS, United Kingdom
525 B Street, Suite 1650, San Diego, CA 92101, United States
50 Hampshire Street, 5th Floor, Cambridge, MA 02139, United States
The Boulevard, Langford Lane, Kidlington, Oxford OX5 1GB, United Kingdom

Copyright © 2023 Elsevier Inc. All rights reserved.

MATLAB® is a trademark of The MathWorks, Inc. and is used with permission.
The MathWorks does not warrant the accuracy of the text or exercises in this book.
This book's use or discussion of MATLAB® software or related products does not constitute
endorsement or sponsorship by The MathWorks of a particular pedagogical approach or
particular use of the MATLAB® software.

No part of this publication may be reproduced or transmitted in any form or by any means,
electronic or mechanical, including photocopying, recording, or any information storage and
retrieval system, without permission in writing from the publisher. Details on how to seek
permission, further information about the Publisher's permissions policies and our arrangements
with organizations such as the Copyright Clearance Center and the Copyright Licensing Agency,
can be found at our website: www.elsevier.com/permissions.

This book and the individual contributions contained in it are protected under copyright by the
Publisher (other than as may be noted herein).

Notices

Knowledge and best practice in this field are constantly changing. As new research and
experience broaden our understanding, changes in research methods, professional practices, or
medical treatment may become necessary.

Practitioners and researchers must always rely on their own experience and knowledge in
evaluating and using any information, methods, compounds, or experiments described herein. In
using such information or methods they should be mindful of their own safety and the safety of
others, including parties for whom they have a professional responsibility.

To the fullest extent of the law, neither the Publisher nor the authors, contributors, or editors,
assume any liability for any injury and/or damage to persons or property as a matter of products
liability, negligence or otherwise, or from any use or operation of any methods, products,
instructions, or ideas contained in the material herein.

ISBN: 978-0-323-99792-8

For information on all Academic Press publications
visit our website at https://www.elsevier.com/books-and-journals

Publisher: Glyn Jones
Editorial Project Manager: Elaine Desamero
Production Project Manager: Fahmida Sultana
Cover Designer: Vicky Pearson Esser

Typeset by VTeX

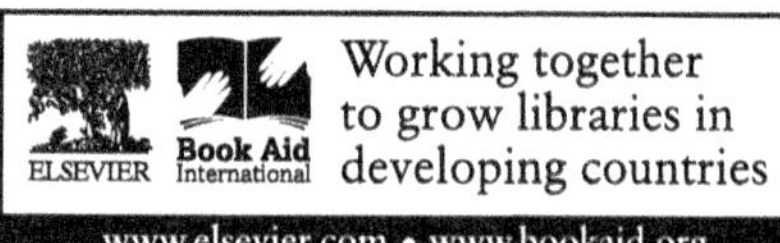

Contents

7. Stochastic delay systems

Preface

It is well known that stability and control analysis of delay systems has received much attention in the fields of mathematics and automatic control. Various mathematical methods as well as stability and controllability concepts are explored to deal with such problems.

In recent decades, there have been few developments in seeking explicit formulas of solutions to delay differential/discrete equations by introducing continuous/discrete delayed exponential matrices. One of the biggest advantages of continuous/discrete delayed exponential matrices is to transfer the classical idea of representing the solution of linear ordinary differential equations to linear delay differential/discrete equations. Stability analysis and mathematical control theory are important areas of research for classical differential delay systems. Various mathematical methods as well as new stability concepts are explored to deal with such problems. However, there is no book that uses delayed exponential matrices to deal with the stability, controllability, and iterative learning control of delay systems such as first order systems, oscillating systems, impulsive systems, fractional systems, difference systems, and stochastic systems. This was the main motivation to write this book.

This book is devoted to the study of the stability, controllability, and iterative learning control of delay systems of several kinds, such as first order systems, oscillating systems, impulsive systems, fractional systems, difference systems, and stochastic systems in the fields of physics, biology, population dynamics, ecology, and economics, which have not been presented in other books on conventional fields. The delayed exponential matrix function approach is widely used to derive the representation of the solutions, stability, and controllability. Iterative learning control designs are also established, which can be regarded as a way to find the control function. The broad variety of achieved results with rigorous proofs and many numerical examples makes this book unique.

This monograph is useful for researchers and graduate students to study stability and control for delay systems. It may also be used as seminars and advanced graduate courses of pure and applied mathematics and related disciplines.

We would like to thank Professors A. Debbouche, J. Diblík, Z.S. Hou, M. Pospíšil, D. O'Regan, X. Ruan, D. Shen, J.J. Trujillo, W. Wei, and Y. Zhou for their support. We also wish to express appreciation to our graduate students C.B. Liang, Z.J. Luo, D.H. Luo, and W.Z. Qiu for their help.

Guiyang and Bratislava

JinRong Wang
Michal Fečkan
Mengmeng Li
2022

Acknowledgments

We acknowledge with gratitude the support of the National Natural Science Foundation of China (12161015), the Training Object of High Level and Innovative Talents of Guizhou Province ((2016)4006), the Major Research Project of Innovation Group of the Guizhou Education Department ([2018]012), the Guizhou Data Driven Modeling Learning and Optimization Innovation Team ([2020]5016), the Major Project of Guizhou Postgraduate Education and Teaching Teform (YJSJGKT[2021]041), the Slovak Research and Development Agency (Contract No. APVV-18-0308), and the Slovak Grant Agency VEGA (No. 1/0358/20 and No. 2/0127/20).

Chapter 1

Introduction

It is well known that delay differential equations arise naturally in economics, physics, and control problems. It is an interesting task to develop the idea of Duhamel's principle in classical linear ordinary differential equations (ODEs) to seek the explicit representation of solutions [1] to delay discrete/differential systems. In fact, it is not an easy task to construct a fundamental matrix for linear differential delay systems, even for a simple first order delay system $\dot{x}(t) = Ax(t) + Bx(t - \tau), t \geq 0$, with initial condition $x(t) = \varphi(t), t \in [-\tau, 0]$, $\tau > 0$, where A, B are suitable constant matrices. Khusainov and Shuklin in [1] introduced the delayed matrix exponential $e_\tau^{B\cdot} : \mathbb{R} \to \mathbb{R}^n$ [1, Definition 0.3] and derived an explicit formula of solutions to such linear differential delay systems with $AB = BA$. Diblík and Khusainov [2] adopted the idea to construct the discrete matrix delayed exponential, which was used to derive an explicit formula of solutions to a discrete delay system.

There is rapid development in seeking explicit formulas of solutions to delay differential/discrete equations by introducing various continuous/discrete delayed exponential matrices (see, for example, [3–18]). One of the biggest advantages of continuous/discrete delayed exponential matrices is the possibility to transfer the classical idea of representing the solution of linear ODEs to linear delay differential/discrete equations. This provides a new idea and approach to study the representation of solutions, asymptotic stability, finite time stability, controllability, and iterative learning control (ILC) for various kinds of linear continuous/discrete delay systems, for example, oscillating delay systems, impulsive delay systems, fractional delay systems, and stochastic delay systems.

Stability analysis is one of the most important issues in control systems. In general, the Lyapunov method is used to deal with the asymptotic stability of trivial solutions. However, there exists a stable system which yields undesirable transient performance at one fixed point in practical applications. For this reason, it is necessary to consider the boundedness of system trajectories over a given finite time interval from the engineering point of view instead of determining the long-time asymptotical behavior for the system trajectory from the mathematical point of view. As a result, the concept of finite time stability (the boundedness of system trajectories) is offered to characterize the behavior of dynamical systems.

The finite time stability concept, introduced by Dorato [19], of delay differential equations arises from the fields of multibody mechanics, automatic engines, and physiological systems. Finite time stability means that the system

Stability and Controls Analysis for Delay Systems. https://doi.org/10.1016/B978-0-32-399792-8.00007-4
Copyright © 2023 Elsevier Inc. All rights reserved.

state does not exceed a certain bound for a given finite time interval and seems more appropriate from practical considerations. It is remarkable that Weiss and Infante [20–22] give sufficient conditions for finite time stability of nonlinear systems by introducing the suitable Lyapunov functionals. A solution of an equation or a state of a system is said to be finite time stable if it does not exceed a certain threshold during a fixed finite time interval. Obviously, finite time stability is very different from the classical exponential stability, which deals with equations operated in the whole infinite time interval. Concerning finite time stability, Ulam's stability and stable manifolds of linear systems, impulsive systems and fractional systems, fundamental matrix methods, linear matrix inequalities, algebraic inequalities, and integral inequalities are often used to deal with this issue. For more recent contributions, one can see [23–37].

The concept of controllability was first introduced by Kalman in the early 1960s. Some very important conclusions about the behavior of linear and nonlinear dynamical systems have been obtained. In the field of engineering technology, there are many real controlled systems with obvious delay effects, such as vehicle active suspension systems, rocket engine combustion systems, and other mechanical systems. The delay effect leads to the evolution of the system state with time depending on not only the current state of the system, but also the state of the system for a certain period of time in the past. Time-delay systems are one of the main tools to study delay effects. Thus, the situation will be much more complicated for time-independent delay differential controlled systems, or even for linear differential systems with pure delay with one input. In fact, in contrast to the theory of classical linear differential/difference controlled systems, when driving these delay systems to rest, one is required to control not only the value of the state at the final time, but also the memory accumulated with an aftereffect. In addition, the representation of the solution is not easy to characterize without knowing the fundamental matrix of a homogeneous delay differential system. As a result, various classes of control methods for delay differential systems were considered and different variants of controllability were developed in the past decades.

Khusainov and Shuklin [38] initially studied relative controllability of linear differential systems with pure delay via the associated delayed matrix exponential. Thereafter, Khusainov et al. [5] studied the Cauchy problem for a second order linear differential equation with pure delay. Next, in [4], relative controllability of linear discrete systems with a single constant delay was initially studied using the so-called discrete delayed matrix exponential, where an equivalent condition was stated for an initial value problem to have a control function. In addition, one of the control functions was found. Moreover, Pospíšil et al. [39] extended the study of controllability to linear discrete pure delay systems with constant coefficients and multiple control functions and derived a representation of solution in the form of a matrix polynomial using the Z-transform [18] to a system of nonhomogeneous linear difference equations with any finite number of constant delays and linear parts given by pairwise permutable matrices. Various criteria of relative controllability for linear discrete pure delay systems are presented and the associated control functions are also constructed.

In 1978, Uchiyama [40] proposed the concept of ILC from the viewpoint of human learning ability to deal with robotic systems. In the past decades, various types of iterative updating laws with uniform trial lengths have been widely used to deal with the issues in robotics, process control, and biological systems. A large amount of literature on ILC has been published for various types of systems such as discrete time systems, fractional differential systems, impulsive hybrid systems, distributed parameter systems, and networked stochastic systems. After reviewing the previous literature, we observe a restriction that the length of process operation and the desired trajectory are invariant in different iterations. Obviously, this condition may limit the application range of ILC. In fact, the case of operation terminating early does exist [41], for example, the functional electrical stimulation for upper limb movement [42]. This case strongly motivated us to consider ILC under randomly iteration-varying length environments. For more recent contributions on this hot topic, one can refer to [43–55] and reference therein. It is remarkable that the ILC problem of a delay spring–mass–damper system, which has received much attention, is very important in many practical mechanical systems [56].

ILC, which has become an important topic in modern data-driven control, is proposed in [57,58]. Note that ILC problems for linear discrete delayed controlled systems have been studied extensively by transferring to analyze a constructed Roesser model. For more related contributions, one can refer to [59–69] and reference therein.

This monograph is devoted to the rapidly developing research area of the stability, controllability, and ILC theory of delay difference/differential controlled systems. Such basic theory should be the starting point for further research concerning the optimal control, numerical analysis, and applications of other types of delay systems. The book is divided into seven chapters. Chapter 1 gives the introduction. In Chapter 2, we present the finite time stability, controllability, and ILC for first order delay differential systems. Chapter 3 is devoted to the study of the same issues in Chapter 2 for second order oscillating delay systems. In Chapter 4, we study asymptotical stability, finite time stability, and controllability for impulsive delay systems. In Chapter 5, we extend some content in Chapter 1 to fractional delay systems by introducing impulsive delay fundamental matrices. Chapter 6 is devoted to establishing the same issues as in Chapter 2 for difference delay systems. Chapter 7 further establishes the related controllability results for first order and second order stochastic delay systems.

Key features of this book are stated here. We present the representation and stability of solutions via the delayed exponential matrix function approach. We give useful sufficient conditions to guarantee controllability. We discuss ILC design and focus on new systems such as first order systems, oscillating systems, impulsive systems, fractional systems, difference systems, and stochastic systems occurring in various fields.

This book is useful for researchers and graduate students for research, seminars, and advanced graduate courses in pure and applied mathematics, engineering, and related disciplines.

Chapter 2

Delay systems

2.1 Finite time stability

2.1.1 Finite time stability for linear delay differential systems

When reviewing the previous literature dealing with finite time stability problems for delay systems, we observe the following facts: (i) Delay system $\dot{x}(t) = Ax(t) + Bx(t - \tau)$, $t > 0$, is considered mostly as an integral system, where A, B are suitable matrices. (ii) A uniform transition matrix associated with A, B is not computed directly and the structure of solution $x(t)$ is not well characterized on every subinterval $[n\tau, (n + 1)\tau]$, $n \in \mathbb{N}$. (iii) The Gronwall inequality, linear matrix inequality, and Lyapunov function methods are used to obtain finite time stability results.

Recently, Khusainov and Shuklin [1] introduced the delayed exponential matrix $e_\tau^{B_1 t}$, $B_1 = e^{-A\tau} B$, to give a representation of a solution of a linear system with $AB = BA$ and one delay term.

Now, we are concerned with the following issue: Can we apply the delayed exponential matrix in [1] to establish a direct method to find the sufficient conditions to guarantee finite time stability? If so, we will provide another direct method to find some conditions to make the desired delay system finite time stable, which is a supplement of the current methods in some sense. Thus, in this part, we study finite time stability of the following linear homogeneous delay differential system:

$$\begin{cases} \dot{x}(t) = Ax(t) + Bx(t - \tau), \ t \in J := [0, T], \\ x(t) = \varphi(t), \ -\tau \leq t \leq 0, \ \tau > 0, \end{cases} \tag{2.1}$$

where $x \in \mathbb{R}^n$ and T denotes the prefixed iteration domain length. Without loss of generality, we set $T = n\tau$ and $n \in \mathbb{N}$, $\varphi \in C_\tau^1 := C^1([-\tau, 0], \mathbb{R}^n)$, and the $n \times n$ matrices A and B satisfy $AB = BA$.

According to [1, Remark 1.3], any solution of system (2.1) has the form

$$x(t) = X_0(t)e^{A\tau}\varphi(-\tau) + \int_{-\tau}^{0} X_0(t - \tau - s)e^{A\tau}[\varphi'(s) - A\varphi(s)]ds, \tag{2.2}$$

where

$$X_0(t) = e^{At}e_\tau^{B_1 t}, \ B_1 = e^{-A\tau} B \tag{2.3}$$

Stability and Controls Analysis for Delay Systems. https://doi.org/10.1016/B978-0-32-399792-8.00008-6
Copyright © 2023 Elsevier Inc. All rights reserved.

is called the fundamental matrix of (2.2). The delayed exponential matrix $e_\tau^{B_1 t}$ is defined by

$$
e_\tau^{B_1 t} = \begin{cases} \Theta, & t < -\tau, \\ I, & -\tau \le t < 0, \\ I + B_1 t + \cdots + B_1^k \dfrac{(t - (k-1)\tau)^k}{k!}, & (k-1)\tau \le t < k\tau, \ k = 1, 2, \cdots, \end{cases}
$$

$$(2.4)$$

where Θ and I denote the zero and identity matrices, respectively.

Let $\mathbb{R}^n$ be the n-dimensional vector space, let $\mathbb{R}^{n \times n}$ denote the $n \times n$ matrix with real value elements, let $X^\top$ denote the transpose of the matrix X, and let $\lambda_{\max}(X)$ denote the maximum eigenvalue of the matrix X. Denote by $C(J, \mathbb{R}^n)$ the Banach space of vector-valued continuous functions from $J \to \mathbb{R}^n$ endowed with the norm $\|x\| = \max\limits_{t \in J} \|x(t)\|$ for a norm $\| \cdot \|$ on $\mathbb{R}^n$. We introduce a space $C^1(J, \mathbb{R}^n) = \{x \in C(J, \mathbb{R}^n) : \dot{x} \in C(J, \mathbb{R}^n)\}$. For $A : \mathbb{R}^n \to \mathbb{R}^n$, we consider its matrix norm $\|A\| = \max\limits_{\|x\|=1} \|Ax\|$ generated by $\| \cdot \|$. In addition, we note $\|\varphi\| = \max\limits_{\theta \in [-\tau, 0]} \|\varphi(\theta)\|$. The λ-norm on $C(J, \mathbb{R}^n)$ is represented as $\|x\|_\lambda = \sup\limits_{t \in J} \left\{ e^{-\lambda t} \|x(t)\| \right\}, \ \lambda > 0.$

Definition 2.1. (see [19] or [70, Definition 1]) System (2.1) is finite time stable with respect to $\{0, J, \alpha, \beta, \tau\}$ if and only if $\varphi^\top(t)\varphi(t) < \alpha, \forall\, t \in [-\tau, 0]$, implies

$$
x^\top(t)x(t) < \beta, \ \forall\, t \in J,
$$

where α is a real positive number and $\beta \in \mathbb{R}$ with $\beta > \alpha$.

Definition 2.2. (see [70, Definition 2]) System (2.1) is finite time stable with respect to $\{0, J, \alpha, \beta, \tau\}$ if and only if $\|\varphi\|^2 < \alpha$ implies

$$
\|x(t)\|^2 < \beta, \ \forall\, t \in J.
$$

Remark 2.1. Definition 2.1 coincides with Definition 2.2 when $\| \cdot \|$ is a Euclidean norm.

Definition 2.3. (see [71] or [72]) The matrix measure (or matrix logarithmic norm) of X is defined as

$$
\mu(X) = \lim_{h \to 0^+} \frac{\|I + hX\| - 1}{h}.
$$

In particular, if $\| \cdot \|$ denotes the 2-norm of a matrix, then

$$
\mu(X) = \frac{1}{2}\lambda_{\max}(X + X^\top).
$$

This norm is called the Lozinski matrix norm and this norm was used to characterize stability and error bounds in the numerical integration of ODEs [73].

Lemma 2.1. *(see [6, Lemma 3]) For all $t \in \mathbb{R}$,*

$$\|e_\tau^{Bt}\| \le e^{\|B\|(t+\tau)}.$$

Lemma 2.2. *(see [1, Lemma 4]) For all $t \in \mathbb{R}$,*

$$\frac{d}{dt}e_\tau^{Bt} = Be_\tau^{B(t-\tau)}.$$

Lemma 2.3. *(see [74, Lemma 1]) For any positive symmetric constant matrix $W \in \mathbb{R}^{n \times n}$, scalars a, b satisfying $a < b$, and a vector function $f : [a,b] \to \mathbb{R}^n$ such that the integrations are well defined, we have*

$$\left(\int_a^b f(s)ds\right)^\top W\left(\int_a^b f(s)ds\right) \le (b-a)\int_a^b f^\top(s)Wf(s)ds.$$

Lemma 2.4. *(see [75] or [76, Lemma 2]) For any real square matrix $W \in \mathbb{R}^{n \times n}$ and scalar variable t, the expression*

$$\lambda_{\max}(e^{Wt}e^{W^\top t}) \le e^{2\mu(W)t}$$

holds, where $\mu(W)$ is the matrix measure of the matrix W defined in Definition 2.3.

Lemma 2.5. *(see [76, Lemma 3]) For any vectors $u, v \in \mathbb{R}^n$ and symmetric positive definite matrix $\mathbb{R}^{n \times n} \ni \Gamma = \Gamma^{-1} > 0$, the following inequalities hold:*

$$2u(t)^\top v(t) \le u^\top(t)\Gamma u(t) + v^\top(t)\Gamma^{-1}v(t)$$

and

$$-2u(t)^\top v(t) \le u^\top(t)\Gamma u(t) + v^\top(t)\Gamma^{-1}v(t).$$

The finite time stability results are presented by providing an approach which is different from previous methods.

2.1.1.1 Finite time stability results based on Definition 2.1

Theorem 2.1. *System (2.1) is finite time stable with respect to $\{0, J, \alpha, \beta, \tau\}$ if*

$$\|X_0(t)\| < \frac{\sqrt{\beta}}{\|e^{A\tau}\|\sqrt{N(\alpha)}}, \quad \forall\, t \in J, \tag{2.5}$$

where $X_0(t)$ is given in (2.3), α, β are defined in Definition 2.1, and

$$N(\alpha) := \alpha + 2\sqrt{\alpha}\int_{-\tau}^0 \|\varphi'(s) - A\varphi(s)\|ds + \left(\int_{-\tau}^0 \|\varphi'(s) - A\varphi(s)\|ds\right)^2. \tag{2.6}$$

Proof. Let x be the solution of system (2.1) satisfying (2.2). Denote $a(t) := X_0(t)e^{A\tau}\varphi(-\tau) \in \mathbb{R}^n$, and $d(t) := \int_{-\tau}^{0} X_0(t-\tau-s)e^{A\tau}[\varphi'(s) - A\varphi(s)]ds \in \mathbb{R}^n$. Clearly, $a^\top(t)d(t) = d^\top(t)a(t)$. For any $t \in J$, one can get

$$
\begin{aligned}
x(t)^\top x(t) &= \left(X_0(t)e^{A\tau}\varphi(-\tau) + \int_{-\tau}^{0} X_0(t-\tau-s)e^{A\tau}[\varphi'(s) - A\varphi(s)]ds \right)^\top \\
&\quad \times \left(X_0(t)e^{A\tau}\varphi(-\tau) + \int_{-\tau}^{0} X_0(t-\tau-s)e^{A\tau}[\varphi'(s) - A\varphi(s)]ds \right) \\
&= (e^{A\tau}\varphi(-\tau))^\top X_0^\top(t) X_0(t) e^{A\tau}\varphi(-\tau) \\
&\quad + a^\top(t)d(t) + d^\top(t)a(t) + d^\top(t)d(t) \\
&\leq \lambda_m(t)(e^{A\tau}\varphi(-\tau))^\top e^{A\tau}\varphi(-\tau) + 2a^\top(t)d(t) + \|d(t)\|^2, \quad (2.7)
\end{aligned}
$$

where $\lambda_m(t) = \max_{t \in J} \rho\{X_0^\top(t)X_0(t)\}$ and $\rho\{Y\}$ denotes the spectrum of Y.

For the term $a^\top(t)d(t)$, we have

$$
\begin{aligned}
&a^\top(t)d(t) \\
&= (e^{A\tau}\varphi(-\tau))^\top X_0^\top(t) \int_{-\tau}^{0} X_0(t-\tau-s)e^{A\tau}[\varphi'(s) - A\varphi(s)]ds \\
&\leq \|\varphi(-\tau)^\top\| \|e^{A\tau}\| \|X_0^\top(t)\| \int_{-\tau}^{0} \|X_0(t-\tau-s)\| \|e^{A\tau}\| \|\varphi'(s) - A\varphi(s)\|ds.
\end{aligned}
$$

$$(2.8)$$

Let $\eta = t - \tau - s$, $s \in [-\tau, 0]$. Then, $\eta \in [t - \tau, t] \subseteq [T - \tau, T] \subseteq J$. Thus,

$$
\|X_0(\eta)\|\,|_{\eta \in [t-\tau,t]} \leq \|X_0(t)\|\,|_{t \in [T-\tau,T]} \leq \|X_0(t)\|\,|_{t \in J}. \quad (2.9)
$$

Linking (2.7) and (2.8) via (2.9) we obtain

$$
\begin{aligned}
x(t)^\top x(t) &\leq \lambda_m(t)(e^{A\tau}\varphi(-\tau))^\top e^{A\tau}\varphi(-\tau) \\
&\quad + 2\|\varphi(-\tau)^\top\| \|e^{A\tau}\| \|X_0^\top(t)\| \|X_0(t)\| \|e^{A\tau}\| \int_{-\tau}^{0} \|\varphi'(s) - A\varphi(s)\|ds \\
&\quad + \|X_0(t)\|^2 \|e^{A\tau}\|^2 \left(\int_{-\tau}^{0} \|\varphi'(s) - A\varphi(s)\|ds \right)^2. \quad (2.10)
\end{aligned}
$$

Using the fact that

$$
\lambda_m(t) \leq \|X_0^\top(t)X_0(t)\| \leq \|X_0^\top(t)\| \|X_0(t)\| = \|X_0(t)\|^2,
$$

by (2.10), we obtain

$$
x(t)^\top x(t) \leq \|X_0(t)\|^2 \|e^{A\tau}\|^2 \varphi^\top(-\tau)\varphi(-\tau)
$$

$$+ 2\|\varphi(-\tau)^\top\|\,\|X_0(t)\|^2\|e^{A\tau}\|^2 \int_{-\tau}^0 \|\varphi'(s) - A\varphi(s)\|ds$$

$$+ \|X_0(t)\|^2\|e^{A\tau}\|^2 \left(\int_{-\tau}^0 \|\varphi'(s) - A\varphi(s)\|ds\right)^2. \qquad (2.11)$$

Next, we choose $\varphi^\top(t)\varphi(t) < \alpha$, $\forall\, t \in [-\tau, 0]$, which implies $\|\varphi^\top(-\tau)\| < \sqrt{\alpha}$. Finally, applying condition (2.5) to the above inequality, (2.11) reduces to

$$x(t)^\top x(t) \le \|X_0(t)\|^2\|e^{A\tau}\|^2 N(\alpha) < \beta,$$

where $N(\alpha)$ is given in (2.6). According to Definition 2.1, system (2.1) is finite time stable. $\qquad\square$

Theorem 2.2. *System (2.1) is finite time stable with respect to* $\{0, J, \alpha, \beta, \tau\}$ *if*

$$e^{\mathfrak{M}(t+\tau)} < \frac{\beta}{2(\alpha + \tau \int_{-\tau}^0 \|\varphi'(s) - A\varphi(s)\|^2 ds)}, \quad t \in J, \qquad (2.12)$$

where $\mathfrak{M} = 2\mu(A^\top) + 2\|B_1\|$ *and* α, β *are defined in Definition 2.1.*

Proof. Let x be the solution of system (2.1) satisfying (2.2). By Lemma 2.5, we have

$$
\begin{aligned}
x(t)^\top x(t) \le\ & \varphi^\top(-\tau)(e_\tau^{B_1 t})^\top (e^{A(t+\tau)})^\top e^{A(t+\tau)} e_\tau^{B_1 t} \varphi(-\tau) \\
& + [e^{A(t+\tau)} e_\tau^{B_1 t} \varphi(-\tau)] I [e^{A(t+\tau)} e_\tau^{B_1 t} \varphi(-\tau)]^\top + \Upsilon \\
& + \int_{-\tau}^0 (e^{A(t-s)} e_\tau^{B_1(t-\tau-s)} [\varphi'(s) - A\varphi(s)])^\top ds \\
& \times \int_{-\tau}^0 e^{A(t-s)} e_\tau^{B_1(t-\tau-s)} [\varphi'(s) - A\varphi(s)] ds,
\end{aligned}
$$

where

$$
\begin{aligned}
\Upsilon =\ & \int_{-\tau}^0 (e^{A(t-s)} e_\tau^{B_1(t-\tau-s)} [\varphi'(s) - A\varphi(s)])^\top ds \\
& \times I^{-1} \int_{-\tau}^0 e^{A(t-s)} e_\tau^{B_1(t-\tau-s)} [\varphi'(s) - A\varphi(s)] ds.
\end{aligned}
$$

Further, by Lemmas 2.3 and 2.4,

$$
\begin{aligned}
x(t)^\top x(t) \le\ & 2\lambda_{\max}((e^{A(t+\tau)})^\top e^{A(t+\tau)})\varphi^\top(-\tau)(e_\tau^{B_1 t})^\top e_\tau^{B_1 t} \varphi(-\tau) \\
& + 2\tau \int_{-\tau}^0 (e^{A(t-s)} e_\tau^{B_1(t-\tau-s)} [\varphi'(s) - A\varphi(s)])^\top e^{A(t-s)} \\
& \times e_\tau^{B_1(t-\tau-s)} [\varphi'(s) - A\varphi(s)] ds
\end{aligned}
$$

$$
\leq 2e^{2\mu(A^\top)(t+\tau)}\varphi^\top(-\tau)(e_\tau^{B_1 t})^\top e_\tau^{B_1 t}\varphi(-\tau)
$$

$$
+ 2\tau \int_{-\tau}^{0} (e^{A(t-s)}e_\tau^{B_1(t-\tau-s)}[\varphi'(s) - A\varphi(s)])^\top e^{A(t-s)}
$$

$$
\times e_\tau^{B_1(t-\tau-s)}[\varphi'(s) - A\varphi(s)]ds
$$

$$
\leq 2e^{2\mu(A^\top)(t+\tau)}\varphi^\top(-\tau)(e_\tau^{B_1 t})^\top e_\tau^{B_1 t}\varphi(-\tau)
$$

$$
+ 2\tau \int_{-\tau}^{0} e^{2\mu(A^\top)(t-s)}([\varphi'(s) - A\varphi(s)])^\top
$$

$$
\times (e_\tau^{B_1(t-\tau-s)})^\top e_\tau^{B_1(t-\tau-s)}[\varphi'(s) - A\varphi(s)]ds. \tag{2.13}
$$

Note that $\|Y^\top\| = \|Y\|$, $\|\varphi(-\tau)\|^2 = \varphi^\top(-\tau)\varphi(-\tau)$. Applying Lemma 2.1 to (2.13), one has

$$
x(t)^\top x(t) \leq 2e^{\mathfrak{M}(t+\tau)}\varphi^\top(-\tau)\varphi(-\tau) + 2\tau \int_{-\tau}^{0} e^{\mathfrak{M}(t-s)}\|\varphi'(s) - A\varphi(s)\|^2 ds
$$

$$
\leq 2e^{\mathfrak{M}(t+\tau)}\alpha + 2\tau e^{\mathfrak{M}(t+\tau)} \int_{-\tau}^{0} \|\varphi'(s) - A\varphi(s)\|^2 ds
$$

$$
\leq 2e^{\mathfrak{M}(t+\tau)}\left[\alpha + \tau \int_{-\tau}^{0} \|\varphi'(s) - A\varphi(s)\|^2 ds\right]. \tag{2.14}
$$

Linking (2.14) and (2.12), we obtain $x^\top(t)x(t) < \beta, \forall\, t \in J$. According to Definition 2.1, system (2.1) is finite time stable. $\qquad\square$

Theorem 2.3. *System* (2.1) *is finite time stable with respect to* $\{0, J, \alpha, \beta, \tau\}$ *if*

$$
4e^{\mathfrak{M}t}\left[1 + \frac{2\tau\lambda_{\max}(A^\top A)}{\mathfrak{M}}(e^{\mathfrak{M}\tau} - 1) + \frac{\tau\lambda_{\max}(B^\top B)}{\mathfrak{M}}(1 - e^{-\mathfrak{M}\tau})\right]
$$

$$
< \frac{\beta}{\alpha},\ t \in J, \tag{2.15}
$$

where $\mathfrak{M} \neq 0$ *and* α, β *are defined in Definition 2.1.*

Proof. According to (2.2) and integration by parts via Lemma 2.2, the solution of system (2.1) can be expressed as

$$
x(t) = e^{At}e_\tau^{B_1(t-\tau)}\varphi(0) - 2 \int_{-\tau}^{0} Ae^{A(t-s)}e_\tau^{B_1(t-\tau-s)}\varphi(s)ds
$$

$$
- \int_{-\tau}^{0} Be^{A(t-\tau-s)}e_\tau^{B_1(t-2\tau-s)}\varphi(s)ds.
$$

By analogy to the proof of Theorem 2.2, we get

$$x^\top(t)x(t) \le (e^{At}e_\tau^{B_1(t-\tau)}\varphi(0))^\top e^{At}e_\tau^{B_1(t-\tau)}\varphi(0)$$
$$- 4(e^{At}e_\tau^{B_1(t-\tau)}\varphi(0))^\top \int_{-\tau}^0 Ae^{A(t-s)}e_\tau^{B_1(t-\tau-s)}\varphi(s)ds$$
$$- 2(e^{At}e_\tau^{B_1(t-\tau)}\varphi(0))^\top \int_{-\tau}^0 Be^{A(t-\tau-s)}e_\tau^{B_1(t-2\tau-s)}\varphi(s)ds$$
$$+ 4\int_{-\tau}^0 (Ae^{A(t-s)}e_\tau^{B_1(t-\tau-s)}\varphi(s))^\top ds$$
$$\times \int_{-\tau}^0 Be^{A(t-\tau-s)}e_\tau^{B_1(t-2\tau-s)}\varphi(s)ds$$
$$+ 4\tau\int_{-\tau}^0 (Ae^{A(t-s)}e_\tau^{B_1(t-\tau-s)}\varphi(s))^\top Ae^{A(t-s)}e_\tau^{B_1(t-\tau-s)}\varphi(s)ds$$
$$+ \tau\int_{-\tau}^0 (Be^{A(t-\tau-s)}e_\tau^{B_1(t-2\tau-s)}\varphi(s))^\top Be^{A(t-\tau-s)}e_\tau^{B_1(t-2\tau-s)}\varphi(s)ds.$$

By Lemmas 2.3 and 2.5, we have

$$x^\top(t)x(t) \le 4e^{2\mu(A^\top)t}\varphi^\top(0)(e_\tau^{B_1(t-\tau)})^\top e_\tau^{B_1(t-\tau)}\varphi(0) + 8\tau P_1 + 4\tau P_2, \quad (2.16)$$

where

$$P_1 = \int_{-\tau}^0 (Ae^{A(t-s)}e_\tau^{B_1(t-\tau-s)}\varphi(s))^\top Ae^{A(t-s)}e_\tau^{B_1(t-\tau-s)}\varphi(s)ds,$$

$$P_2 = \int_{-\tau}^0 (Be^{A(t-\tau-s)}e_\tau^{B_1(t-2\tau-s)}\varphi(s))^\top Be^{A(t-\tau-s)}e_\tau^{B_1(t-2\tau-s)}\varphi(s)ds.$$

By Lemma 2.4,

$$P_1 \le \lambda_{\max}(A^\top A)\int_{-\tau}^0 (e^{A(t-s)}e_\tau^{B_1(t-\tau-s)}\varphi(s))^\top e^{A(t-s)}e_\tau^{B_1(t-\tau-s)}\varphi(s)ds$$
$$\le \lambda_{\max}(A^\top A)$$
$$\times \int_{-\tau}^0 \lambda_{\max}[(e^{A(t-s)})^\top(e^{A(t-s)})](e_\tau^{B_1(t-\tau-s)}\varphi(s))^\top e_\tau^{B_1(t-\tau-s)}\varphi(s)ds$$
$$\le \lambda_{\max}(A^\top A)\int_{-\tau}^0 e^{2\mu(A^\top)(t-s)}(e_\tau^{B_1(t-\tau-s)}\varphi(s))^\top e_\tau^{B_1(t-\tau-s)}\varphi(s)ds$$
$$\le \lambda_{\max}(A^\top A)\int_{-\tau}^0 e^{2\mu(A^\top)(t-s)}e^{\|B_1\|(t-s)}\varphi(s)^\top\varphi(s)ds$$
$$\le \lambda_{\max}(A^\top A)\alpha\int_{-\tau}^0 e^{\mathfrak{M}(t-s)}ds$$

$$\leq \frac{\lambda_{\max}(A^\top A)\alpha}{\mathfrak{M}} e^{\mathfrak{M}t}(e^{\mathfrak{M}\tau} - 1), \tag{2.17}$$

where $\mathfrak{M} = 2\mu(A^\top) + 2\|B_1\|$. Similarly, we can get

$$P_2 \leq \frac{\lambda_{\max}(B^\top B)\alpha}{\mathfrak{M}} e^{\mathfrak{M}t}(1 - e^{-\mathfrak{M}\tau}). \tag{2.18}$$

From (2.15)–(2.18), we get

$$
\begin{aligned}
x^\top(t)x(t) &\leq 4e^{2\mu(A^\top)t} e^{2\|B_1\|t}\alpha + \frac{8\tau\lambda_{\max}(A^\top A)\alpha}{\mathfrak{M}} e^{\mathfrak{M}t}(e^{\mathfrak{M}\tau} - 1) \\
&\quad + \frac{4\tau\lambda_{\max}(B^\top B)\alpha}{\mathfrak{M}} e^{\mathfrak{M}t}(1 - e^{-\mathfrak{M}\tau}) \\
&\leq 4e^{\mathfrak{M}t}\alpha\left[1 + \frac{2\tau\lambda_{\max}(A^\top A)\alpha}{\mathfrak{M}}(e^{\mathfrak{M}\tau} - 1) \right. \\
&\quad \left. + \frac{\tau\lambda_{\max}(B^\top B)\alpha}{\mathfrak{M}}(1 - e^{-\mathfrak{M}\tau})\right] < \beta.
\end{aligned}
$$

According to Definition 2.1, system (2.1) is finite time stable. $\qquad\square$

2.1.1.2 Finite time stability results based on Definition 2.2

Theorem 2.4. *System* (2.1) *is finite time stable with respect to* $\{0, J, \alpha, \beta, \tau\}$ *if*

$$e^{(\|A\|+\|B_1\|)(t+\tau)} < \frac{\sqrt{\beta}}{\sqrt{\alpha} + \int_{-\tau}^{0} \|\varphi'(s) - A\varphi(s)\|ds}, \quad t \in J, \tag{2.19}$$

holds, where α, β *are defined in Definition 2.2.*

Proof. Linking to Eq. (2.2) and Lemma 2.1, it is easy to get

$$
\begin{aligned}
\|x(t)\| &\leq \|e^{A(t+\tau)}e_\tau^{B_1 t}\varphi(-\tau)\| + \int_{-\tau}^{0} \|e^{A(t-s)}\|\|e_\tau^{B_1(t-\tau-s)}\|\|\varphi'(s) - A\varphi(s)\|ds \\
&\leq e^{(\|A\|+\|B_1\|)(t+\tau)}\|\varphi(-\tau)\| + e^{(\|A\|+\|B_1\|)t} \\
&\quad \times \int_{-\tau}^{0} e^{-(\|A\|+\|B_1\|)s}\|\varphi'(s) - A\varphi(s)\|ds \\
&\leq e^{(\|A\|+\|B_1\|)(t+\tau)}\|\varphi(-\tau)\| + e^{(\|A\|+\|B_1\|)(t+\tau)}\int_{-\tau}^{0} \|\varphi'(s) - A\varphi(s)\|ds \\
&\leq e^{(\|A\|+\|B_1\|)(t+\tau)}\sqrt{\alpha} + e^{(\|A\|+\|B_1\|)(t+\tau)}\int_{-\tau}^{0} \|\varphi'(s) - A\varphi(s)\|ds \\
&\leq e^{(\|A\|+\|B_1\|)(t+\tau)}\left[\sqrt{\alpha} + \int_{-\tau}^{0} \|\varphi'(s) - A\varphi(s)\|ds\right].
\end{aligned}
$$

Using condition (2.19), one can get $\|x(t)\| < \sqrt{\beta}$, $\forall\, t \in J$. According to Definition 2.2,

$$\|x(t)\|^2 < \beta, \ \forall\, t \in J.$$

Thus, system (2.1) is finite time stable. $\qquad\qquad\qquad\qquad\qquad\square$

Theorem 2.5. *System (2.1) is finite time stable with respect to $\{0, J, \alpha, \beta, \tau\}$ if*

$$e^{2Nt}\left[1 + (2\|A\| + \|B\|)\frac{1}{N}(1 - e^{-N\tau})\right]^2 < \frac{\beta}{\alpha}, \ \forall\, t \in J, \qquad (2.20)$$

where $N = \|A\| + \|B_1\| \neq 0$ and α, β are given in Definition 2.2.

Proof. According to (2.2), the solution of system (2.1) can be expressed as

$$x(t) = e^{A(t+\tau)}e_\tau^{B_1 t}\varphi(-\tau) + \int_{-\tau}^{0} e^{A(t-s)}e_\tau^{B_1(t-\tau-s)}\varphi'(s)ds$$

$$- \int_{-\tau}^{0} e^{A(t-s)}e_\tau^{B_1(t-\tau-s)}A\varphi(s)ds. \qquad (2.21)$$

Integrating by parts via Lemma 2.2, one can change expression (2.21) to the form

$$x(t) = e^{At}e_\tau^{B_1(t-\tau)}\varphi(0) - 2\int_{-\tau}^{0} Ae^{A(t-s)}e_\tau^{B_1(t-\tau-s)}\varphi(s)ds$$

$$- \int_{-t}^{0} Be^{A(t-\tau-s)}e_\tau^{B_1(t-2\tau-s)}\varphi(s)ds. \qquad (2.22)$$

Taking the norm on both sides of (2.22) and using Lemma 2.1, we have

$$\|x(t)\| \le e^{Nt}\|\varphi\| + 2\|\varphi\|\int_{-\tau}^{0}\|A\|e^{N(t-s)}ds + \|\varphi\|\int_{-\tau}^{0}\|B\|e^{N(t-\tau-s)}ds$$

$$\le \|\varphi\|e^{Nt}\left[1 + (2\|A\| + \|B\|)\frac{1}{N}(1 - e^{-N\tau})\right]. \qquad (2.23)$$

By condition (2.20) and (2.23), we have

$$\|x(t)\|^2 \le \alpha e^{2Nt}\left[1 + (2\|A\| + \|B\|)\frac{1}{N}(1 - e^{-N\tau})\right]^2 < \beta.$$

Thus, system (2.1) is finite time stable. $\qquad\qquad\qquad\qquad\qquad\square$

2.1.1.3 Numerical examples and discussion

In this section, we give numerical examples to demonstrate the validity of our method and provide some discussion of the inspiring literature [70].

Let us consider the following linear delay system:

$$\begin{cases} \dot{x}(t) = Ax(t) + Bx(t-0.2), & x(t) \in \mathbb{R}^2, \ t \in [0, 1.4], \\ \varphi(t) = (0.3, 0.4)^\top, & -0.2 \leq t \leq 0, \end{cases} \tag{2.24}$$

where we set $T = 1.4$, $\tau = 0.2$, and

$$A = \begin{pmatrix} 1 & 0 \\ 0 & 1 \end{pmatrix}, \quad B = \begin{pmatrix} 0.2 & 0 \\ 0 & 0.2 \end{pmatrix}, \quad AB = BA.$$

It is obvious that $\varphi^\top(t)\varphi(t) = 0.25 < \alpha := 0.3$, $t \in [-0.2, 0]$. By elementary calculation, one has

$$B_1 = \begin{pmatrix} 0.1637 & 0 \\ 0 & 0.1637 \end{pmatrix}, \quad e^{A\tau} = \begin{pmatrix} 1.2214 & 0 \\ 0 & 1.2214 \end{pmatrix}, \quad \|e^{A\tau}\| = 1.2214.$$

Next, $\mu(A^\top) = 1$, $\|B_1\| = 0.1637$,

$$\int_{-0.2}^{0} \|\varphi'(s) - A\varphi(s)\|^2 ds = 0.05, \quad e^{2(\mu(A^\top)+\|B_1\|)(T+\tau)} = 6.4361.$$

Moreover, $\mathfrak{M} = 2.3274$, $\lambda_{\max}(A^\top A) = 1$, $\lambda_{\max}(B^\top B) = 0.04$, $4e^{\mathfrak{M}T} = 16.1625$, $e^{\mathfrak{M}\tau} = 1.5928$, and $e^{-\mathfrak{M}\tau} = 0.6278$.

Fig. 2.1 shows the state $x(t)$ of (2.24) when $T = 1.4$. Fig. 2.2 tells us the corresponding finite stability time when choosing $\beta = 150$. According to Theorem 2.5, one can conclude t firstly achieves maximum stability at 1.19 s. In

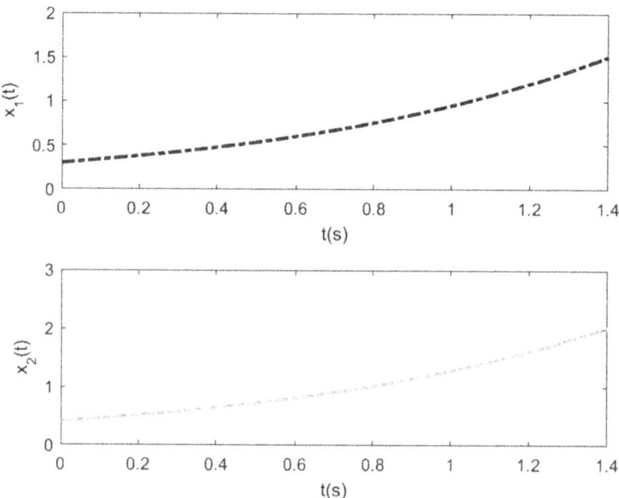

FIGURE 2.1 The state response $x(t)$ of (2.24) when $T = 1.4$.

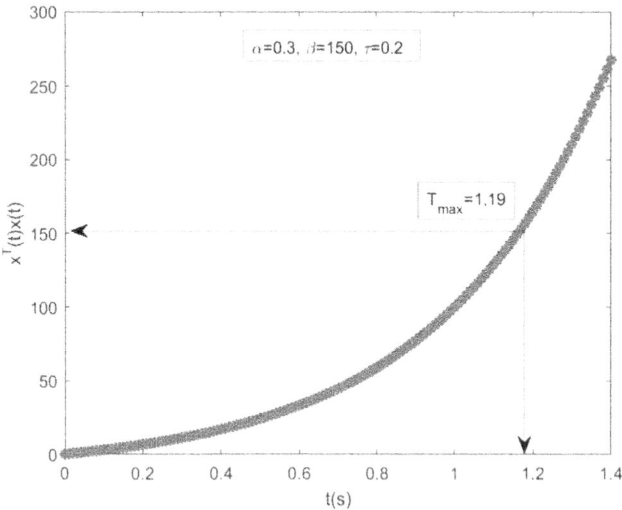

FIGURE 2.2   The square norm of the state vector of (2.24) when $T = 1.4$.

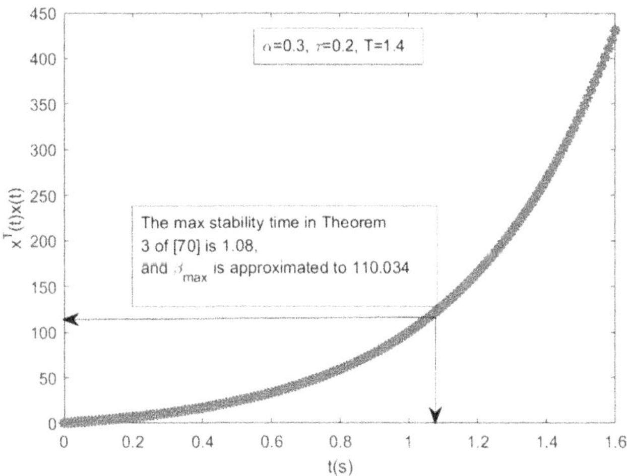

FIGURE 2.3   The square norm of the state vector of (2.24) when $T = 1.4$.

addition, one can check Theorems 2.1–2.5 by elementary computation and simulation. It is not a difficult task, and we omit the process. According to [70, Theorem 3], one knows the maximum finite stability time is 1.08 s, which is less than the stability time in our Theorem 2.1. Fig. 2.3 shows the finite stability time in line with [70, Theorem 3].

By virtue of the delayed exponential matrix corresponding to the linear delay continuous system, like the transition matrix for classical linear continuous systems, we provide another method different from the Gronwall inequality,

Lyapunov function, and linear matrix inequality (LMI) approaches to study finite time stability of linear delay continuous systems. New criteria to guarantee finite time stability are derived by using a direct formula solution.

(i) Theorem 2.1 shows that one can make (2.1) finite time stable by putting restrictions on the fundamental delay matrix $X_0(t)$ directly.

(ii) Theorem 2.2 shows that one can make (2.1) finite time stable by putting conditions associated with matrix measures for $A^\top$ and $\|B_1\|$.

(iii) Theorem 2.3 shows that one can make (2.1) finite time stable by putting conditions associated with matrix measures for $A^\top$ and $\|B_1\|$ and the maximum eigenvalue for $A^\top A$. Thus, the conditions in Theorem 2.3 seem a little stronger than Theorem 2.2.

(iv) Theorems 2.4 and 2.5 show that one can make (2.1) finite time stable by putting conditions associated with matrix norms $\|A\|$ and $\|B_1\|$. Theorems 2.1, 2.4, and 2.5 are related; however, their presentations are different.

2.1.1.4 Conclusion

We give sufficient conditions for the finite time stability results derived based on the delayed matrix exponential approach and Jensen's and Coppel's inequalities. We demonstrate the validity of the designed method and provide some discussion by using a numerical example. The results in this part are motivated from [77].

2.1.2 Finite time stability for semilinear delay differential systems

In this part, we study finite time stability of the following two problems:
(i) equations with a nonlinearity independent of delay:

$$\begin{cases} \dot{x}(t) = Ax(t) + Bx(t-\tau) + f(x(t)), \ t \in J := [0, T], \\ x(t) = \varphi(t), \ -\tau \le t \le 0, \ \tau > 0; \end{cases} \tag{2.25}$$

(ii) equations with a nonlinearity depending on delay:

$$\begin{cases} \dot{x}(t) = Ax(t) + Bx(t-\tau) + F(x(t), x(t-\tau)), \ t \in J, \\ x(t) = \varphi(t), \ -\tau \le t \le 0, \ \tau > 0, \end{cases} \tag{2.26}$$

where $x \in C^1([-\tau, T], \mathbb{R}^n)$, $f \in C(\mathbb{R}^n, \mathbb{R}^n)$, and $F \in C(\mathbb{R}^n \times \mathbb{R}^n, \mathbb{R}^n)$. According to [1, Corollary 2.2], if $AB = BA$, then any solution of (2.25) has the form

$$x(t) = X_0(t)e^{A\tau}\varphi(-\tau) + \int_{-\tau}^{0} X_0(t-\tau-s)e^{A\tau}[\varphi'(s) - A\varphi(s)]ds$$

$$+ \int_{0}^{t} X_0(t-\tau-s)e^{A\tau}f(x(s))ds, \tag{2.27}$$

and any solution of (2.26) has the form

$$x(t) = X_0(t)e^{A\tau}\varphi(-\tau) + \int_{-\tau}^{0} X_0(t-\tau-s)e^{A\tau}[\varphi'(s) - A\varphi(s)]ds$$

$$+ \int_{0}^{t} X_0(t-\tau-s)e^{A\tau} F(x(s), x(s-\tau))ds, \tag{2.28}$$

where $X_0(t) = e^{At}e_{\tau}^{B_1 t}$, $B_1 = e^{-A\tau} B$ is called the fundamental matrix of (2.25) and (2.26), and the delayed exponential matrix $e_{\tau}^{B_1 t}$ is defined in (2.4).

Integral inequalities play an important role in finite time stability analysis of the solutions to differential equations.

Next, the finite time stability results are presented by virtue of the delayed exponential matrix $e_{\tau}^{B_1 t}$, $B_1 = e^{-A\tau} B$, and the above two powerful Gronwall inequalities.

We give the following hypotheses:

$[H_0]$: A and B are permutation matrices, that is, $AB = BA$.

$[H_1]$: For $f \in C(\mathbb{R}^n, \mathbb{R}^n)$, there exists $P > 0$ such that $\|f(x)\| \le P\|x\|$ for all $x \in \mathbb{R}^n$.

$[H_2]$: For $F \in C(\mathbb{R}^n \times \mathbb{R}^n, \mathbb{R}^n)$, there exists $Q > 0$ such that $\|F(x, y)\| \le Q(\|x\| + \|y\|)$ for all $x, y \in \mathbb{R}^n$.

2.1.2.1   Finite time stability results for system (2.25)

Theorem 2.6. *The solution of* (2.25) *is finite time stable with respect to* $\{0, J, \alpha, \beta, \tau\}$ *if* $[H_0]$, $[H_1]$, *and the inequality*

$$e^{(N+P)t} < \frac{\sqrt{\beta}}{e^{N\tau}\sqrt{\alpha} + \int_{-\tau}^{0} e^{-Ns}\|\varphi'(s) - A\varphi(s)\|ds}, \quad \forall t \in J, \tag{2.29}$$

hold, where $N = \|A\| + \|B_1\|$ *and* α, β *are defined in Definition 2.2.*

Proof. Note that by $[H_0]$, the solution of (2.25) can be formulated as (2.27). Taking the norm for (2.27) via Lemma 2.1, we obtain

$$\|x(t)\| \le e^{\|A\|(t+\tau)}e^{\|B_1\|(t+\tau)}\|\varphi(-\tau)\|$$

$$+ \int_{-\tau}^{0} e^{\|A\|(t-s)}e^{\|B_1\|(t-s)}\|\varphi'(s) - A\varphi(s)\|ds$$

$$+ \int_{0}^{t} e^{\|A\|(t-s)}e^{\|B_1\|(t-s)}\|f(x(s))\|ds. \tag{2.30}$$

By assumption $[H_1]$ and inequality (2.30), we obtain

$$\|x(t)\| \le e^{N(t+\tau)}\|\varphi(-\tau)\| + \int_{-\tau}^{0} e^{N(t-s)}\|\varphi'(s) - A\varphi(s)\|ds$$

$$+ \int_0^t e^{N(t-s)} P \|x(s)\| ds. \tag{2.31}$$

If we denote $u(t) = e^{-Nt} \|x(t)\|$, (2.31) can be simplified to

$$u(t) \le e^{N\tau} \|\varphi(-\tau)\| + \int_{-\tau}^0 e^{-Ns} \|\varphi'(s) - A\varphi(s)\| ds + \int_0^t Pu(s)ds,$$

$$\le a(t) + \int_0^t Pu(s)ds, \ t \in J, \tag{2.32}$$

where

$$a(t) = e^{N\tau} \|\varphi\| + \int_{-\tau}^0 e^{-Ns} \|\varphi'(s) - A\varphi(s)\| ds, \ t \in J. \tag{2.33}$$

It is obvious that $a(t) \ge 0$. In addition,

$$u(t) \le e^{N\tau} \|x(t)\| = e^{N\tau} \|\varphi(t)\| \le e^{N\tau} \|\varphi\| \le a(t), \ t \in [-\tau, 0). \tag{2.34}$$

Now we set $\varphi(t) = a(t)$. Obviously, $a(0) = \varphi(0)$. Observing (2.32) and (2.34), applying the Gronwall inequality (see [78]), one can get

$$u(t) \le a(t)e^{Pt}.$$

By (2.29), we infer that

$$\|x(t)\| \le a(t)e^{(N+P)t}$$

$$= \left[e^{N\tau} \|\varphi\| + \int_{-\tau}^0 e^{-Ns} \|\varphi'(s) - A\varphi(s)\| ds \right] e^{(N+P)t}$$

$$< \left[e^{N\tau} \sqrt{\alpha} + \int_{-\tau}^0 e^{-Ns} \|\varphi'(s) - A\varphi(s)\| ds \right] e^{(N+P)t}$$

$$< \sqrt{\beta}, \ t \in J.$$

That is, $\|x(t)\|^2 < \beta$, $\forall \, t \in J$. The proof is completed. $\qquad\square$

Remark 2.2. In Theorem 2.6, condition (2.29) involves φ', which seems a slightly stronger requirement than φ. In order to weaken (2.29), we expect to obtain a general formula for (2.27) without φ'. By using integration by parts and Lemma 2.2, via $[H_0]$, we have

$$\int_{-\tau}^0 e^{A(t-\tau-s)} e_\tau^{B_1(t-\tau-s)} e^{A\tau} \varphi'(s)ds$$

$$= e^{At} e_\tau^{B_1(t-\tau)} \varphi(0) - e^{A(t+\tau)} e_\tau^{B_1 t} \varphi(-\tau)$$

$$+ \int_{-\tau}^0 [Ae^{A(t-s)} e_\tau^{B_1(t-\tau-s)} + Be^{A(t-\tau-s)} e_\tau^{B_1(t-2\tau-s)}]\varphi(s)ds. \tag{2.35}$$

Combining (2.35) and (2.27), we obtain

$$x(t) = e^{At} e_\tau^{B_1(t-\tau)} \varphi(0) + \int_{-\tau}^{0} B e^{A(t-\tau-s)} e_\tau^{B_1(t-2\tau-s)} \varphi(s) ds$$

$$+ \int_{0}^{t} e^{A(t-\tau-s)} e_\tau^{B_1(t-\tau-s)} e^{A\tau} f(x(s)) ds, \ t \geq 0. \qquad (2.36)$$

Clearly, (2.36) does not need φ' to exist.

Theorem 2.7. *The solution of* (2.25) *is finite time stable with respect to* $\{0, J, \alpha, \beta, \tau\}$ *if* $[H_0]$, $[H_1]$, *and the inequality*

$$\gamma e^{(N+P)t} < \sqrt{\frac{\beta}{\alpha}}, \ \forall\, t \in J, \qquad (2.37)$$

hold, where

$$\gamma = \max \left\{ e^{N\tau}, 1 + \frac{\|B\|}{N}(1 - e^{-N\tau}) \right\}. \qquad (2.38)$$

Proof. Taking the norm for (2.36) via Lemma 2.1 and $[H_1]$ again, one has

$$\|x(t)\| \leq e^{Nt} \|\varphi(0)\| + \int_{-\tau}^{0} \|B\| e^{N(t-\tau-s)} \|\varphi(s)\| ds + \int_{0}^{t} e^{N(t-s)} P \|x(s)\| ds$$

$$\leq e^{Nt} \|\varphi\| + \int_{\tau}^{0} \|B\| e^{N(t-\tau-s)} \|\varphi\| ds + \int_{0}^{t} e^{N(t-s)} P \|x(s)\| ds,$$

Similar to the procedure of Theorem 2.6,

$$\|u(t)\| \leq \|\varphi\| + \int_{-\tau}^{0} \|B\| e^{-N(\tau+s)} \|\varphi\| ds + \int_{0}^{t} P \|u(s)\| ds$$

$$= \|\varphi\| \left[1 + \frac{\|B\|}{N}(1 - e^{-N\tau}) \right] + \int_{0}^{t} P \|u(s)\| ds$$

$$\leq \tilde{a}(t) \|\varphi\| e^{Pt}, \qquad (2.39)$$

where $\tilde{a}(t) = \gamma, t \in J$.

It is obvious that $\tilde{a}(t) \geq 0$. In addition, $u(t) \leq e^{N\tau} \|x(t)\| \leq e^{N\tau} \|\varphi\| \leq \tilde{a}(t) \|\varphi\|, t \in [-\tau, 0)$. Now we set $\varphi(t) = \tilde{a}(t) \|\varphi\|$. Obviously, $a(0) = \varphi(0)$. Then (2.39) becomes

$$\|x(t)\| \leq \tilde{a}(t) \|\varphi\| e^{(P+N)t} < \sqrt{\alpha} \gamma e^{(P+N)t} < \sqrt{\beta}, \ t \in J,$$

due to (2.37). That is, $\|x(t)\|^2 < \beta, \forall\, t \in J$. The proof is completed. $\qquad \square$

2.1.2.2 Finite time stability results for system (2.26)

Theorem 2.8. *The solution of* (2.26) *is finite time stable with respect to* $\{0, J, \alpha,$ $\beta, \tau\}$ *if* $[H_0]$, $[H_2]$, *and the inequality*

$$e^{(Q+Qe^{-N\tau}+N)t} < \frac{\sqrt{\beta}}{e^{N\tau}\sqrt{\alpha} + \int_{-\tau}^{0} e^{-Ns}\|\varphi'(s) - A\varphi(s)\|ds}, \quad \forall\, t \in J, \quad (2.40)$$

hold.

Proof. Note that by $[H_0]$, the solution of (2.26) can be formulated as (2.28). Taking the norm for (2.28) via Lemma 2.1 we obtain

$$\|x(t)\| \leq e^{\|A\|(t+\tau)}e^{\|B_1\|(t+\tau)}\|\varphi(-\tau)\|$$

$$+ \int_{-\tau}^{0} e^{\|A\|(t-s)}e^{\|B_1\|(t-s)}\|\varphi'(s) - A\varphi(s)\|ds$$

$$+ \int_{0}^{t} e^{\|A\|(t-s)}e^{\|B_1\|(t-s)}\|F(x(s), x(s-\tau))\|ds. \quad (2.41)$$

By $[H_2]$, inequality (2.41) becomes

$$\|x(t)\| \leq e^{N(t+\tau)}\|\varphi(-\tau)\| + \int_{-\tau}^{0} e^{N(t-s)}\|\varphi'(s) - A\varphi(s)\|ds$$

$$+ \int_{0}^{t} e^{N(t-s)}Q(\|x(s)\| + \|x(s-\tau)\|)ds. \quad (2.42)$$

Setting $u(t) = e^{-Nt}\|x(t)\|$ again, (2.42) can be simplified to

$$u(t) \leq e^{N\tau}\|\varphi(-\tau)\| + \int_{-\tau}^{0} e^{-Ns}\|\varphi'(s) - A\varphi(s)\|ds$$

$$+ \int_{0}^{t} Q(u(s) + e^{-N\tau}u(s-\tau))ds$$

$$\leq a(t) + Q\int_{0}^{t} u(s)ds + Qe^{-N\tau}\int_{0}^{t} u(s-\tau)ds, \quad (2.43)$$

where $a(t)$ is given in (2.33).

Denote $b(t) = Q, c(t) = Qe^{-N\tau}$. Obviously, $u(t) \leq e^{N\tau}\|\varphi\| < a(t) := \varphi(t)$, $t \in [-\tau, 0)$. Applying [79, Lemma 2.3], one can get

$$u(t) \leq a(t)e^{(Q+Qe^{-N\tau})t}, \ t \in J.$$

This yields

$$\|x(t)\| \leq a(t)e^{(Q+Qe^{-N\tau}+N)t}$$

$$< \left[e^{N\tau} \sqrt{\alpha} + \int_{-\tau}^{0} e^{-Ns} \|\varphi'(s) - A\varphi(s)\| ds \right] e^{(Q+Qe^{-N\tau}+N)t}.$$

According to (2.40), we have $\|x(t)\| < \sqrt{\beta}$. That is, $\|x(t)\|^2 < \beta, \forall t \in J$. $\square$

Theorem 2.9. *The solution of (2.26) is finite time stable with respect to $\{0, J, \alpha, \beta, \tau\}$ if $[H_0]$, $[H_2]$, and*

$$\gamma e^{(Q+Qe^{-N\tau}+N)t} < \sqrt{\frac{\beta}{\alpha}}, \ \forall t \in J, \tag{2.44}$$

hold, where γ is defined in (2.38).

Proof. According to (2.36), it is not difficult to obtain

$$\|x(t)\| \leq e^{Nt} \|\varphi\| + \int_{-\tau}^{0} \|B\| e^{N(t-\tau-s)} \|\varphi\| ds$$
$$+ \int_{0}^{t} e^{N(t-s)} Q \left(\|x(s)\| + \|x(s-\tau)\| \right) ds.$$

Set $a(t) = \varphi(t) = \gamma$. Similar to the proof of Theorem 2.8, based on [79, Lemma 2.3], we get

$$e^{-Nt} \|x(t)\| \leq \|\varphi\| + \int_{-\tau}^{0} \|B\| e^{-N(\tau+s)} \|\varphi\| ds$$
$$+ \int_{0}^{t} Q e^{-Ns} \|x(s)\| ds + \int_{0}^{t} Q e^{-N\tau} e^{-N(s-\tau)} \|x(s-\tau)\| ds$$
$$< \gamma \sqrt{\alpha} + Q \int_{0}^{t} e^{-Ns} \|x(s)\| ds + Q e^{-N\tau} \int_{0}^{t} e^{-N(s-\tau)} \|x(s-\tau)\| ds$$
$$< \gamma \sqrt{\alpha} e^{(Q+Qe^{-N\tau})t}, \tag{2.45}$$

which implies that

$$\|x(t)\| < \sqrt{\alpha} \gamma e^{(Q+Qe^{-N\tau}+N)t}.$$

According to (2.44), we get $\|x(t)\| < \sqrt{\beta}$. That is, $\|x(t)\|^2 < \beta, \forall t \in J$. $\square$

In addition, we can take the methods and techniques in [6] to obtain finite time stability, which are similar to our above Theorems 2.8 and 2.9.

Theorem 2.10. *The solution of (2.26) is finite time stable with respect to $\{0, J, \alpha, \beta, \tau\}$ if $[H_0]$, $[H_2]$, and*

$$c e^{(c_1+2c_2+N)t} < \sqrt{\beta}, \ \forall t \in J, \tag{2.46}$$

hold, where

$$c = e^{N\tau}\sqrt{\alpha} + \int_{-\tau}^{0} e^{-Ns}\|\varphi'(s) - A\varphi(s)\|ds, \ c_1 = Q, \ c_2 = Qe^{-N\tau}.$$

Proof. Since the proof is very similar to the proof of Theorem 2.8, here we only give the differences. Following (2.43), one has

$$u(t) \leq a(t) + Q\int_0^t u(s)ds + Qe^{-N\tau}\int_0^t u(s-\tau)ds, \ t \in J,$$

where $a(t)$ is given in (2.33).

Set

$$g(t) = c + c_1\int_0^t u(s)ds + c_2\int_0^t u(s-\tau)ds, \ t \in J.$$

Obviously, $u(t) \leq g(t)$ and the following inequality holds (see [6, (3.4)] for details):

$$u(t-\tau) \leq \sup_{0\leq\sigma\leq t} u(\sigma-\tau) \leq \sup_{0\leq\sigma\leq\tau} u(\sigma-\tau) + \sup_{\tau\leq\sigma\leq t} u(\sigma-\tau) \leq 2g(t).$$

Now we consider the following Cauchy problem:

$$\begin{cases} g'(t) = c_1 u(t) + c_2 u(t-\tau) \leq (c_1 + 2c_2)g(t), \ t \in J, \\ g(0) = c. \end{cases} \tag{2.47}$$

Solving (2.47) for $g(t)$, we obtain an estimate

$$u(t) \leq g(t) \leq ce^{(c_1+2c_2)t},$$

i.e.,

$$\|x(t)\| \leq ce^{(c_1+2c_2+N)t} < \sqrt{\beta}, \ t \in J,$$

due to (2.46). The proof is completed. $\qquad\square$

Theorem 2.11. *The solution of* (2.26) *is finite time stable with respect to* $\{0, J, \alpha, \beta, \tau\}$ *if* $[H_0]$, $[H_2]$, *and*

$$\left[1 + \frac{\|B\|}{N}(1 - e^{-N\tau})\right]e^{(c_1+2c_2+N)t} < \sqrt{\frac{\beta}{\alpha}}, \ \forall t \in J, \tag{2.48}$$

hold, where c_1, c_2 *are defined in Theorem 2.10.*

Proof. Since the proof is very similar to the proof of Theorem 2.9, here we only give the differences.

It follows from inequality (2.45) that

$$u(t) \le \|\varphi\|\left[1 + \frac{\|B\|}{N}(1 - e^{-N\tau})\right]$$
$$+ Q\int_0^t e^{-Ns}\|x(s)\|ds + Qe^{-N\tau}\int_0^t e^{-N(s-\tau)}\|x(s-\tau)\|ds.$$

Set

$$g(t) = \tilde{c} + c_1\int_0^t u(s)ds + c_2\int_0^t u(s-\tau)ds, \ t \ge 0,$$

where

$$\tilde{c} = \sqrt{\alpha}\left[1 + \frac{\|B\|}{N}(1 - e^{-N\tau})\right].$$

Solving (2.47) with $g(0) = \tilde{c}$, we obtain an estimate

$$u(t) \le g(t) \le \tilde{c}e^{(c_1+2c_2)t},$$

i.e.,

$$\|x(t)\| \le \tilde{c}e^{(c_1+2c_2+N)t}, \ t \in J.$$

According to (2.48), it is easy to see that $\|x(t)\| < \sqrt{\beta}, \ t \in J$. The proof is finished. $\qquad\square$

2.1.2.3 *Numerical examples and discussion*

First, we give some necessary explanation for systems (2.25) and (2.26) from the point of view of practical application.

For example, in a combustion chamber or reactor, we consider the change process of pressure. Based on our knowledge, the movement process of pressure depends not only on the current state value, but also on the past state value and control. From this point of view, systems (2.25) and (2.26) have a practical significance. The nonlinear term $f(x(t))$ of system (2.25) and $F(x(t), x(t-\tau))$ of system (2.26) can be considered as state feedback control or interference dependent on the state. According to the actual needs, people often hope that the pressure of the combustion chamber cannot exceed the expected threshold. In other words, this problem reduces to the finite time stability of systems (2.25) and (2.26).

In addition, we would like to remind the reader that Lazarević [80] studied a Newcastle robot model, which is equivalent to a fractional differential system with time delay. If the order of a fractional differential system is equal to 1, then our model is a special case of the model with state interference in [80]. From this point of view, it was also shown that the offered delay systems arise from the practical engineering problem. Moreover, we provide a direct method to study the finite time stability of delay differential systems, which is different from the results in [80].

In what follows, we give numerical examples to demonstrate the validity of our method and provide a discussion. The norm that we use in this section is the standard Euclidean norm.

Example 2.1. Consider a delay differential equation with a nonlinearity independent of delay:

$$\begin{cases} \dot{x}(t) = Ax(t) + Bx(t - 0.2) + f(x(t)), \ x(t) \in \mathbb{R}^2, \ t \in J = [0, 1.6], \\ \varphi(t) = (0.6, 0.8)^\top, \ -0.2 \le t \le 0, \end{cases} \tag{2.49}$$

where we set $T = 1.6$, $\tau = 0.2$, and

$$A = \begin{bmatrix} 1 & 0 \\ 0 & 1 \end{bmatrix}, \ B = \begin{bmatrix} -0.4 & 0.5 \\ 0.3 & 0 \end{bmatrix}, \ f(x(t)) = \begin{pmatrix} 0.1 \sin |x_1(t)| \\ 0.1 \sin |x_2(t)| \end{pmatrix}. \tag{2.50}$$

It is obvious that $AB = BA$ and

$$\|\varphi\|^2 = 1 < \alpha := 1.2, \ t \in [-0.2, 0].$$

By elementary calculation, one has

$$e^{-A\tau} = \begin{bmatrix} 0.8187 & 0 \\ 0 & 0.8187 \end{bmatrix}, \ B_1 = \begin{bmatrix} -0.3275 & 0.4094 \\ 0.2456 & 0 \end{bmatrix},$$
$$N = 1.5492, \quad e^{N\tau} = 1.3632.$$

Next,

$$e^{-N\tau} = 0.7336, \ \int_{-0.2}^0 e^{-Ns} \|\varphi'(s) - A\varphi(s)\| ds = 0.2344,$$

and

$$e^{N\tau} \sqrt{\alpha} + \int_{-0.2}^0 e^{-Ns} \|\varphi'(s) - A\varphi(s)\| ds = 1.7277.$$

Moreover, by calculation,

$$\|f(x)\| = \sqrt{0.02(\sin^2 |x_1| + \sin^2 |x_2|)} \le 0.1\sqrt{2}\sqrt{|x_1|^2 + |x_2|^2} \doteq 0.1414 \|x\|.$$

So we can set $P = 0.1414$ such that $\|f(x)\| \le P\|x\|$.

By elementary computation, one can choose a suitable $\beta \ge 668.116$ satisfying (2.29) and $\beta \ge 498.653$ satisfying (2.37). Both of them are greater than the actual value 50.924, which means that Theorems 2.6 and 2.7 are effective. Thus,

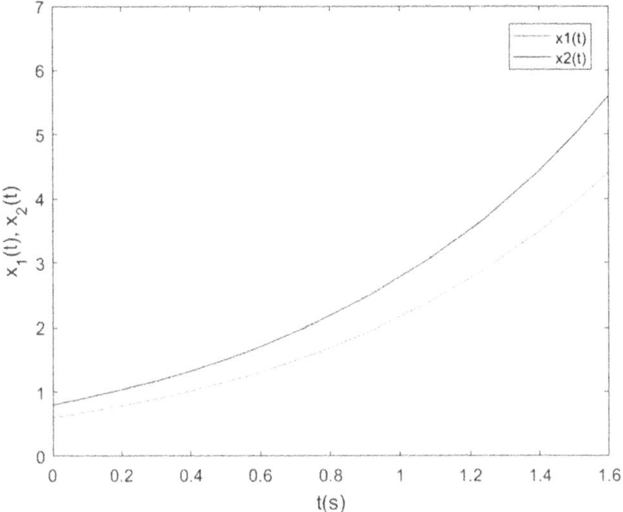

FIGURE 2.4   The state response $x(t)$ of (2.49).

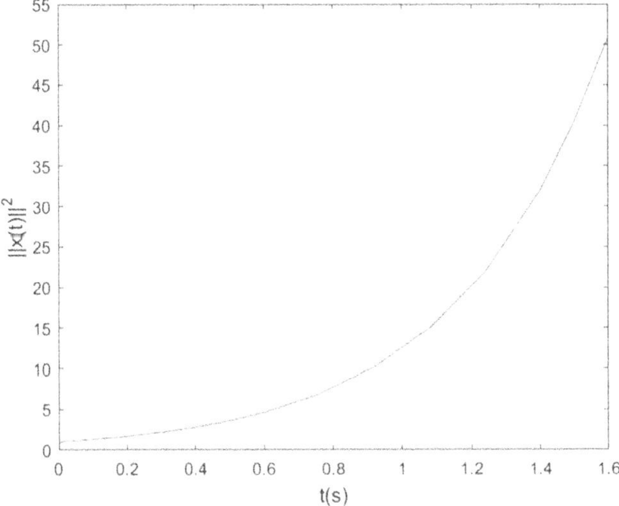

FIGURE 2.5   The square norm of the state vector of (2.49).

the trivial solution of (2.49) is finite time stable with respect to $\{0, J, \alpha, \beta, \tau\}$ (see Fig. 2.4 and Fig. 2.5) by Theorems 2.6 and 2.7.

Finally, we list tables to show finite time stability related to the stability time interval J and explicit stability bounds β.

As for the nonlinear delay system (2.49), we fixed $\beta = 28$, $\alpha = 1.2$ and observed the stability time interval J. By computer simulation, we obtain the

TABLE 2.1 Stability time interval J with β fixed.

Theorem	$\|\varphi\|$	J	τ	α	$\|x(t)\|^2$	β
2.6	1	[0, 0.60]	0.2	1.2	4.615	28
2.7	1	[0, 1.08]	0.2	1.2	14.975	28

TABLE 2.2 β with the stability time interval J fixed.

Theorem	$\|\varphi\|$	J	τ	α	$\|x(t)\|^2$	β
2.6	1	[0, 0.8]	0.2	1.2	7.069	44.657
2.7	1	[0, 0.8]	0.2	1.2	7.069	33.346

different stability times according to Theorems 2.6 and 2.7. From Table 2.1, one knows that the range of the stability time interval of Theorem 2.7 is wider than that of Theorem 2.6.

As for the nonlinear delay system (2.49), we fixed the stability time interval $J = [0, 0.8]$, $\alpha = 1.2$ and observed the threshold β. By computer simulation, we obtain the different thresholds β according to Theorems 2.6 and 2.7. From Table 2.2, one knows that the threshold β of Theorem 2.7 is smaller than that of Theorem 2.6.

Based on the above results, we claim that Theorem 2.7 is more effective than Theorem 2.6.

Example 2.2. Consider the following delay differential equation with a nonlinearity depending on delay:

$$\begin{cases} \dot{x}(t) = Ax(t) + Bx(t - 0.2) + F(x(t), x(t - 0.2)), \ x(t) \in \mathbb{R}^2, \ t \in J, \\ \varphi(t) = (0.3, 0.4)^\top, \ -0.2 \leq t \leq 0, \end{cases}$$

$$(2.51)$$

where we set $T = 1.6$, $\tau = 0.2$, $J = [0, 1.6]$, A, B are given in (2.50), $\alpha := 0.3$, and

$$F(x(t), x(t - 0.2)) = \begin{pmatrix} 0.1 \sin \|x_1(t)\| - |x_1(t - 0.2)|\| \\ 0.1 \sin \|x_2(t)\| - |x_2(t - 0.2)|\| \end{pmatrix}.$$

Similar to Example 2.1, we only need to find Q for the nonlinear term F. By calculation,

$$\|F(x, y)\| \leq 0.1\sqrt{2}(\|x\| + \|y\|) \doteq 0.1414(\|x\| + \|y\|).$$

Then, we can set $Q = 0.1414$ such that $\|F(x, y)\| \leq Q(\|x\| + \|y\|)$.

By elementary computation, one can choose a suitable $\beta \geq 246.611$ satisfying (2.40), $\beta \geq 184.258$ satisfying (2.44), $\beta \geq 324.009$ satisfying (2.46), and $\beta \geq 162.074$ satisfying (2.48). Both of them greater than the actual value 36.0722, which means that Theorems 2.8–2.11 are effective. Thus, the trivial solution of (2.51) is finite time stable with respect to $\{0, J, \alpha, \beta, \tau\}$ (see Fig. 2.6 and Fig. 2.7) by Theorems 2.8–2.11.

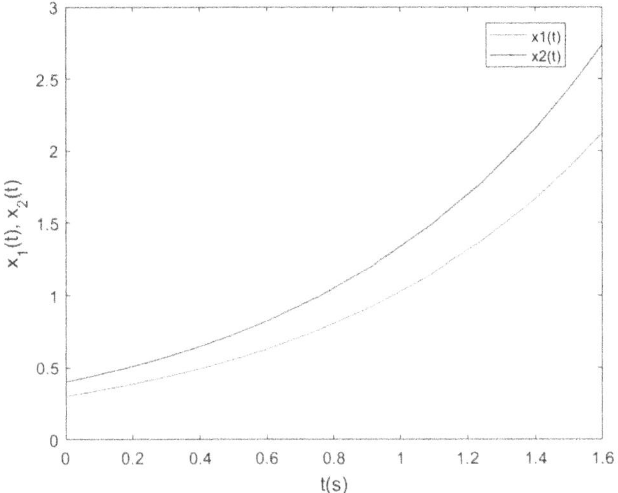

FIGURE 2.6 The state response $x(t)$ of (2.51).

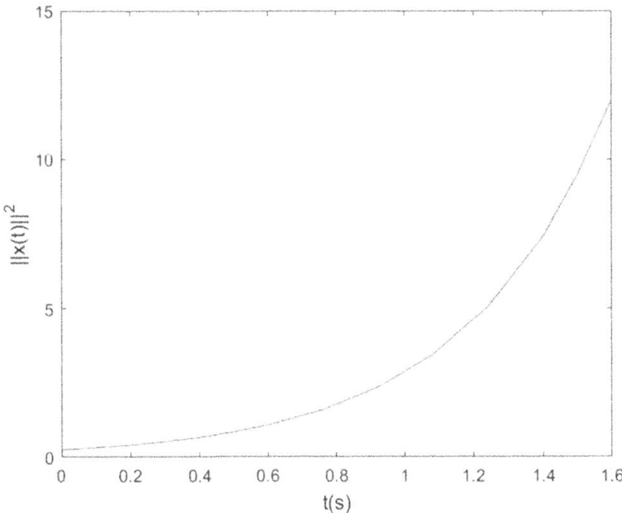

FIGURE 2.7 The square norm of the state vector of (2.51).

As for the nonlinear delay system (2.51), we fixed threshold $\beta = 70, \alpha = 0.3$ and observed the stability time interval J. By computer simulation, we obtain

TABLE 2.3 Stability time interval J with β fixed.

Theorem	$\|\varphi\|$	J	τ	α	$\|x(t)\|^2$	β
2.8	0.25	[0, 1.245]	0.2	0.3	5.149	70
2.9	0.25	[0, 1.329]	0.2	0.3	5.539	70
2.10	0.25	[0, 1.194]	0.2	0.3	3.624	70
2.11	0.25	[0, 1.378]	0.2	0.3	5.594	70

TABLE 2.4 β with the stability time interval J fixed.

Theorem	$\|\varphi\|$	J	τ	α	$\|x(t)\|^2$	β
2.8	0.25	[0, 1]	0.2	0.3	3.224	28.010
2.9	0.25	[0, 1]	0.2	0.3	3.224	20.928
2.10	0.25	[0, 1]	0.2	0.3	3.224	33.221
2.11	0.25	[0, 1]	0.2	0.3	3.224	16.617

the different stability times according to Theorems 2.8–2.11. From Table 2.3, one knows that the range of the stability time interval of Theorem 2.11 is wider than those of other theorems.

As for the nonlinear delay system (2.51), we fixed the stability time interval $J = [0, 1]$, $\alpha = 0.3$ and observed the threshold β. By computer simulation, we obtain the different thresholds β by Theorems 2.8–2.11. From Table 2.4, we know that the threshold β of Theorem 2.11 is smaller than those of others. Based on the above results, one knows that Theorem 2.11 is more effective than other theorems.

2.1.2.4 Conclusion

Following our work on the finite time stability of linear delay continuous systems via the delayed exponential matrix approach, we continue to study the finite time stability of nonlinear delay differential equations under mild linear growth conditions for the nonlinear terms. Based on the constant variation formula to present the solution structure for nonlinear delay systems involving delayed exponential matrices, sufficient conditions are derived to ensure the system states are confined in a certain range for a given bound on the initial conditions. The results in this part are motivated from [81].

2.2 Controllability

2.2.1 Relative controllability for delay differential systems

In this section, we deal with relative controllability of the following semilinear delay differential systems with linear parts defined by pairwise permutable matrices:

$$\begin{cases} \dot{x}(t) = Ax(t) + Bx(t-\tau) + f(t, x(t)) + Cu(t), \ t \in J, \ \tau \geq 0, \\ x(t) = \varphi(t), \ -\tau \leq t \leq 0, \end{cases} \tag{2.52}$$

where $A, B, C \in \mathbb{R}^{n \times n}$, $AB = BA$, $f : J \times \mathbb{R}^n \to \mathbb{R}^n$, and $\varphi \in C^1([-\tau, 0], \mathbb{R}^n)$. The state $x(\cdot)$ takes values from $\mathbb{R}^n$ and the control function $u(\cdot)$ takes values from $L^2(J, \mathbb{R}^n)$.

Concerning the relative controllability of system (2.52), we would like to address the following difficulties:

(i) The formula of the solution of system (2.52) is much more complicated since the representation of the solution involves a more complicated delayed matrix exponential $e_\tau^{e^{-A\tau}Bt}$, $t \in J$.

(ii) Since both linear parts A and B appeared, it is debatable whether we can establish the similar Kalman criterion since the Cayley formula for the delayed matrix exponential $e_\tau^{e^{-A\tau}Bt}$, $t \in J$, is not an easy task.

Apart from the Kalman criterion to judge whether a linear controlled system is controllable, the Grammian matrix criterion is an alternative method which has been widely used to deal with linear controlled systems. The controllability problem for the linear differential systems with control

$$\begin{cases} \dot{x}(t) = Ax(t) + Cu(t), \ t \in J, \\ x(0) = x_0 \in \mathbb{R}^n \end{cases}$$

is known to be solved if and only if the Grammian matrix of the form

$$W_c[0, t_1] = \int_0^{t_1} e^{-As} CC^\top e^{A^\top s} ds$$

is nonsingular. By checking the proof of [38, Theorem 4] via the control function

$$u(t) = b^\top e_\tau^{B^\top(t_1 - \tau - t)} \left[\int_0^{t_1} e_\tau^{B(t_1 - \tau - s)} bb^\top e_\tau^{B^\top(t_1 - \tau - s)} ds \right]^{-1} \eta$$

and

$$\eta = x_1 - e_\tau^{Bt_1} \varphi(-\tau) - \int_{-\tau}^0 e_\tau^{B(t_1 - \tau - s)} \varphi'(s) ds,$$

and the vector $x_1 \in \mathbb{R}^n$ is prefixed. One can directly conclude that a sufficient condition for system

$$\begin{cases} \dot{x}(t) = Bx(t-\tau) + bu(t), \ t \in J, \ \tau \geq 0, \\ x(t) = \varphi(t), \ -\tau \leq t \leq 0, \end{cases}$$

is relatively controllable when the following Grammian-type matrix

$$\int_0^{t_1} e_\tau^{B(t_1-\tau-s)} bb^\top e_\tau^{B^\top (t_1-\tau-s)} ds$$

is nonsingular, where $x : [-\tau, t_1] \to \mathbb{R}^n$ is continuous differentiable on $J = [0, t_1]$, $t_1 > 0$, b is a constant vector, $u : J \to \mathbb{R}$ is an input, and $\varphi : [-\tau, 0] \to \mathbb{R}^n$ is a given function.

This implies that it is possible to adopt the Grammian-type matrix method to address the relative controllability of system (2.52) although it is not easy to verify the necessity.

First, the explicit formula of solutions to nonhomogeneous linear delay systems is collected.

Lemma 2.6. *(see [1, Corollary 2.2]) Let $f : J \to \mathbb{R}^n$ be a continuous vector-valued function. A solution $x \in C([-\tau, t_1], \mathbb{R}^n)$ of the system*

$$\begin{cases} \dot{x}(t) = Ax(t) + Bx(t-\tau) + f(t), \ t \in J, \ \tau \ge 0, \\ x(t) = \varphi(t), \ -\tau \le t \le 0, \ \tau > 0, \end{cases} \tag{2.53}$$

can be formulated by an integral equation of the type

$$x(t) = e^{A(t+\tau)} e_\tau^{B_1 t} \varphi(-\tau) + \int_{-\tau}^0 e^{A(t-s)} e_\tau^{B_1 (t-\tau-s)} [\varphi'(s) - A\varphi(s)] ds$$

$$+ \int_0^t e^{A(t-s)} e_\tau^{B_1 (t-\tau-s)} f(s) ds, \ t \in [-\tau, t_1],$$

where $e_\tau^{B\cdot}$ is defined in (2.4) and $B_1 = e^{-A\tau} B$.

By Lemma 2.6, a solution $x \in C([-\tau, t_1], \mathbb{R}^n)$ of system (2.52) can be formulated by

$$x(t) = e^{A(t+\tau)} e_\tau^{B_1 t} \varphi(-\tau) + \int_{-\tau}^0 e^{A(t-s)} e_\tau^{B_1 (t-\tau-s)} [\varphi'(s) - A\varphi(s)] ds$$

$$+ \int_0^t e^{A(t-s)} e_\tau^{B_1 (t-\tau-s)} [f(s, x(s)) + Cu(s)] ds, \ t \in [-\tau, t_1]. \tag{2.54}$$

We note that the main motivation for having permutable matrices is that the above formula now is simple while in general it is rather awkward. In fact, $e^{At} e_\tau^{B_1 t} (B_1 = e^{-A\tau} B)$ is an explicit form of the fundamental matrix of $\dot{x}(t) = Ax(t) + Bx(t-\tau)$, $t > 0$, with $AB = BA$. This admits one can apply the classical ideas of the variation of constants method via the principle of superposition for ODEs to seek an explicit simple formula of solutions to (2.53). The simple form of (2.54) allows to find a simple criterion of relative controllability. Moreover, we also give some details to show that a model of population dynamics with delayed birthrates and delayed logistic terms can be turned into delay differential equations with permutable matrices.

Definition 2.4. (see [38, Definition 4])System (2.52) is called relatively controllable if for an arbitrary initial vector function $\varphi \in C^1([-\tau, 0], \mathbb{R}^n)$, the final state of the vector $x_1 \in \mathbb{R}^n$, and time t_1, there exists a control $u \in L^2(J, \mathbb{R}^n)$ such that system (2.52) has a solution $x \in C([-\tau, t_1], \mathbb{R}^n)$ that satisfies the boundary conditions $x(t) = \varphi(t)$, $-\tau \le t \le 0$, and $x(t_1) = x_1$.

2.2.1.1 Relative controllability for linear delay differential systems

Let $f(t, x(t)) \equiv \mathbf{0} = \underbrace{(0, \cdots, 0)}_{n}^{\top}$, $t \in J$, i.e., system (2.52) reduces to the following linear delay controlled system with linear parts defined by pairwise permutable matrices:

$$\begin{cases} \dot{x}(t) = Ax(t) + Bx(t-\tau) + Cu(t), \ t \in J, \\ x(t) = \varphi(t), \ -\tau \le t \le 0. \end{cases} \tag{2.55}$$

Next, we introduce a notation of a delay Grammian matrix, an extension of the classical Grammian matrix for linear differential systems, as follows:

$$W_\tau[0, t_1] = \int_0^{t_1} e^{A(t_1-s)} e_\tau^{B_1(t_1-\tau-s)} CC^\top e_\tau^{B_1^\top(t_1-\tau-s)} e^{A^\top(t_1-s)} ds. \tag{2.56}$$

Theorem 2.12. *System (2.55) is relatively controllable if and only if $W_\tau[0, t_1]$ defined in (2.56) is nonsingular.*

Proof. **Sufficiency.** Since $W_\tau[0, t_1]$ is nonsingular, its inverse $W_\tau^{-1}[0, t_1]$ is well defined. One can select a control function as follows:

$$u(t) = C^\top e_\tau^{B_1^\top(t_1-\tau-t)} e^{A^\top(t_1-t)} W_\tau^{-1}[0, t_1]\eta, \tag{2.57}$$

where

$$\eta = x_1 - e^{A(t_1+\tau)} e_\tau^{B_1 t_1} \varphi(-\tau) - \int_{-\tau}^0 e^{A(t_1-s)} e_\tau^{B_1(t_1-\tau-s)} [\varphi'(s) - A\varphi(s)] ds, \tag{2.58}$$

where the vector $x_1 \in \mathbb{R}^n$ is arbitrary before it is chosen.

Inserting (2.57) in (2.54) (with $f(\cdot, x) = \mathbf{0}$), we have

$$x(t_1) = e^{A(t_1+\tau)} e_\tau^{B_1 t_1} \varphi(-\tau) + \int_{-\tau}^0 e^{A(t_1-s)} e_\tau^{B_1(t_1-\tau-s)} [\varphi'(s) - A\varphi(s)] ds$$

$$+ \int_0^{t_1} e^{A(t_1-s)} e_\tau^{B_1(t_1-\tau-s)} CC^\top e_\tau^{B_1^\top(t_1-\tau-s)} e^{A^\top(t_1-s)} ds \cdot W_\tau^{-1}[0, t_1]\eta. \tag{2.59}$$

Linking (2.56) and (2.59) via (2.58), it is not difficult to derive

$$x(t_1) = e^{A(t_1+\tau)} e_\tau^{B_1 t_1} \varphi(-\tau) + \int_{-\tau}^{0} e^{A(t_1-s)} e_\tau^{B_1(t_1-\tau-s)} [\varphi'(s) - A\varphi(s)]ds + \eta$$

$$= x_1. \tag{2.60}$$

The boundary condition $x(t) = \varphi(t)$, $-\tau \leq t \leq 0$, holds by Lemma 2.6. Combining with formula (2.60) via Definition 2.4, system (2.52) is relatively controllable.

Necessity. We prove our result by contradiction. Assume $W_\tau[0, t_1]$ is singular, i.e., there exists at least one nonzero state $\tilde{x} \in \mathbb{R}^n$ such that

$$\tilde{x}^\top W_\tau[0, t_1]\tilde{x} = 0.$$

Furthermore, we obtain

$$
\begin{aligned}
0 &= \tilde{x}^\top W_\tau[0, t_1]\tilde{x} \\
&= \int_0^{t_1} \tilde{x}^\top e^{A(t_1-s)} e_\tau^{B_1(t_1-\tau-s)} CC^\top e_\tau^{B_1^\top(t_1-\tau-s)} e^{A^\top(t_1-s)} \tilde{x}\, ds \\
&= \int_0^{t_1} \left[\tilde{x}^\top e^{A(t_1-s)} e_\tau^{B_1(t_1-\tau-s)} C\right]\left[\tilde{x}^\top e^{A(t_1-s)} e_\tau^{B_1(t_1-\tau-s)} C\right]^\top ds \\
&= \int_0^{t_1} \left\| \tilde{x}^\top e^{A(t_1-s)} e_\tau^{B_1(t_1-\tau-s)} C \right\|^2 ds,
\end{aligned}
$$

which implies that

$$\tilde{x}^\top e^{A(t_1-s)} e_\tau^{B_1(t_1-\tau-s)} C = \mathbf{0}^\top, \ \forall\, s \in J. \tag{2.61}$$

Since system (2.55) is relatively controllable, according to Definition 2.4, there exists a control $u_1(t)$ that drives the initial state to zero at t_1, that is,

$$x(t_1) = e^{A(t_1+\tau)} e_\tau^{B_1 t_1} \varphi(-\tau) + \int_{-\tau}^{0} e^{A(t_1-s)} e_\tau^{B_1(t_1-\tau-s)} [\varphi'(s) - A\varphi(s)]ds$$

$$+ \int_0^{t_1} e^{A(t_1-s)} e_\tau^{B_1(t_1-\tau-s)} Cu_1(s)ds = \mathbf{0}. \tag{2.62}$$

Similarly, there also exists a control $\tilde{u}(t)$ that drives the initial state to the state $\tilde{x}$ at t_1, i.e.,

$$x(t_1) = e^{A(t_1+\tau)} e_\tau^{B_1 t_1} \varphi(-\tau) + \int_{-\tau}^{0} e^{A(t_1-s)} e_\tau^{B_1(t_1-\tau-s)} [\varphi'(s) - A\varphi(s)]ds$$

$$+ \int_0^{t_1} e^{A(t_1-s)} e_\tau^{B_1(t_1-\tau-s)} C\tilde{u}(s)ds$$

$$= \tilde{x}. \tag{2.63}$$

Then by (2.62) and (2.63), we have

$$\tilde{x} = \int_0^{t_1} e^{A(t_1-s)} e_\tau^{B_1(t_1-\tau-s)} C[\tilde{u}(s) - u_1(s)]ds. \qquad (2.64)$$

Multiplying both sides of (2.64) by $\tilde{x}^{\mathrm{T}}$, we obtain

$$\tilde{x}^{\top}\tilde{x} = \int_0^{t_1} \tilde{x}^{\top} e^{A(t_1-s)} e_\tau^{B_1(t_1-\tau-s)} C[\tilde{u}(s) - u_1(s)]ds.$$

Note that by (2.61), one can obtain $\tilde{x}^{\top}\tilde{x} = 0$, i.e., $\tilde{x} = \mathbf{0}$, which conflicts with $\tilde{x}$ being nonzero. Thus, the delay Grammian matrix $W_\tau[0, t_1]$ is nonsingular. The proof is finished. $\qquad\qquad\square$

Remark 2.3. In Theorem 2.12, if $B = \Theta$, then $B_1 = \Theta$ and therefore $e_\tau^{B_1\cdot} = I$. Then, the delay Grammian matrix $W_\tau[0, t_1]$ given in (2.56) reduces to the classical Grammian matrix for linear systems. Hence Theorem 2.12 contains the classical controllability result for linear systems. Concerning on the case of $A = \Theta$, some related interesting results for linear continuous delay systems have been reported in [82, Theorem 3] and some results for linear discrete delay systems have been reported in [2,3,17]. It is remarkable that a criterion of relative controllability, expressed using a defining equation, has been obtained in [83,84]. From a practical point of view, it is more convenient than Theorem 2.12, which presents other theoretical criteria to guarantee (2.55) is relatively controllable and also enriches the results in this field.

2.2.1.2 *Relative controllability for semilinear delay differential systems*

Consider $f(t, x(t)) \not\equiv \mathbf{0}$, $t \in J$. We need the following hypothesis:

$[H_1]$: The operator $W : L^2(J, \mathbb{R}^n) \to \mathbb{R}^n$ defined by

$$Wu = \int_0^{t_1} e^{A(t_1-s)} e_\tau^{B_1(t_1-\tau-s)} Cu(s)ds$$

has an inverse operator W^{-1} which takes values in $L^2(J, \mathbb{R}^n)/\ker W$.

Then we set

$$M = \|W^{-1}\|_{L_b(\mathbb{R}^n, L^2(J,\mathbb{R}^n)/\ker W)}.$$

Remark 2.4. Clearly, W must be surjective to satisfy $[H_1]$. On the other hand, if W is surjective, then we can define an inverse $W^{-1} : \mathbb{R}^n \to L^2(J, \mathbb{R}^n)/\ker W$. We propose its natural construction as follows. Let $(\cdot, \cdot)$ denote the Euclidean scalar product in $\mathbb{R}^n$. Then $L^2(J, \mathbb{R}^n)$ and $\mathbb{R}^n$ are Hilbert spaces. So we can use $\ker W = \mathrm{im}\, W^{*\perp}$ and $\mathrm{im}\, W = \ker W^{*\perp}$. We need to find $W^{\star}$. Set $W(\cdot) =$

$e^{A(t_1-\cdot)}e_\tau^{B_1(t_1-\tau-\cdot)}C$. For any $w \in \mathbb{R}^n$ and $u \in L^2(J, \mathbb{R}^n)$ we derive

$$(Wu, w) = \left(\int_0^{t_1} W(s)u(s)ds, w\right) = \int_0^{t_1} (u(s), W(s)^\top w)ds,$$

which gives $W^* w = W(s)^\top w$. Thus $\ker W^* = \{\mathbf{0}\}$ if and only if

$$\int_0^{t_1} \|W(s)^\top w\|^2 ds \neq 0$$

for any $\mathbf{0} \neq w \in \mathbb{R}^n$. But

$$\int_0^{t_1} \|W(s)^\top w\|^2 ds = \int_0^{t_1} (W(s)^\top w, W(s)^\top w)ds$$

$$= \int_0^{t_1} (W(s)W(s)^\top w, w)ds = (W_\tau[0, t_1]w, w). \quad (2.65)$$

So the surjectivity of W is equivalent to the regularity of $W_\tau[0, t_1]$, and we assume this. To solve $Wu = v$, $u \in \ker W^\perp = \operatorname{im} W^*$, we take $u(t) = W(t)^\top w$ and then solve

$$v = W(W(\cdot)^\top w) = \int_0^{t_1} W(s)W(s)^\top w\, ds = W_\tau[0, t_1]w,$$

which gives $w = W_\tau[0, t_1]^{-1}v$, and this implies

$$u(t) = W^{-1}w = W(t)^\top W_\tau[0, t_1]^{-1}w,$$

where we consider $L_b(\mathbb{R}^n, L^2(J, \mathbb{R}^n))/\ker W) = \ker W^\perp$. Moreover, by (2.65), we derive

$$\int_0^{t_1} \|u(s)\|^2 ds = \int_0^{t_1} \|W(s)^\top W_\tau[0, t_1]^{-1}w\|^2 ds$$

$$= \int_0^{t_1} \left(W(s)^\top W_\tau[0, t_1]^{-1}w, W(s)^\top W_\tau[0, t_1]^{-1}w\right)ds$$

$$= \int_0^{t_1} \left((W_\tau[0, t_1]^{-1})^\top W(s)W(s)^\top W_\tau[0, t_1]^{-1}w, w\right)ds$$

$$= \left((W_\tau[0, t_1]^{-1})^\top \int_0^{t_1} W(s)W(s)^\top ds\, W_\tau[0, t_1]^{-1}w, w\right)$$

$$= \left((W_\tau[0, t_1]^{-1})^\top w, w\right) = \left(w, W_\tau[0, t_1]^{-1}w\right),$$

which gives

$$M = \sqrt{\|W_\tau[0, t_1]^{-1}\|}. \quad (2.66)$$

We note that (2.65) also implies

$$\|W\| = \|W^*\| = \sqrt{\|W_\tau[0, t_1]\|}.$$

$[H_2]$: The function $f : J \times \mathbb{R}^n \to \mathbb{R}^n$ is continuous and there exists a constant $q > 1$ and $L_f(\cdot) \in L^q(J, \mathbb{R}^+)$ such that

$$\|f(t, x_1) - f(t, x_2)\| \leq L_f(t)\|x_1 - x_2\|, \ x_i \in \mathbb{R}^n, \ i = 1, 2.$$

In view of $[H_1]$, for arbitrary $x(\cdot) \in \mathcal{C}$, consider a control function $u_x(t)$ given by

$$u_x(t) = W^{-1}\Bigg[x_1 - e^{A(t_1+\tau)}e_\tau^{B_1 t_1}\varphi(-\tau)$$
$$- \int_{-\tau}^0 e^{A(t_1-s)}e_\tau^{B_1(t_1-\tau-s)}(\varphi'(s) - A\varphi(s))ds$$
$$- \int_0^{t_1} e^{A(t_1-s)}e_\tau^{B_1(t_1-\tau-s)} f(s, x(s))ds\Bigg](t), \ t \in J. \tag{2.67}$$

Now we state our main idea to prove our main result via the fixed point method. We firstly show that, using control (2.67), the operator $\mathcal{P} : \mathcal{C} \to \mathcal{C}$ defined by

$$(\mathcal{P}x)(t) = e^{A(t+\tau)}e_\tau^{B_1 t}\varphi(-\tau) + \int_{-\tau}^0 e^{A(t-s)}e_\tau^{B_1(t-\tau-s)}[\varphi'(s) - A\varphi(s)]ds$$
$$+ \int_0^t e^{A(t-s)}e_\tau^{B_1(t-\tau-s)} f(s, x(s))ds$$
$$+ \int_0^t e^{A(t-s)}e_\tau^{B_1(t-\tau-s)}Cu_x(s)ds$$

has a fixed point x, which is just a solution of system (2.52). Secondly, we check that $(\mathcal{P}x)(t_1) = x_1$ and $(\mathcal{P}x)(0) = \varphi(0) = x_0$, which means that u_x steers system (2.52) from x_0 to x_1 in finite time t_1. This implies system (2.52) is relatively controllable on J.

For each positive number r, define $\mathcal{B}_r = \{x \in \mathcal{C} : \|x\|_C \leq r\}$. Then, for each r, $\mathcal{B}_r$ is obviously a bounded, closed, and convex set of $\mathcal{C}$. For brevity, we set $R_f = \sup_{t \in J}\|f(t, 0)\|$ and $N = \|A\| + \|B_1\|$.

In what follows, we apply Krasnoselskii's fixed point theorem (see [82]) to derive the relative controllability result for system (2.52).

Theorem 2.13. *Suppose that $[H_1]$ and $[H_2]$ are satisfied. Then system (2.52) is relatively controllable provided that*

$$M_2\Bigg[1 + \frac{M}{N}(e^{Nt_1} - 1)\|C\|\Bigg] < 1, \tag{2.68}$$

where $M_2 = [\frac{1}{Np}(e^{Npt_1} - 1)]^{\frac{1}{p}} \|L_f\|_{L^q(J,\mathbb{R}^+)}$, $\frac{1}{p} + \frac{1}{q} = 1$, *p, q > 1, and M is given by* (2.66).

Proof. To examine the conditions for Krasnoselskii's fixed point theorem (see [82]), we divide our proof into several steps.

Step 1. We prove that there exists a positive number r such that $\mathcal{P}(\mathcal{B}_r) \subseteq \mathcal{B}_r$. In light of $[H_2]$ and the Hölder inequality, we obtain

$$\int_0^t e^{N(t-s)} L_f(s)ds \leq \left(\int_0^t e^{Np(t-s)}ds \right)^{\frac{1}{p}} \left(\int_0^t L_f^q(s)ds \right)^{\frac{1}{q}}$$

$$\leq \left[\frac{1}{Np}(e^{Npt} - 1) \right]^{\frac{1}{p}} \|L_f\|_{L^q(J,\mathbb{R}^+)}$$

and

$$\int_0^t e^{N(t-s)} \|f(s,0)\|ds \leq R_f \int_0^t e^{N(t-s)}ds \leq \frac{R_f}{N}(e^{Nt} - 1).$$

Taking into account (2.67), using $[H_1]$, $[H_2]$, and Lemma 2.1, we have

$$\|u_x(t)\| \leq \|W^{-1}\|_{L(\mathbb{R}^n, L^2(J,\mathbb{R}^n)/\ker W)} \left(\|x_1\| + e^{N(t_1+\tau)}\|\varphi(-\tau)\| \right.$$

$$\left. + \int_{-\tau}^0 e^{N(t_1-s)}\|\varphi'(s) - A\varphi(s)\|ds + \int_0^{t_1} e^{N(t_1-s)}\|f(s,x(s))\|ds \right)$$

$$\leq M\|x_1\| + Me^{N(t_1+\tau)}\|\varphi(-\tau)\| + M\int_{-\tau}^0 e^{N(t_1-s)}\|\varphi'(s) - A\varphi(s)\|ds$$

$$+ M\int_0^{t_1} e^{N(t_1-s)}L_f(s)\|x(s)\|ds + M\int_0^{t_1} e^{N(t_1-s)}\|f(s,0)\|ds$$

$$\leq M\|x_1\| + Me^{N(t_1+\tau)}\|\varphi(-\tau)\| + M\int_{-\tau}^0 e^{N(t_1-s)}\|\varphi'(s) - A\varphi(s)\|ds$$

$$+ M\left[\frac{1}{Np}(e^{Npt_1} - 1) \right]^{\frac{1}{p}} \|L_f\|_{L^q(J,\mathbb{R}^+)}\|x\| + \frac{MR_f}{N}(e^{Nt_1} - 1)$$

$$\leq M\|x_1\| + Ma + MM_2r,$$

where

$$a = e^{N(t_1+\tau)}\|\varphi\| + \int_{-\tau}^0 e^{N(t_1-s)}\|\varphi'(s) - A\varphi(s)\|ds + \frac{R_f}{N}(e^{Nt_1} - 1)$$

and M_2 is defined above.

From $[H_1]$ and $[H_2]$ we have

$$\|(\mathcal{P}x)(t)\| \leq e^{N(t+\tau)}\|\varphi(-\tau)\| + \int_{-\tau}^0 e^{N(t-s)}\|\varphi'(s) - A\varphi(s)\|ds$$

$$+ \int_0^t e^{N(t-s)} \| f(s, x(s)) \| ds + \int_0^t e^{N(t-s)} \| C \| \| u_x(s) \| ds$$

$$\leq e^{N(t+\tau)} \| \varphi(-\tau) \| + \int_{-\tau}^0 e^{N(t-s)} \| \varphi'(s) - A\varphi(s) \| ds$$

$$+ \int_0^t e^{N(t-s)} L_f(s) \| x(s) \| ds + \int_0^t e^{N(t-s)} \| f(s, 0) \| ds$$

$$+ \int_0^t e^{N(t-s)} \| C \| (M \| x_1 \| + Ma + MM_2 \| x \|) ds$$

$$\leq e^{N(t+\tau)} \| \varphi(-\tau) \| + \int_{-\tau}^0 e^{N(t-s)} \| \varphi'(s) - A\varphi(s) \| ds$$

$$+ \left[\frac{1}{Np}(e^{Npt} - 1) \right]^{\frac{1}{p}} \| L_f \|_{L^{\frac{1}{q}}(J, \mathbb{R}^+)} \| x \| + \frac{R_f}{N}(e^{Nt} - 1)$$

$$+ \| C \| \| x_1 \| M \int_0^t e^{N(t-s)} ds + M \| C \| a \int_0^t e^{N(t-s)} ds$$

$$+ MM_2 \| C \| \| x \| \int_0^t e^{N(t-s)} ds$$

$$\leq a \left[1 + \frac{M}{N}(e^{Nt_1} - 1) \| C \| \right] + \frac{M}{N}(e^{Nt_1} - 1) \| C \| \| x_1 \|$$

$$+ M_2 \left[1 + \frac{M}{N}(e^{Nt_1} - 1) \| C \| \right] r = r$$

for

$$r = \frac{a \left[1 + \frac{M}{N}(e^{Nt_1} - 1) \| C \| \right] + \frac{M}{N}(e^{Nt_1} - 1) \| C \| \| x_1 \|}{1 - M_2 \left[1 + \frac{M}{N}(e^{Nt_1} - 1) \| C \| \right]}.$$

Hence, we obtain $\mathcal{P}(\mathcal{B}_r) \subseteq \mathcal{B}_r$ for such an r.

Now, we divide $\mathcal{P}$ into two operators $\mathcal{P}_1$ and $\mathcal{P}_2$ on $\mathcal{B}_r$ as

$$(\mathcal{P}_1 x)(t) = e^{A(t+\tau)} e_\tau^{B_1 t} \varphi(-\tau) + \int_{-\tau}^0 e^{A(t-s)} e_\tau^{B_1(t-\tau-s)} [\varphi'(s) - A\varphi(s)] ds$$

$$+ \int_0^t e^{A(t-s)} e_\tau^{B_1(t-\tau-s)} C u_x(s) ds,$$

$$(\mathcal{P}_2 x)(t) = \int_0^t e^{A(t-s)} e_\tau^{B_1(t-\tau-s)} f(s, x(s)) ds,$$

for $t \in J$, respectively.

Step 2. We show that $\mathcal{P}_1$ is a contraction mapping.

Let $x, y \in \mathcal{B}_r$. In view of $[H_1]$ and $[H_2]$, for each $t \in J$, we have

$$
\begin{aligned}
\|u_x(t) - u_y(t)\| &\leq M \int_0^t e^{N(t-s)} L_f(s) \|x(s) - y(s)\| ds \\
&\leq M \int_0^t e^{N(t-s)} L_f(s) ds \|x - y\| \\
&\leq M \left[\frac{1}{Np} (e^{Npt} - 1) \right]^{\frac{1}{p}} \|L_f\|_{L^q(J,\mathbb{R}^+)} \|x - y\| \\
&\leq M M_2 \|x - y\|.
\end{aligned}
$$

Using the above, we derive

$$
\begin{aligned}
\|(\mathcal{P}_1 x)(t) - (\mathcal{P}_1 y)(t)\| &\leq \int_0^t e^{N(t-s)} \|C\| \|u_x(s) - u_y(s)\| ds \\
&\leq \|C\| \int_0^t e^{N(t-s)} ds \, M M_2 \|x - y\| \\
&\leq \frac{\|C\|}{N} (e^{Nt_1} - 1) M M_2 \|x - y\|,
\end{aligned}
$$

which gives

$$
\|\mathcal{P}_1 x - \mathcal{P}_1 y\| \leq L \|x - y\|, \quad L := \frac{M M_2}{N} (e^{Nt_1} - 1) \|C\|.
$$

In view of (2.68), we conclude that $L < 1$, which implies $\mathcal{P}_1$ is a contraction.

Step 3. We show that $\mathcal{P}_2$ is a compact and continuous operator.

Let $x_n \in \mathcal{B}_r$ with $x_n \to x$ in $\mathcal{B}_r$. Denote $F_n(\cdot) = f(\cdot, x_n(\cdot))$ and $F(\cdot) = f(\cdot, x(\cdot))$. Using $[H_2]$, we have $F_n \to F$ in C and thus

$$
\|(\mathcal{P}_2 x_n)t - (\mathcal{P}_2 x)(t)\| \leq \int_0^t e^{N(t-s)} \|F_n(s) - F(s)\| ds \to 0 \quad \text{as } n \to \infty
$$

uniformly for $t \in J$, which implies that $\mathcal{P}_2$ is continuous on $\mathcal{B}_r$.

To check the compactness of $\mathcal{P}_2$, we prove that $\mathcal{P}_2(\mathcal{B}_r) \subset C$ is equicontinuous and bounded.

In fact, for any $x \in \mathcal{B}_r$, $t_1 \geq t + h \geq t > 0$, we have

$$
\begin{aligned}
(\mathcal{P}_2 x)(t + h) &- (\mathcal{P}_2 x)(t) \\
&= \int_0^{t+h} e^{A(t+h-s)} e_\tau^{B_1(t+h-\tau-s)} F(s) ds - \int_0^t e^{A(t-s)} e_\tau^{B_1(t-\tau-s)} F(s) ds \\
&= I_1 + I_2 + I_3,
\end{aligned}
$$

where

$$
I_1 = \int_t^{t+h} e^{A(t+h-s)} e_\tau^{B_1(t+h-\tau-s)} F(s) ds,
$$

$$I_2 = \int_0^t e^{A(t+h-s)} \left[e_\tau^{B_1(t+h-\tau-s)} - e_\tau^{B_1(t-\tau-s)} \right] F(s)\,ds,$$

$$I_3 = \int_0^t \left[e^{A(t+h-s)} - e^{A(t-s)} \right] e_\tau^{B_1(t-\tau-s)} F(s)\,ds.$$

From the above, we derive

$$\|(\mathcal{P}_2 x)(t+h) - (\mathcal{P}_2 x)(t)\| \leq \|I_1\| + \|I_2\| + \|I_3\|.$$

Next, we check $\|I_i\| \to 0$ as $h \to 0$, $i = 1, 2, 3$, uniformly for t.
For I_1, using $[H_2]$,

$$\|I_1\| \leq \int_t^{t+h} e^{N(t+h-s)} \|F(s)\|\,ds$$

$$\leq \int_t^{t+h} e^{N(t+h-s)} (\|f(s, x(s)) - f(s, 0)\| + \|f(s, 0)\|)\,ds$$

$$\leq \int_t^{t+h} e^{N(t+h-s)} L_f(s)\|x(s)\|\,ds + \int_t^{t+h} e^{N(t+h-s)} \|f(s, 0)\|\,ds$$

$$\leq \|x\|_C \int_t^{t+h} e^{N(t+h-s)} L_f(s)\,ds + R_f \int_t^{t+h} e^{N(t+h-s)}\,ds$$

$$\leq \left[\frac{1}{Np}(e^{pNh} - 1) \right]^{\frac{1}{p}} \|L_f\|_{L^q(J,\mathbb{R}^+)} r + \frac{R_f}{N}(e^{Nh} - 1) \to 0 \text{ as } h \to 0.$$

One can apply $[H_2]$ to derive

$$\|I_2\| \leq \int_0^t e^{\|A\|(t+h-s)} \|e_\tau^{B_1(t+h-\tau-s)} - e_\tau^{B_1(t-\tau-s)}\| \|F(s)\|\,ds$$

$$\leq e^{\|A\|t_1} \int_0^t \|e_\tau^{B_1(t+h-\tau-s)} - e_\tau^{B_1(t-\tau-s)}\| L_f(s)\,ds\,\|x\|$$

$$+ e^{\|A\|t_1} \int_0^t \|e_\tau^{B_1(t+h-\tau-s)} - e_\tau^{B_1(t-\tau-s)}\| \|f(s, 0)\|\,ds$$

$$\leq e^{\|A\|t_1} \|L_f\|_{L^q(J,\mathbb{R}^+)} r \left(\int_0^t \|e_\tau^{B_1(t+h-\tau-s)} - e_\tau^{B_1(t-\tau-s)}\|^p\,ds \right)^{\frac{1}{p}}$$

$$+ e^{\|A\|t_1} R_f \int_0^t \|e_\tau^{B_1(t+h-\tau-s)} - e_\tau^{B_1(t-\tau-s)}\|\,ds$$

$$\leq e^{\|A\|t_1} \|L_f\|_{L^q(J,\mathbb{R}^+)} r \left(\int_{-\tau}^{t_1-\tau} \|e_\tau^{B_1(s+h)} - e_\tau^{B_1 s}\|^p\,ds \right)^{\frac{1}{p}}$$

$$+ e^{\|A\|t_1} R_f \int_{-\tau}^{t_1-\tau} \|e_\tau^{B_1(s+h)} - e_\tau^{B_1 s}\|\,ds \to 0 \text{ as } h \to 0,$$

by using Lebesgue's dominated convergence theorem.

For I_3, it is easy to get

$$\|I_3\| \le \int_0^t \|e^{A(t+h-s)} - e^{A(t-s)}\| e^{\|B_1\|(t-s)} \|F(s)\| ds$$

$$\le \int_0^t \|e^{A(t+h-s)} - e^{A(t-s)}\| e^{\|B_1\|(t-s)} (L_f(s)\|x\| + \|f(s,0)\|) ds$$

$$\le \int_0^t \|e^{Ah} - I\| e^{N(t-s)} (L_f(s)r + R_f) ds$$

$$\le \|e^{Ah} - I\| \left[r \int_0^t e^{N(t-s)} L_f(s) ds + R_f \int_0^t e^{N(t-s)} ds \right]$$

$$\le \|e^{Ah} - I\| \left[\left(\frac{1}{Np}(e^{Npt} - 1) \right)^{\frac{1}{p}} \|L_f\|_{L^q(J,\mathbb{R}^+)} r + \frac{R_f}{N}(e^{Nt} - 1) \right] \to 0,$$

as $h \to 0$.

From the above, we immediately obtain

$$\|(\mathcal{P}_2 x)(t+h) - (\mathcal{P}_2 x)(t)\| \to 0, \ h \to 0,$$

uniformly for all t and $x \in \mathcal{B}_r$. Therefore, $\mathcal{P}_2(\mathcal{B}_r) \subset \mathcal{C}$ is equicontinuous. Next, repeating the above computations, we have

$$\|(\mathcal{P}_2 x)(t)\| \le \int_0^t e^{N(t-s)} (L_f(s)r + R_f) ds$$

$$\le \left(\frac{1}{Np}(e^{Npt_1} - 1) \right)^{\frac{1}{p}} \|L_f\|_{L^q(J,\mathbb{R}^+)} r + \frac{R_f}{N}(e^{Nt_1} - 1).$$

Hence $\mathcal{P}(\mathcal{B}_r)$ is bounded. By the Arzela–Ascoli theorem, $\mathcal{P}_2(\mathcal{B}_r) \subset \mathcal{C}$ is relatively compact in $\mathcal{C}$.

Thus, $\mathcal{P}_2$ is a compact operator. Hence, using Krasnoselskii's fixed point theorem (see [82]), $\mathcal{P}$ has a fixed point x on $\mathcal{B}_r$. Obviously, x is a solution of system (2.52) satisfying $x(t_1) = x_1$. The boundary condition $x(t) = \varphi(t)$, $-\tau \le t \le 0$, holds by Lemma 2.6. The proof is completed. $\qquad\square$

2.2.1.3 Numerical examples and discussion

Firstly, we give some reasonable explanations for system (2.52) from the point of view of practical application. We also note that relative controllability for linear systems of neutral differential equations with a delay is studied in [85] and an example of competing biological systems is used to illustrate the theoretical results.

For instance, in a combustion chamber or reactor, we consider the change process of pressure. Based on our knowledge, the movement process of pressure depends not only on the current state value, but also on the past state value and control. From this point of view, system (2.52) has a practical significance, and $f(t, x(t))$ of system (2.52) can be considered as an interference which depends on the state. According to the actual needs, people often hope that the pressure of the combustion chamber cannot exceed the expected threshold and it can be controlled. Therefore, we study the controllability of system (2.52), (see also [85, Section 4]).

Delay differential equations have been widely used to model the population growth of certain species [86] in the past decades. Now, we present a model of population dynamics with delayed birthrates and delayed logistic terms given by the system

$$
\begin{cases}
\dot{z}_1(t) = z_1(t)\big[-a_1 - b_1 z_2(t) - c(z_1(t-\tau) + z_2(t-\tau)) \big] \\
\qquad + d_1 z_1(t-\tau) + \varphi_1(t), \\
\dot{z}_2(t) = z_2(t)\big[-a_2 + b_2 z_1(t) - c(z_1(t-\tau) + z_2(t-\tau)) \big] \\
\qquad + d_2 z_2(t-\tau) + \varphi_2(t),
\end{cases}
\tag{2.69}
$$

where $t > 0$, $\tau > 0$, $a_2 > a_1 > 0$, $b_1, b_2, d_1, d_2 > 0$, and φ_1, φ_2 are given functions. We put

$$
Z(t) = \begin{pmatrix} z_1(t) \\ z_2(t) \end{pmatrix}, \quad
A = \begin{bmatrix} -a_1 & 0 \\ 0 & -a_2 \end{bmatrix}, \quad
B = \begin{bmatrix} d_1 & 0 \\ 0 & d_2 \end{bmatrix},
$$

and $F : \mathbb{R}^2 \times \mathbb{R}^2 \to \mathbb{R}^2$ is defined by

$$
F(t, Z(t), Z(t-\tau))
$$
$$
= \begin{pmatrix} -b_1 z_1(t) z_2(t) - c z_1(t) z_1(t-\tau) - c z_1(t) z_2(t-\tau) + \varphi_1(t) \\ b_2 z_1(t) z_2(t) - c z_2(t) z_1(t-\tau) - c z_2(t) z_2(t-\tau) + \varphi_2(t) \end{pmatrix}.
$$

Then, (2.69) can be turned into $\dot{Z}(t) = AZ(t) + BZ(t-\tau) + F(t, Z(t), Z(t-\tau))$. This may be an explanation from the practical point of view for studying systems with permutable matrices. Concerning the controllability of such systems, we can understand that people find a strategy to attain a target growth of populations by adjusting one of them.

Secondly, we give numerical examples to demonstrate the validity of our method and provide some discussion. We use in this section the Euclidean norm.

Example 2.3. Set $t_1 = 1$, $\tau = 0.2$, $J_1 := [0, 1]$, and $u \in L^2(J_1, \mathbb{R}^2)$. Consider the following semilinear delay differential controlled system:

$$
\begin{cases}
\dot{x}(t) = Ax(t) + Bx(t-0.2) + f(t, x(t)) + Cu(t), \; x(t) \in \mathbb{R}^2, \; t \in J_1, \\
\varphi(t) = (0.4, 0.3)^\top, \; -0.2 \le t \le 0.
\end{cases}
$$
$$
\tag{2.70}
$$

For the sake of simplicity, we set

$$A = \begin{bmatrix} 0.3 & 0 \\ 0 & 0.3 \end{bmatrix}, \quad B = \begin{bmatrix} 0.2 & 0 \\ 0 & 0.2 \end{bmatrix}, \quad f(t, x(t)) = \begin{pmatrix} \frac{1}{10}(t + 0.1)x_1(t) \\ \frac{1}{10}(t + 0.1)x_2(t) \end{pmatrix}, \quad C = I.$$

Obviously, $AB = BA = \begin{bmatrix} 0.06 & 0 \\ 0 & 0.06 \end{bmatrix}$ since A, B are diagonal matrices. By elementary calculation, we have

$$B_1 = e^{-A\tau}B = \begin{bmatrix} 0.1884 & 0 \\ 0 & 0.1884 \end{bmatrix}, \quad N = \|A\| + \|B_1\| = 0.4884, \quad (2.71)$$

and

$$e^{A(1-s)} = \begin{bmatrix} e^{0.3(1-s)} & 0 \\ 0 & e^{0.3(1-s)} \end{bmatrix}, \quad e^{A(t_1+\tau)} = \begin{bmatrix} 1.4333 & 0 \\ 0 & 1.4333 \end{bmatrix},$$

$$e_\tau^{B_1 t_1} = \begin{bmatrix} 1.2 & 0 \\ 0 & 1.2 \end{bmatrix}.$$

Now we use (2.66) to estimate M. For this purpose, we need to obtain $W_\tau[0, t_1]$ and then derive $W_\tau[0, t_1]^{-1}$. Obviously, $A = A^\top$, $B = B^\top$, $B_1 = B_1^\top$, and $C = C^\top = I$. Hence, the delay Grammian matrix (2.56) has the following explicit form:

$$\begin{aligned}
W_\tau[0, t_1] &= \int_0^{t_1} e^{A(t_1-s)} e_\tau^{B_1(t_1-\tau-s)} CC e_\tau^{B_1(t_1-\tau-s)} e^{A(t_1-s)} ds \\
&= \int_0^{t_1} e^{A(t_1-s)} e_\tau^{B_1(t_1-\tau-s)} e_\tau^{B_1(t_1-\tau-s)} e^{A(t_1-s)} ds \\
&= \int_0^1 e^{A(1-s)} e_{0.2}^{B_1(0.8-s)} e_{0.2}^{B_1(0.8-s)} e^{A(1-s)} ds \\
&= W_1 + W_2 + W_3 + W_4 + W_5,
\end{aligned}$$

where

$$\begin{aligned}
W_1 = \int_0^{0.2} e^{A(1-s)} &\left[I + B_1(0.8 - s) + B_1^2 \frac{(0.6-s)^2}{2!} \right. \\
&\left. + B_1^3 \frac{(0.4-s)^3}{3!} + B_1^4 \frac{(0.2-s)^4}{4!} \right]^2 e^{A(1-s)} ds,
\end{aligned}$$

$$\begin{aligned}
W_2 = \int_{0.2}^{0.4} e^{A(1-s)} &\left[I + B_1(0.8-s) + B_1^2 \frac{(0.6-s)^2}{2!} + B_1^3 \frac{(0.4-s)^3}{3!} \right]^2 \\
&\times e^{A(1-s)} ds,
\end{aligned}$$

$$W_3 = \int_{0.4}^{0.6} e^{A(1-s)} \left[I + B_1(0.8 - s) + B_1^2 \frac{(0.6 - s)^2}{2!} \right]^2 e^{A(1-s)} ds,$$

$$W_4 = \int_{0.6}^{0.8} e^{A(1-s)} \left[I + B_1(0.8 - s) \right]^2 e^{A(1-s)} ds,$$

$$W_5 = \int_{0.8}^{1} e^{A(1-s)} I^2 e^{A(1-s)} ds.$$

By computation, one can get

$$W_1 = \begin{bmatrix} 0.4438 & 0 \\ 0 & 0.4438 \end{bmatrix}, \quad W_2 = \begin{bmatrix} 0.8099 & 0 \\ 0 & 0.8099 \end{bmatrix},$$

$$W_3 = \begin{bmatrix} 1.1118 & 0 \\ 0 & 1.1118 \end{bmatrix}, \quad W_4 = \begin{bmatrix} 1.3559 & 0 \\ 0 & 1.3559 \end{bmatrix},$$

$$W_5 = \begin{bmatrix} 1.3702 & 0 \\ 0 & 1.3702 \end{bmatrix}.$$

Therefore, we obtain

$$W_\tau[0, 1] = \begin{bmatrix} 5.0916 & 0 \\ 0 & 5.0916 \end{bmatrix},$$

and we further derive

$$W_\tau[0, 1]^{-1} = \begin{bmatrix} 0.1964 & 0 \\ 0 & 0.1964 \end{bmatrix}. \tag{2.72}$$

Consequently, we obtain

$$M = \sqrt{\|W_\tau[0, 1]^{-1}\|} = \sqrt{0.1964} = 0.4432.$$

Hence, W satisfies assumption $[H_1]$.

Further, it is easy to see that for any $x(t), y(t) \in \mathbb{R}^2$ and $t \in J_1$,

$$\|f(t, x) - f(t, y)\| = \frac{1}{10}(t + 0.1)\sqrt{(x_1(t) - y_1(t))^2 + (x_2(t) - y_2(t))^2}$$

$$\leq \frac{1}{10}(t + 0.1)\|x - y\|.$$

Hence, f satisfies assumption $[H_2]$, where we set $L_f(\cdot) = \frac{\cdot + 0.1}{10} \in L^q(J_1, \mathbb{R}^+)$.
Obviously, $\|L_f\|_{L^q(J_1, \mathbb{R}^+)} = \frac{1}{10}(\frac{1.1^{q+1} - 0.1^{q+1}}{q+1})^{\frac{1}{q}}$ and $R_f = \sup_{t \in J_1} \|f(t, 0)\| = $
0. Next, $\|C\| = 1$, $\|L_f\|_{L^2(J_1, \mathbb{R}^+)} = 0.0661$, and

$M_2 = [\frac{1}{2N}(e^{2N} - 1)]^{\frac{1}{2}} \|L_f\|_{L^2(J_1, \mathbb{R}^+)} = 0.0942$ when we choose $p = q = 2$. Hence,

$$
\gamma = M_2 \left[1 + \frac{M}{N}(e^{Nt_1} - 1)\|C\| \right] = 0.0942 \left[1 + \frac{0.4432}{0.4884}(e^{0.4884} - 1) \right]
$$
$$
= 0.1480 < 1,
$$

which guarantees that condition (2.68) holds.

Thus all conditions of Theorem 2.13 are satisfied. Hence, system (2.70) is relatively controllable on [0, 1].

Example 2.4. Consider the relative controllability of system (2.70) (with $f(\cdot, x) \equiv 0$) on J_1, where A, B, B_1, C are defined in Example 2.3.

According to Theorem 2.12, we can know that system (2.70) is relatively controllable when $f(\cdot, x) = 0$. Further, keeping in mind (2.58), one can get

$$
\eta = x_1 - \begin{bmatrix} -0.6486 \\ -0.4865 \end{bmatrix}, \quad x_1 \in \mathbb{R}^2.
$$

By using the selection form of the control in (2.57), we arrive at

$$
u(t) = C^\top e_\tau^{B_1^\top (t_1 - \tau - t)} e^{A^\top (t_1 - t)} W_\tau[0, t_1]^{-1} \eta
$$
$$
= e_{0.2}^{B_1(0.8 - t)} e^{A(1 - t)} W_\tau[0, 1]^{-1} \eta
$$

$$
= \begin{cases}
\left[I + B_1(0.8 - t) + B_1^2 \dfrac{(0.6 - t)^2}{2!} + B_1^3 \dfrac{(0.4 - t)^3}{3!} + B_1^4 \dfrac{(0.2 - t)^4}{4!} \right] \\
\quad \times e^{A(1-t)} W_\tau[0, 1]^{-1}\eta, \ 0 \leq t < 0.2, \\[4pt]
\left[I + B_1(0.8 - t) + B_1^2 \dfrac{(0.6 - t)^2}{2!} + B_1^3 \dfrac{(0.4 - t)^3}{3!} \right] \\
\quad \times e^{A(1-t)} W_\tau[0, 1]^{-1}\eta, \ 0.2 \leq t < 0.4, \\[4pt]
\left[I + B_1(0.8 - t) + B_1^2 \dfrac{(0.6 - t)^2}{2!} \right] \\
\quad \times e^{A(1-t)} W_\tau[0, 1]^{-1}\eta, \ 0.4 \leq t < 0.6, \\[4pt]
\left[I + B_1(0.8 - t) \right] \times e^{A(1-t)} W_\tau[0, 1]^{-1}\eta, \ 0.6 \leq t < 0.8, \\[4pt]
I \times e^{A(1-t)} W_\tau[0, 1]^{-1}\eta, \ 0.8 \leq t < 1,
\end{cases}
$$

where B_1 is given in (2.71) and $W_\tau[0, 1]^{-1}$ is given in (2.72).

2.2.1.4   Conclusions

The purpose of this contribution is to develop a controllability method for linear and semilinear delay controlled systems with linear parts defined by permutable matrices. In order to achieve this purpose, a representation of solutions is used with the help of a delayed matrix exponential. Such an approach leads to new criteria for the relative controllability of our issues by constructing a delay Grammian matrix and applying the fixed point method. The results in this part are motivated from [87].

2.3   Iterative learning control

2.3.1   Iterative learning control for delay differential systems

In this section, we discuss ILC for time-delay systems via initial state learning. More precisely, we study the following linear controlled systems with pure delay:

$$\begin{cases} \dot{x}_k(t) = Ax_k(t) + Bx_k(t - \tau) + u_k(t), \ t \in [0, T], \\ x_k(t) = \varphi(t), \ -\tau \le t \le 0, \ \tau > 0, \\ y_k(t) = Cx_k(t) + Du_k(t). \end{cases} \tag{2.73}$$

Let $\varphi \in C_\tau^1 := C^1([-\tau, 0], \mathbb{R}^n)$, let A and B be two $n \times n$ matrices such that $AB = BA$, let C and D be two $m \times n$ matrices, let k denote the k-th learning iteration, and let the variables x_k, $u_k \in \mathbb{R}^n$, and $y_k \in \mathbb{R}^m$ denote the state, input, and output, respectively. By [1, Corollary 2.2], we derive that the state $x_k(\cdot)$ has the form

$$x_k(t) = e^{A(t+\tau)}e_\tau^{B_1 t}\varphi(-\tau) + \int_{-\tau}^0 e^{A(t-\tau-s)}e_\tau^{B_1(t-\tau-s)}e^{A\tau}[\varphi'(s) - A\varphi(s)]ds$$

$$+ \int_0^t e^{A(t-\tau-s)}e_\tau^{B_1(t-\tau-s)}e^{A\tau}u_k(s)ds, \tag{2.74}$$

where e_τ^{Bt} is defined in (2.4) and $B_1 = e^{-A\tau}B$.

Let y_d be a desired trajectory and set $e_k = y_d - y_k$, which denotes the output error, and $\delta u_k = u_{k+1} - u_k$.

For system (2.73), we consider the open-loop P-type ILC updating law

$$u_{k+1}(t) = u_k(t) + P_o e_k(t). \tag{2.75}$$

For system (2.73) with $D = \Theta$, we consider the open-loop D-type ILC updating law

$$u_{k+1}(t) = u_k(t) + D_o \dot{e}_k(t), \tag{2.76}$$

where P_o and $D_o \in \mathbb{R}^{n \times m}$ are learning gain matrices.

The main objective is to use a delayed exponential matrix to generate the control input u_k. Then the time-delay system output y_k is tracking the iteratively varying reference trajectories y_d by adopting P-type ILC and D-type ILC as $k \to \infty$ uniformly on $[0, T]$ in the λ-norm sense.

Here we point out that our method is different from the method given in the previous reference. However, we obtain the same convergence results. Our method relies on a direct formula solution, so it is constructive.

Lemma 2.7. *(see [88, Chapter 5.6]) For a matrix $A \in \mathbb{R}^{n \times n}$ and $\forall \epsilon > 0$, there is a norm $\| \cdot \|$ on $\mathbb{R}^{n \times n}$ such that*

$$\|A\| \leq \rho(A) + \epsilon,$$

where $\rho(A)$ denotes the spectral radius of matrix A.

Next, we give an alternative formula to compute the solution of the linear system with pure delay, which is a direct corollary of [1, Corollary 2.2].

Lemma 2.8. *Let $f : J \to \mathbb{R}^n$ be a continuous function. The solution $x \in \mathcal{C}^1(J, \mathbb{R}^n)$ of*

$$\begin{cases} \dot{x}(t) = Ax(t) + Bx(t - \tau) + f(t), \ t > 0, \ \tau \geq 0, \\ x(t) = \varphi(t), \ -\tau \leq t \leq 0, \end{cases} \tag{2.77}$$

has the form

$$x(t) = e^{A(t+\tau)} e_\tau^{B_1 t} \varphi(-\tau)$$

$$+ \int_{-\tau}^{0} e^{A(t-\tau-s)} \sum_{l=0}^{j-1} B_1^l \frac{(t - l\tau - s)^l}{l!} e^{A\tau} [\varphi'(s) - A\varphi(s)] ds$$

$$+ \int_{-\tau}^{t-j\tau} e^{A(t-\tau-s)} B_1^j \frac{(t - j\tau - s)^j}{j!} e^{A\tau} [\varphi'(s) - A\varphi(s)] ds$$

$$+ \sum_{i=0}^{j-1} \int_{0}^{t-i\tau} e^{A(t-\tau-s)} B_1^i \frac{(t - i\tau - s)^i}{i!} e^{A\tau} f(s) ds, \tag{2.78}$$

where $B_1 = e^{-A\tau} B$ and $(j - 1)\tau \leq t < j\tau$, $j = 1, 2, \cdots n$.

Proof. According to [1, (21)], we know the solution of system (2.77) has the form

$$x(t) = e^{A(t+\tau)} e_\tau^{B_1 t} \varphi(-\tau)$$

$$+ \int_{-\tau}^{0} e^{A(t-\tau-s)} e_\tau^{B_1(t-\tau-s)} e^{A\tau} [\varphi'(s) - A\varphi(s)] ds$$

$$+ \int_{0}^{t} e^{A(t-\tau-s)} e_\tau^{B_1(t-\tau-s)} e^{A\tau} f(s) ds. \tag{2.79}$$

Without loss of generality, we consider $(j-1)\tau \le t < j\tau$, $j = 1, 2, \cdots n$.

Next, we combine the formula of the delayed matrix exponential (2.4) with (2.79) to prove the result. We divide our proof into two steps.

Step 1. We prove that

$$\int_{-\tau}^{0} e^{A(t-\tau-s)} e_{\tau}^{B_1(t-\tau-s)} e^{A\tau} [\varphi'(s) - A\varphi(s)] ds$$

$$= \int_{-\tau}^{0} e^{A(t-\tau-s)} \sum_{l=0}^{j-1} B_1^l \frac{(t-l\tau-s)^l}{l!} e^{A\tau} [\varphi'(s) - A\varphi(s)] ds$$

$$+ \int_{-\tau}^{t-j\tau} e^{A(t-\tau-s)} B_1^j \frac{(t-j\tau-s)^j}{j!} e^{A\tau} [\varphi'(s) - A\varphi(s)] ds. \qquad (2.80)$$

Due to the fact that $-\tau < s < 0$, we obtain $t - \tau < t - \tau - s < t$ and $(j-2)\tau < t - \tau < t - \tau - s < t < j\tau$. When $-\tau < s < t - j\tau$, we have $(j-1)\tau < t - \tau - s < t < j\tau$. When $t - j\tau < s < 0$, we have $(j-2)\tau < t - \tau < t - \tau - s < (j-1)\tau$. Hence

$$\int_{-\tau}^{0} e^{A(t-\tau-s)} e_{\tau}^{B_1(t-\tau-s)} e^{A\tau} [\varphi'(s) - A\varphi(s)] ds$$

$$= \int_{-\tau}^{t-j\tau} e^{A(t-\tau-s)} \sum_{l=0}^{j} B_1^l \frac{(t-l\tau-s)^l}{l!} e^{A\tau} [\varphi'(s) - A\varphi(s)] ds$$

$$+ \int_{t-j\tau}^{0} e^{A(t-\tau-s)} \sum_{l=0}^{j-1} B_1^l \frac{(t-l\tau-s)^l}{l!} e^{A\tau} [\varphi'(s) - A\varphi(s)] ds$$

$$= \int_{-\tau}^{0} e^{A(t-\tau-s)} \sum_{l=0}^{j-1} B_1^l \frac{(t-l\tau-s)^l}{l!} e^{A\tau} [\varphi'(s) - A\varphi(s)] ds$$

$$+ \int_{-\tau}^{t-j\tau} e^{A(t-\tau-s)} B_1^j \frac{(t-j\tau-s)^j}{j!} e^{A\tau} [\varphi'(s) - A\varphi(s)] ds.$$

Step 2. We check that

$$\int_{0}^{t} e^{A(t-\tau-s)} e_{\tau}^{B_1(t-\tau-s)} e^{A\tau} f(s) ds$$

$$= \sum_{i=0}^{j-1} \int_{0}^{t-i\tau} e^{A(t-\tau-s)} B_1^i \frac{(t-i\tau-s)^i}{i!} e^{A\tau} f(s) ds. \qquad (2.81)$$

Due to the fact that $0 < s < t$, we obtain $-\tau < t - \tau - s < t - \tau$. When $t - (i+1)\tau < s < t - i\tau$, we have $(i-1)\tau < t - \tau - s < i\tau$, $i = 0, 1, \cdots, j-2$. When

$0 < s < t - (j-1)\tau$, we have $(j-2)\tau < t - \tau - s < t - \tau < (j-1)\tau$. Hence

$$\int_0^t e^{A(t-\tau-s)} e_\tau^{B_1(t-\tau-s)} e^{A\tau} f(s) ds$$

$$= \sum_{i=0}^{j-2} \int_{t-(i+1)\tau}^{t-i\tau} e^{A(t-\tau-s)} e_\tau^{B_1(t-\tau-s)} e^{A\tau} f(s) ds$$

$$+ \int_0^{t-(j-1)\tau} e^{A(t-\tau-s)} e_\tau^{B_1(t-\tau-s)} e^{A\tau} f(s) ds$$

$$= \sum_{i=0}^{j-2} \int_{t-(i+1)\tau}^{t-i\tau} e^{A(t-\tau-s)} \sum_{l=0}^{i} B_1^l \frac{(t-l\tau-s)^l}{l!} e^{A\tau} f(s) ds$$

$$+ \int_0^{t-(j-1)\tau} e^{A(t-\tau-s)} \sum_{l=0}^{j-1} B_1^l \frac{(t-l\tau-s)^l}{l!} e^{A\tau} f(s) ds$$

$$= \sum_{i=0}^{j-2} \sum_{l=0}^{i} \int_{t-(i+1)\tau}^{t-i\tau} e^{A(t-\tau-s)} B_1^l \frac{(t-l\tau-s)^l}{l!} e^{A\tau} f(s) ds$$

$$+ \sum_{l=0}^{j-1} \int_0^{t-(j-1)\tau} e^{A(t-\tau-s)} B_1^l \frac{(t-l\tau-s)^l}{l!} e^{A\tau} f(s) ds$$

$$= \sum_{l=0}^{j-2} \int_{t-(j-1)\tau}^{t-l\tau} e^{A(t-\tau-s)} B_1^l \frac{(t-l\tau-s)^l}{l!} e^{A\tau} f(s) ds$$

$$+ \sum_{l=0}^{j-1} \int_0^{t-(j-1)\tau} e^{A(t-\tau-s)} B_1^l \frac{(t-l\tau-s)^l}{l!} e^{A\tau} f(s) ds$$

$$= \sum_{i=0}^{j-2} \int_0^{t-i\tau} e^{A(t-\tau-s)} B_1^i \frac{(t-i\tau-s)^i}{i!} e^{A\tau} f(s) ds$$

$$+ \int_0^{t-(j-1)\tau} e^{A(t-\tau-s)} B_1^{j-1} \frac{(t-(j-1)\tau-s)^{j-1}}{(j-1)!} e^{A\tau} f(s) ds$$

$$= \sum_{i=0}^{j-1} \int_0^{t-i\tau} e^{A(t-\tau-s)} B_1^i \frac{(t-i\tau-s)^i}{i!} e^{A\tau} f(s) ds.$$

Linking (2.79), (2.80), and (2.81), one can get the result (2.78). The proof is finished. $\qquad\square$

2.3.1.1 Convergence analysis of P-type

In this section, we give the first convergence result of P-type.

Theorem 2.14. *Let $y_d(t)$, $t \in [0, T]$, be a desired trajectory for system (2.73). If $\rho(I - DP_o) < 1$, then the P-type ILC law (2.75) guarantees $\lim_{k \to \infty} y_k(t) = y_d(t)$ uniformly on $[0, T]$.*

Proof. Without loss of generality, we consider $(j - 1)\tau \le t < j\tau$, $j = 1, 2, \cdots, n$. Linking (2.73) and (2.74), we have

$$e_{k+1}(t) - e_k(t) = y_k(t) - y_{k+1}(t)$$

$$= -C \int_0^t e^{A(t-\tau-s)} e_\tau^{B_1(t-\tau-s)} e^{A\tau} \delta u_k(s) ds - D\delta u_k(t).$$

According to (2.75) and Lemma 2.8, we have

$$e_{k+1}(t) = (I - DP_o)e_k(t)$$

$$- C \sum_{i=0}^{j-1} \int_0^{t-i\tau} e^{A(t-\tau-s)} B_1^i \frac{(t - i\tau - s)^i}{i!} e^{A\tau} \delta u_k(s) ds. \qquad (2.82)$$

Further, by Lemma 2.7 we know that for a given $\epsilon > 0$, there is a norm $\| \cdot \|$ on $\mathbb{R}^n$ such that

$$\|I - DP_o\| \le \rho(I - DP_o) + \epsilon.$$

Using (2.82), we have

$$\|e_{k+1}(t)\| \le \left[\rho(I - DP_o) + \epsilon \right] \|e_k(t)\|$$

$$+ \|C\| \sum_{i=0}^{j-1} \int_0^{t-i\tau} \|e^{A(t-\tau-s)}\| \|B_1^i\| \frac{(t - i\tau - s)^i}{i!} \|e^{A\tau}\| \|\delta u_k(s)\| ds.$$

Hence, we obtain

$$\|e_{k+1}(t)\| \le \left[\rho(I - DP_o) + \epsilon \right] \|e_k(t)\|$$

$$+ \|C\| \sum_{i=0}^{j-1} \int_0^{t-i\tau} e^{\|A\|(t-s)} \|B_1\|^i \frac{(t - i\tau - s)^i}{i!} \|\delta u_k(s)\| ds. \qquad (2.83)$$

For fixed i, $i = 0, 1, \cdots, j - 1$, we have

$$\int_0^{t-i\tau} e^{\|A\|(t-s)} \|B_1\|^i \frac{(t - i\tau - s)^i}{i!} \|\delta u_k(s)\| ds$$

$$= \frac{\|B_1\|^i}{i!} \int_0^{t-i\tau} e^{\|A\|(t-s)} (t - i\tau - s)^i \|\delta u_k(s)\| ds$$

$$\leq \frac{\|B_1\|^i}{i!} e^{\|A\|t} \int_0^{t-i\tau} e^{(\lambda-\|A\|)s}(t-i\tau-s)^i ds \|\delta u_k\|_\lambda. \tag{2.84}$$

For any $\lambda > \|A\|$, we apply integration by parts via mathematical induction to derive

$$\int_0^{t-i\tau} e^{(\lambda-\|A\|)s}(t-i\tau-s)^i ds = \frac{1}{\lambda-\|A\|} \int_0^{t-i\tau} (t-i\tau-s)^i de^{(\lambda-\|A\|)s}$$

$$= -\frac{1}{\lambda-\|A\|}(t-i\tau)^i + \frac{i}{(\lambda-\|A\|)^2} \int_0^{t-i\tau} (t-i\tau-s)^{i-1} de^{(\lambda-\|A\|)s}$$

$$= -\frac{1}{\lambda-\|A\|}(t-i\tau)^i - \frac{i}{(\lambda-\|A\|)^2}(t-i\tau)^{i-1}$$

$$+ \frac{i(i-1)}{(\lambda-\|A\|)^3} \int_0^{t-i\tau} (t-i\tau-s)^{i-2} de^{(\lambda-\|A\|)s}$$

$$\vdots$$

$$= -\frac{1}{\lambda-\|A\|}(t-i\tau)^i - \frac{i}{(\lambda-\|A\|)^2}(t-i\tau)^{i-1} - \cdots$$

$$- \frac{i!}{(\lambda-\|A\|)^i}(t-i\tau) + \frac{i!}{(\lambda-\|A\|)^{i+1}}\left(e^{(\lambda-\|A\|)(t-i\tau)} - 1\right)$$

$$= -\sum_{p=0}^{i} \frac{i!(t-i\tau)^{i-p}}{(i-p)!(\lambda-\|A\|)^{p+1}} + \frac{i!e^{(\lambda-\|A\|)(t-i\tau)}}{(\lambda-\|A\|)^{i+1}}. \tag{2.85}$$

Linking (2.83), (2.84), and (2.85), it is not difficult to get

$$\|e_{k+1}(t)\| \leq \left[\rho(I-DP_o)+\epsilon\right]\|e_k(t)\| + \|C\| \sum_{i=0}^{j-1} \frac{\|B_1\|^i}{i!} e^{\|A\|t}$$

$$\times \left[-\sum_{p=0}^{i} \frac{i!(t-i\tau)^{i-p}}{(i-p)!(\lambda-\|A\|)^{p+1}} + \frac{i!e^{(\lambda-\|A\|)(t-i\tau)}}{(\lambda-\|A\|)^{i+1}}\right]\|\delta u_k\|_\lambda. \tag{2.86}$$

By (2.86) and (2.75), we have

$$\|e_{k+1}(t)\|e^{-\lambda t} \leq \left[\rho(I-DP_o)+\epsilon\right]\|e_k(t)\|e^{-\lambda t} + \|C\|\|P_o\| \sum_{i=0}^{j-1} \frac{\|B_1\|^i}{i!} e^{(\|A\|-\lambda)t}$$

$$\times \left[-\sum_{p=0}^{i} \frac{i!(t-i\tau)^{i-p}}{(i-p)!(\lambda-\|A\|)^{p+1}} + \frac{i!e^{(\lambda-\|A\|)(t-i\tau)}}{(\lambda-\|A\|)^{i+1}}\right]\|e_k\|_\lambda,$$

and taking the λ-norm, we arrive at

$$\|e_{k+1}\|_\lambda \leq \left[\rho(I - DP_o) + \epsilon\right]\|e_k\|_\lambda + \|C\|\|P_o\| \sum_{i=0}^{j-1} \frac{\|B_1\|^i}{i!}$$

$$\times \left[\sum_{p=0}^{i} \frac{i!\max\{1, T^p\}}{(i-p)!(\lambda - \|A\|)^{p+1}} + \frac{i!}{(\lambda - \|A\|)^{i+1}}\right]\|e_k\|_\lambda.$$

Since $\rho(I - DP_o) < 1$, for any $\epsilon \in \left(0, \frac{1-\rho(I-DP_o)}{4}\right)$ and $\lambda > \|A\|$ sufficiently large, we have

$$\|e_{k+1}\|_\lambda \leq (\rho(I - DP_o) + 2\epsilon)\|e_k\|_\lambda,$$

which implies $\lim_{k\to\infty} \|e_k\|_\lambda = 0$, since $\rho(I - DP_o) + 2\epsilon < 1$. In addition, $\|e_k\| \leq e^{\lambda T}\|e_k\|_\lambda$. Hence, $\lim_{k\to\infty} \|e_k\| = 0$. The proof is completed. $\qquad\square$

Remark 2.5. We use the delayed matrix exponential method to obtain convergence of the P-type ILC algorithm. Next, we applied the norm $\|\cdot\|_\lambda$ just for a technical reason to get uniform convergence only under condition $\rho(I - DP_o) < 1$ in the end of the above proof. Moreover, fixing $\epsilon \in \left(0, \frac{1-\rho(I-DP_o)}{4}\right)$, the smallest suitable $\lambda > \|A\|$ is given by an equation

$$\|C\|\|P_o\| \sum_{i=0}^{N-1} \frac{\|B_1\|^i}{i!} \left[\sum_{p=0}^{i} \frac{i!\max\{1, T^p\}}{(i-p)!(\lambda - \|A\|)^{p+1}} + \frac{i!}{(\lambda - \|A\|)^{i+1}}\right] = \epsilon,$$

which is rather awkward to solve.

2.3.1.2 Convergence analysis of D-type

In this section, we discuss the ILC convergence of D-type.

Theorem 2.15. *Let $y_d(t)$, $t \in [0, T]$, be a desired trajectory for system (2.73) with $D = \Theta$. If $\rho(I - CD_o) < 1$ and $e_k(0) = 0$, $k = 1, 2, \cdots$, then the D-type ILC law (2.76) guarantees $\lim_{k\to\infty} y_k(t) = y_d(t)$ uniformly on $[0, T]$.*

Proof. First, we consider $(j - 1)\tau \leq t < j\tau$, $j = 2, 3, \cdots, N$. By (2.74) and (2.76), we have

$$\dot{e}_{k+1}(t) - \dot{e}_k(t)$$

$$= C[\dot{x}_k(t) - \dot{x}_{k+1}(t)]$$

$$= CA[x_k(t) - x_{k+1}(t)] + CB[x_k(t - \tau) - x_{k+1}(t - \tau)] + C[u_k(t) - u_{k+1}(t)]$$

$$= -CA \int_0^t e^{A(t-\tau-s)} e_\tau^{B_1(t-\tau-s)} e^{A\tau} \delta u_k(s)ds$$

$$
- CB \int_0^{t-\tau} e^{A(t-2\tau-s)} e_\tau^{B_1(t-2\tau-s)} e^{A\tau} \delta u_k(s) ds - CD_o \dot{e}_k(t).
$$

So we have

$$
\dot{e}_{k+1}(t) = (I - CD_o)\dot{e}_k(t) - CA \int_0^t e^{A(t-\tau-s)} e_\tau^{B_1(t-\tau-s)} e^{A\tau} \delta u_k(s) ds
$$

$$
- CB \int_0^{t-\tau} e^{A(t-2\tau-s)} e_\tau^{B_1(t-2\tau-s)} e^{A\tau} \delta u_k(s) ds.
$$

Similar to the proof of Theorem 2.14, one can derive

$$
\dot{e}_{k+1}(t) = (I - CD_o)\dot{e}_k(t)
$$

$$
- CA \sum_{i=1}^j \int_0^{t-(i-1)\tau} e^{A(t-\tau-s)} B_1^{i-1} \frac{(t-(i-1)\tau-s)^{i-1}}{(i-1)!} e^{A\tau} \delta u_k(s) ds
$$

$$
- CB \sum_{l=2}^j \int_0^{t-(l-1)\tau} e^{A(t-2\tau-s)} B_1^{l-2} \frac{(t-(l-1)\tau-s)^{l-2}}{(l-2)!} e^{A\tau} \delta u_k(s) ds.
$$

Obviously, we have

$$
\|\dot{e}_{k+1}(t)\| e^{-\lambda t} \le \left[\rho(I - CD_o) + \epsilon \right] \|\dot{e}_k(t)\| e^{-\lambda t} + \|CA\| \sum_{i=1}^j \frac{\|B_1\|^{i-1}}{(i-1)!} e^{(\|A\|-\lambda)t}
$$

$$
\times \int_0^{t-(i-1)\tau} e^{\|A\|(\lambda-s)} (t-(i-1)\tau-s)^{i-1} ds \|\delta u_k\|_\lambda
$$

$$
+ \|CB\| \sum_{l=2}^j \frac{\|B_1\|^{l-2}}{(l-2)!} e^{(\|A\|-\lambda)t}
$$

$$
\times \int_0^{t-(l-1)\tau} e^{\|A\|(\lambda-\tau-s)} (t-(l-1)\tau-s)^{l-2} ds \|\delta u_k\|_\lambda.
$$

$$
\tag{2.87}
$$

By analogy to the computation in (2.84)–(2.86), inequality (2.87) becomes

$$
\|\dot{e}_{k+1}(t)\| e^{-\lambda t} \le \left[\rho(I - CD_o) + \epsilon \right] \|\dot{e}_k(t)\| e^{-\lambda t} + \left(W_1 + W_2 \right) \|D_o\| \|\dot{e}_k\|_\lambda,
$$

$$
\tag{2.88}
$$

where

$$
W_1 = \|CA\| \sum_{i=1}^j \frac{\|B_1\|^{i-1}}{(i-1)!} \left[\sum_{p=1}^i \frac{(i-1)! T^{i-p}}{(i-p)!(\lambda-\|A\|)^p} + \frac{(i-1)!}{(\lambda-\|A\|)^i} \right]
$$

and

$$W_2 = \|CB\| \sum_{l=2}^{j} \frac{\|B_1\|^{l-2}}{(l-2)!} \left[\sum_{q=2}^{l} \frac{(l-2)!T^{l-q}}{(l-q)!(\lambda - \|A\|)^{q-1}} + \frac{(l-2)!}{(\lambda - \|A\|)^{l-1}} \right].$$

If $j = 1$, which means $0 \leq t < \tau$, then by (2.74) and (2.76), we have

$$\dot{e}_{k+1}(t) - \dot{e}_k(t)$$
$$= C[\dot{x}_k(t) - \dot{x}_{k+1}(t)]$$
$$= CA[x_k(t) - x_{k+1}(t)] + CB[x_k(t - \tau) - x_{k+1}(t - \tau)] + C[u_k(t) - u_{k+1}(t)]$$
$$= -CA \int_0^t e^{A(t-\tau-s)} e_\tau^{B_1(t-\tau-s)} e^{A\tau} \delta u_k(s)ds - CD_o \dot{e}_k(t).$$

We can repeat the above arguments to arrive at (2.88) with $W_2 = 0$ and $W_1 = \frac{2\|CA\|}{\lambda - \|A\|}$. Hence (2.88) holds on the whole $[0, T]$, which implies

$$\|\dot{e}_{k+1}\|_\lambda \leq \left[\rho(I - CD_o) + \epsilon + \left(W_1 + W_2 \right) \|D_o\| \right] \|\dot{e}_k\|_\lambda.$$

Note that by $\rho(I - CD_o) < 1$, we obtain $\lim_{k \to \infty} \|\dot{e}_k\|_\lambda = 0$. Thus, $\|\dot{e}_k\| \leq \dot{e}^{\lambda T} \|\dot{e}_k\|_\lambda$. So $\|\dot{e}_k\| \to 0$ as $k \to \infty$. Due to the fact that $e_k(0) = 0$, we get $\|e_k\| \leq T\|\dot{e}_k\|$, and consequently we find that $\|e_k\| \to 0$ as $k \to \infty$. The proof is completed. $\square$

Remark 2.6. Since

$$e_k(0) = y_d(0) - y_k(0) = y_d(0) - Cx_k(0) = y_d(0) - C\varphi(0),$$

we need $y_d(0) = C\varphi(0)$, which gives a compatibility condition for φ. Since $y_d(0)$ is arbitrary, we need C to be surjective.

2.3.1.3   Numerical examples and discussion

In this part, four numerical examples are presented to demonstrate the validity of the designed method. Next, noting the fact that $\|e_k\| \leq e^{\lambda T} \|e_k\|_\lambda$, we obtain $\|e_k\| \to 0$ as $k \to \infty$, i.e., $\lim_{k \to \infty} \sup_{t \in J} \|e_k(t)\| = 0$, which yields $\lim_{k \to \infty} \|e_k(t)\| = 0$ for $\forall t \in J$. In order to simulate the tracking error trajectory of every moment of each time, we adopt the vector 2-norm in our simulations.

Example 2.5. Consider the following system:

$$\begin{cases} \dot{x}_k(t) = x_k(t) + x_k(t - 0.5) + u_k(t), \ x(t) \in \mathbb{R}, \ t \geq 0, \\ x_k(t) = t, \ -0.5 \leq t \leq 0, \\ y_k(t) = x_k(t) + 0.3u_k(t). \end{cases} \tag{2.89}$$

The P-type ILC is set as

$$u_{k+1}(t) = u_k(t) + e_k(t).$$

The original reference trajectory is set as

$$y_d(t) = 2\sin(4\pi t), \quad t \in [0, 1],$$

where $t \in [0, 1]$, $\tau = 0.5$, $\varphi(t) = t$. Set $n = m = 1$, $A = 1$, $B = 1$, $C = 1$, and $D = 0.3$. It is not difficult to find that $B_1 = e^{-A\tau}B = e^{-0.5}$ and

$$e_{0.5}^t = \begin{cases} 1 + e^{-0.5}t, & t \in [0, 0.5], \\ 1 + e^{-0.5}t + e^{-1}\dfrac{(t - 0.5)^2}{2!}, & t \in [0.5, 1]. \end{cases}$$

Next, setting $P_o = 1$, one has $\rho(1 - DP_o) = 0.7 < 1$. Thus, all conditions of Theorem 2.14 are satisfied, and the output $y_k(t)$ uniformly converges to $y_d(t)$ for $t \in [0, 1]$. The upper figure of Fig. 2.8 shows system (2.89)'s 10th output y_k and the reference trajectory y_d. The lower figure of Fig. 2.8 shows the tracking error in each iteration.

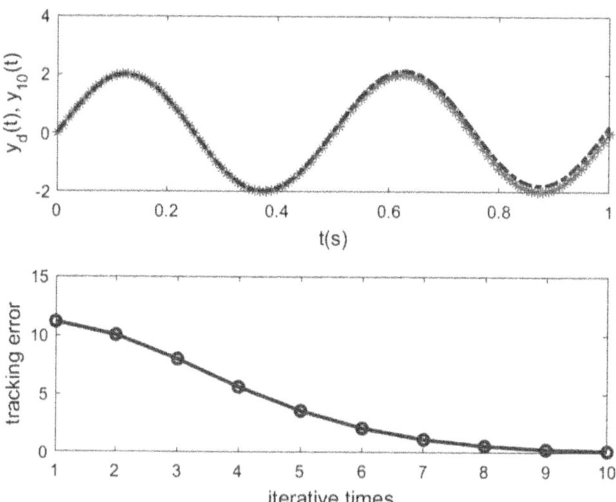

FIGURE 2.8   The system's 10th output and the tracking error for (2.89).

Example 2.6. Consider the following system:

$$\begin{cases} \dot{x}_k(t) = x_k(t) + x_k(t - 0.5) + u_k(t), & x(t) \in \mathbb{R}, \ t \geq 0, \\ x_k(t) = t, & -0.5 \leq t \leq 0, \\ y_k(t) = x_k(t). \end{cases} \tag{2.90}$$

The D-type ILC is set as

$$u_{k+1}(t) = u_k(t) + 0.3\dot{e}_k(t).$$

The original reference trajectory is the same as in Example 2.5. Set $t \in [0, 1]$, $\tau = 0.5$, $\varphi(t) = t$, and $n = m = 1$, $A = 1$, $B = 1$, $C = 1$, and $D = 0$. It is not difficult to see that B_1 and $e'_{0.5}$ are the same as in Example 2.5. Next, setting $D_o = 0.3$, it is easy to check $\rho(1 - CD_o) = 0.7 < 1$. Thus, all conditions of Theorem 2.15 are satisfied. The upper figure of Fig. 2.9 shows system (2.90)'s 10th output y_k and the reference trajectory y_d. The lower figure of Fig. 2.9 shows the tracking error in each iteration.

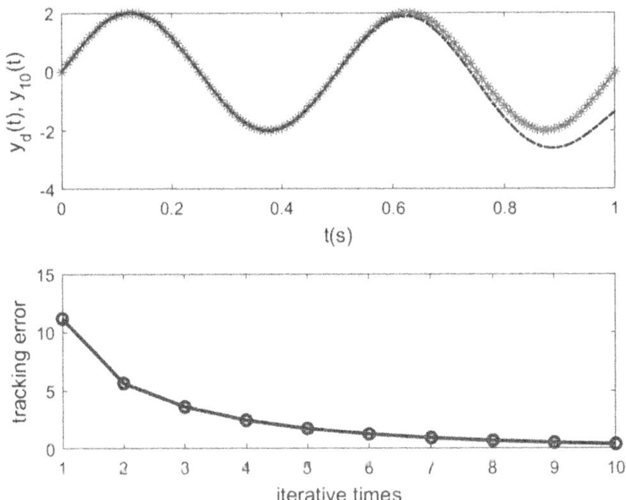

FIGURE 2.9 The system's 10th output and the tracking error for (2.90).

Example 2.7. Consider the following system:

$$\begin{cases} \dot{x}_k(t) = x_k(t) + x_k(t - 0.5) + u_k(t), \ x_k(t), u_k(t) \in \mathbb{R}^2, \ t \geq 0, \\ x_k(t) = (e^t, \ e^t)^\top, \ -0.5 \leq t \leq 0, \\ y_k(t) = (1, \ 2)x_k(t) + (2, \ 1)u_k(t). \end{cases} \tag{2.91}$$

The P-type ILC is set as

$$u_{k+1}(t) = u_k(t) + \begin{pmatrix} 0.5 \\ -0.4 \end{pmatrix} e_k(t).$$

The original reference trajectory is

$$y_d(t) = 2\cos(4\pi t) + 3t^2, \quad t \in [0, 1].$$

Set $t \in [0, 1]$, $\tau = 0.5$, $\varphi(t) = (e^t, \ e^t)^\top$, and $n = 2$, $m = 1$, $A = B = I$, $C = (1, \ 2)$, and $D = (2, \ 1)$. It is not difficult to see that

$$B_1 = e^{-A\tau} B = \begin{pmatrix} 0.6065 & 0 \\ 0 & 0.6065 \end{pmatrix}$$

and

$$e_{0.5}^t = \begin{cases} I, & t \in [-0.5, 0], \\ I + B_1 t, & t \in [0, 0.5], \\ I + B_1 t + B_1^2 \dfrac{(t - 0.5)^2}{2!}, & t \in [0.5, 1]. \end{cases}$$

Next, setting $P_o = (0.5, -0.4)^\top$, one has $\rho(I - DP_o) = 0.8 < 1$. Thus, all conditions of Theorem 2.14 are satisfied. Then the output $y_k(t)$ uniformly converges to $y_d(t)$ for $t \in [0, 1]$. The upper figure of Fig. 2.10 shows system (2.91)'s 10th output y_k and the reference trajectory y_d. The lower figure of Fig. 2.10 shows the tracking error in each iteration.

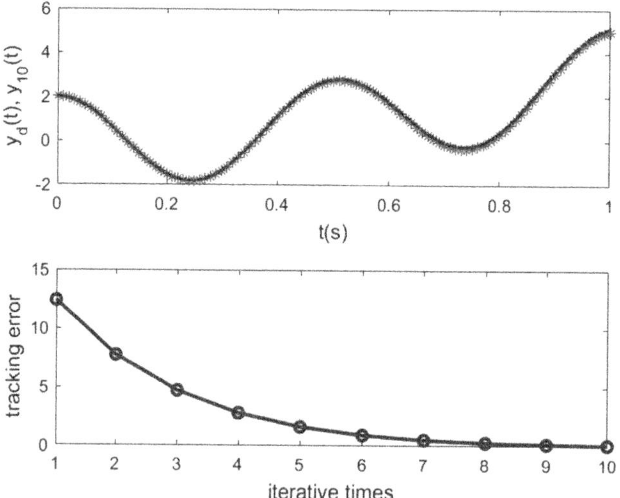

FIGURE 2.10   The system's 10th output and the tracking error for (2.91).

Example 2.8. Consider the following system:

$$\begin{cases} \dot{x}_k(t) = x_k(t) + x_k(t - 0.5) + u_k(t), & x_k(t), u_k(t) \in \mathbb{R}^2, \ t \geq 0, \\ x_k(t) = (t, \ t)^\top, & -0.5 \leq t \leq 0, \\ y_k(t) = (1, \ 2)x_k(t). \end{cases} \tag{2.92}$$

The D-type ILC is set as

$$u_{k+1}(t) = u_k(t) + \begin{pmatrix} 1 \\ -0.4 \end{pmatrix} \dot{e}_k(t).$$

To satisfy the condition $e_k(0) = 0$, the original reference trajectory is chosen as

$$y_d(t) = 2\sin(8\pi t) + 3t^2, \quad t \in [0, 1].$$

It is easy to check that the conditions of Theorem 2.15 are satisfied. Then the output $y_k(t)$ uniformly converges to $y_d(t)$ for $t \in [0, 1]$. The upper figure of Fig. 2.11 shows system (2.92)'s 10th output y_k and the reference trajectory y_d. The lower figure of Fig. 2.11 shows the tracking error in each iteration.

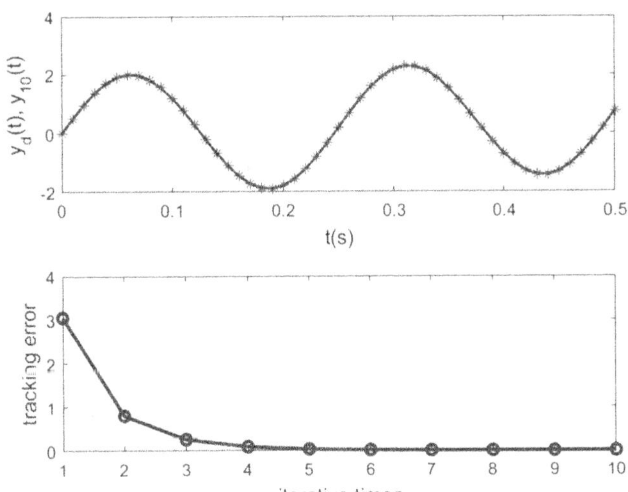

FIGURE 2.11   The system's 10th output and the tracking error for (2.92).

2.3.1.4   Conclusions

We present open-loop P- and D-type asymptotic convergence results for time-delay controlled systems via delayed matrix exponentials in the λ-norm sense by virtue of the spectral matrix radius. Moreover, the simulations show that our ILC strategies are effective. The results in this section are motivated from [15].

Chapter 3

Oscillating delay systems

3.1 Finite time stability

3.1.1 Finite time stability of oscillating delay systems

In this chapter, we study finite time stability of the following second order linear differential equations with pure delay term:

$$
\begin{cases}
\ddot{x}(t) + \Omega^2 x(t - \tau) = 0, \ \tau > 0, \ t \in J := [0, T], \\
x(t) \equiv \varphi(t), \ \dot{x}(t) \equiv \dot{\varphi}(t), \ -\tau \leq t \leq 0,
\end{cases}
\tag{3.1}
$$

where $x \in \mathbb{R}^n$, τ is the time delay, φ is an arbitrary twice continuously differentiable vector function, T is a prefixed positive number, and Ω is an $n \times n$ nonsingular matrix.

Note that Khusainov et al. [5] give a new representation of the solution for (3.1) as follows:

$$
\begin{aligned}
x(t) = \cos_\tau \Omega t \varphi(-\tau) + \Omega^{-1} \sin_\tau \Omega t \dot{\varphi}(-\tau) \\
+ \Omega^{-1} \int_{-\tau}^{0} \sin_\tau \Omega (t - \tau - s) \ddot{\varphi}(s) ds,
\end{aligned}
\tag{3.2}
$$

where $\cos_\tau \Omega t$ is called the delayed matrix cosine of polynomial degree $2k$ ([5, Definition 1]) on the intervals $(k-1)\tau \leq t < k\tau$ formulated by

$$
\cos_\tau \Omega t =
\begin{cases}
\Theta, & -\infty < t < -\tau, \\
I, & -\tau \leq t < 0, \\
I - \Omega^2 \frac{t^2}{2!}, & 0 \leq t < \tau, \\
\ \vdots & \ \vdots \\
I - \Omega^2 \frac{t^2}{2!} + \ldots + (-1)^k \Omega^{2k} \frac{[t-(k-1)\tau]^{2k}}{(2k)!}, & (k-1)\tau \leq t < k\tau
\end{cases}
\tag{3.3}
$$

Stability and Controls Analysis for Delay Systems. https://doi.org/10.1016/B978-0-32-399792-8.00009-8

Copyright © 2023 Elsevier Inc. All rights reserved.

and $\sin_\tau \Omega t$ is called the delayed matrix sine of polynomial degree $2k + 1$ ([5, Definition 2]) on the intervals $(k - 1)\tau \leq t < k\tau$ formulated by

$$
\sin_\tau \Omega t = \begin{cases}
\Theta, & -\infty < t < -\tau, \\
\Omega(t + \tau), & -\tau \leq t < 0, \\
\Omega(t + \tau) - \Omega^3 \frac{t^3}{3!}, & 0 \leq t < \tau, \\
\ \vdots & \vdots \\
\Omega(t + \tau) - \ldots + (-1)^k \Omega^{2k+1} \frac{[t - (k-1)\tau]^{2k+1}}{(2k+1)!}, & (k - 1)\tau \leq t < k\tau.
\end{cases}
\tag{3.4}
$$

Obviously, when $\tau = 0$, $\cos_\tau \Omega t$ and $\sin_\tau \Omega t$ reduce to the matrix cosine function $\cos \Omega t$ and the matrix sine function $\sin \Omega t$, respectively, which are given by the following formal matrix series:

$$
\cos \Omega t = 1 - \Omega^2 \frac{t^2}{2!} + \cdots + (-1)^k \Omega^{2k} \frac{t^{2k}}{(2k)!} + \cdots
$$

and

$$
\sin \Omega t = \Omega \frac{t}{1!} - \Omega^3 \frac{t^3}{3!} + \cdots + (-1)^k \Omega^{2k+1} \frac{t^{2k+1}}{(2k + 1)!} + \cdots .
$$

By Gantmakher [89, p. 123],

$$
x(t) = x_0 \cos \Omega t + \Omega^{-1} \dot{x}_0 \sin \Omega t, \quad \text{provided } \Omega^{-1} \text{ exists},
$$

is a solution of the second order differential system $\ddot{x}(t) + \Omega^2 x(t) = 0$, $t \geq 0$, $x(0) = x_0 \in \mathbb{R}^n$, $\dot{x}(0) = \dot{x}_0 \in \mathbb{R}^n$.

Set $J = [0, t_1]$, $t_1 > 0$. Denote $C^2(J, \mathbb{R}^n) = \{x \in C(J, \mathbb{R}^n) : \ddot{x} \in C(J, \mathbb{R}^n)\}$ endowed with the norm $\|x\|_{C^2(J)} = \max_{t \in J}\{\|x(t)\|, \|\dot{x}(t)\|, \|\ddot{x}(t)\|\}$. Let X, Y be two Banach spaces and let $L_b(X, Y)$ be the space of bounded linear operators from X to Y. Now, $L^p(J, Y)$ denotes the Banach space of functions $f : J \to Y$ which are Bochner integrable normed by $\|f\|_{L^p(J,Y)}$ for some $1 < p < \infty$.

In addition, we let $\|\varphi\| = \max_{s \in [-\tau, 0]} \|\varphi(s)\|$, $\|\dot{\varphi}\| = \max_{s \in [-\tau, 0]} \|\dot{\varphi}(s)\|$, and $\|\ddot{\varphi}\| = \max_{s \in [-\tau, 0]} \|\ddot{\varphi}(s)\|$.

We need the following rules of differentiation for the delayed matrix cosine of polynomial degree $2k$ on the interval $[(k - 1)\tau, k\tau)$ and sine of polynomial degree $2k + 1$ on the interval $[(k - 1)\tau, k\tau)$ defined in (3.3) and (3.4), respectively.

Lemma 3.1. *([5, Lemmas 1 and 2]) The following rules of differentiation are true for the matrix functions (3.3) and (3.4):*

$$
\frac{d}{dt} \cos_\tau \Omega t = -\Omega \sin_\tau \Omega(t - \tau), \quad \frac{d}{dt} \sin_\tau \Omega t = \Omega \cos_\tau \Omega t.
$$

Remark 3.1. For simplification of the next computation, one can divide the term $\int_{-\tau}^{0} \sin_\tau \Omega(t - \tau - s)\ddot{\varphi}(s)ds$ in (3.2) into the following form according to the subintervals $[(k - 1)\tau, k\tau)$:

$$\int_{-\tau}^{0} \sin_\tau \Omega(t - \tau - s)\ddot{\varphi}(s)ds$$

$$= \int_{-\tau}^{t-k\tau} \sin_\tau \Omega(t - \tau - s)\ddot{\varphi}(s)ds + \int_{t-k\tau}^{0} \sin_\tau \Omega(t - \tau - s)\ddot{\varphi}(s)ds.$$

Obviously, $\sin_\tau \Omega(t - \tau - s)$ has different formulas in different subintervals $[(k - 1)\tau, k\tau)$ by (3.4).

By Remark 3.1, the solution (3.2) of system (3.1) can be expressed in the following form:

$$x(t) = \cos_\tau \Omega t \varphi(-\tau) + \Omega^{-1} \sin_\tau \Omega t \dot{\varphi}(-\tau)$$

$$+ \Omega^{-1} \int_{-\tau}^{t-k\tau} \sin_\tau \Omega(t - \tau - s)\ddot{\varphi}(s)ds$$

$$+ \Omega^{-1} \int_{t-k\tau}^{0} \sin_\tau \Omega(t - \tau - s)\ddot{\varphi}(s)ds \qquad (3.5)$$

for $(k - 1)\tau \le t \le k\tau$.

Observing the solution (3.2) involves $\ddot{\varphi}$, which seems to be a slightly stronger requirement than initial conditions.

Remark 3.2. In order to obtain some alternative formulas, one can apply integration by parts via Lemma 3.1 to derive

$$\int_{-\tau}^{0} \sin_\tau \Omega(t - \tau - s)\ddot{\varphi}(s)ds$$

$$= \sin_\tau \Omega(t - \tau)\dot{\varphi}(0) - \sin_\tau \Omega t \dot{\varphi}(-\tau) + \Omega \int_{-\tau}^{0} \cos_\tau \Omega(t - \tau - s)\dot{\varphi}(s)ds.$$

Then the solution (3.2) can be expressed as

$$x(t) = \cos_\tau \Omega t \varphi(-\tau) + \Omega^{-1} \sin_\tau \Omega(t - \tau)\dot{\varphi}(0)$$

$$+ \int_{-\tau}^{0} \cos_\tau \Omega(t - \tau - s)\dot{\varphi}(s)ds. \qquad (3.6)$$

If we use integration by parts again for the integral part of (3.6), then we have

$$\int_{-\tau}^{0} \cos_\tau \Omega(t - \tau - s)\dot{\varphi}(s)ds$$

$$= \cos_\tau \Omega(t - \tau)\varphi(0) - \cos_\tau \Omega t \varphi(-\tau) - \Omega \int_{-\tau}^{0} \sin_\tau \Omega(t - 2\tau - s)\varphi(s)ds,$$

which implies that (3.6) can be expressed as

$$x(t) = \cos_\tau \Omega(t - \tau)\varphi(0) + \Omega^{-1}\sin_\tau \Omega(t - \tau)\dot{\varphi}(0)$$
$$- \Omega \int_{-\tau}^{0} \sin_\tau \Omega(t - 2\tau - s)\varphi(s)ds. \tag{3.7}$$

Definition 3.1. (see [24, Definition 2.1]) System (3.1) satisfying initial conditions $x(t) \equiv \varphi(t)$ and $\dot{x}(t) \equiv \dot{\varphi}(t)$ for $-\tau \leq t \leq 0$ is finite time stable with respect to $\{0, J, \delta, \epsilon, \tau\}$ if and only if

$$\gamma < \delta, \tag{3.8}$$

which implies

$$\|x(t)\| < \epsilon, \ \forall \, t \in J,$$

where $\gamma = \max\{\|\varphi\|, \|\dot{\varphi}\|, \|\ddot{\varphi}\|\}$ denotes the initial time of observation of the system. In addition, δ, ϵ are real positive numbers.

Using the form of $\cos_\tau \Omega t$ and $\sin_\tau \Omega t$ one can prove the following two lemmas, which will be widely used in the sequel.

Lemma 3.2. *For any $t \in [(k - 1)\tau, k\tau)$, $k = 0, 1, \cdots, n$, the following formula is true:*

$$\|\cos_\tau \Omega t\| \leq \cosh(\|\Omega\|t).$$

Proof. Using the form of (3.3), one can calculate that

$$\|\cos_\tau \Omega t\| \leq 1 + \|\Omega\|^2\frac{t^2}{2!} + \|\Omega\|^4\frac{(t - \tau)^4}{4!} + \cdots + \|\Omega\|^{2k}\frac{[t - (k - 1)\tau]^{2k}}{(2k)!}$$
$$\leq 1 + \|\Omega\|^2\frac{t^2}{2!} + \|\Omega\|^4\frac{t^4}{4!} + \cdots + \|\Omega\|^{2k}\frac{t^{2k}}{(2k)!}$$
$$\leq \sum_{k=0}^{\infty} \frac{(\|\Omega\|t)^{2k}}{(2k)!} = \cosh(\|\Omega\|t).$$

The proof is completed. $\qquad\qquad\square$

Lemma 3.3. *For any $t \in [(k - 1)\tau, k\tau)$, $k = 0, 1, \cdots, n$, the following formula is true:*

$$\|\sin_\tau \Omega t\| \leq \sinh[\|\Omega\|(t + \tau)].$$

Proof. Using the form of (3.4), we get

$$\|\sin_\tau \Omega t\| \leq \|\Omega\|(t + \tau) + \|\Omega\|^3\frac{t^3}{3!} + \cdots + \|\Omega\|^{2k+1}\frac{[t - (k - 1)\tau]^{2k+1}}{(2k + 1)!}$$

$$\le \|\Omega\|(t+\tau) + \|\Omega\|^3 \frac{(t+\tau)^3}{3!} + \cdots + \|\Omega\|^{2k+1}\frac{(t+\tau)^{2k+1}}{(2k+1)!}$$

$$\le \sum_{k=0}^{\infty} \frac{[\|\Omega\|(t+\tau)]^{2k+1}}{(2k+1)!} = \sinh[\|\Omega\|(t+\tau)].$$

The proof is finished. $\qquad\qquad\square$

Remark 3.3. When $t \in (-\infty, -\tau)$, we can get $\|\cos_\tau \Omega t\| = \|\sin_\tau \Omega t\| = 0$ by the form (3.3) and (3.4).

As is well known, the exponential forms of hyperbolic functions $\cosh t$ and $\sinh t$ are defined as follows:

$$\cosh t = \frac{e^t + e^{-t}}{2}, \quad \sinh t = \frac{e^t - e^{-t}}{2}, \ t \in \mathbb{R}.$$

Then for all $t \in \mathbb{R}$, $\sinh t \le \cosh t$ holds and the larger the value of t, the closer $\cosh t$ and $\sinh t$. When $t \to +\infty$, we get $\cosh t = \sinh t$. In addition, both $\cosh t$ and $\sinh t$ are nonnegative, monotone increasing functions for $t \ge 0$.

Obviously, the derivatives of $\cosh(\cdot)$ and $\sinh(\cdot)$ are

$$\frac{d}{dt}\cosh t = \sinh t, \quad \frac{d}{dt}\sinh t = \cosh t. \tag{3.9}$$

Next, we give an example to verify the results of Lemmas 3.2 and 3.3 and also show the images of the delayed cosine function and the delayed sine function.

Example 3.1. Set $\tau = 0.3$, $\Omega = 3$, $\Omega \in \mathbb{R}^1$. By (3.3) and (3.4) we derive that $\cos_{0.3} 3t$ and $\sin_{0.3} 3t$ are as follows:

$$\cos_{0.3} 3t = \begin{cases} 1, \ t \in [-0.3, 0), \\[4pt] 1 - 3^2\frac{t^2}{2}, \ t \in [0, 0.3), \\[4pt] 1 - 3^2\frac{t^2}{2} + 3^4\frac{(t-0.3)^4}{4!}, \ t \in [0.3, 0.6), \\[4pt] 1 - 3^2\frac{t^2}{2} + 3^4\frac{(t-0.3)^4}{4!} - 3^6\frac{(t-0.6)^6}{6!}, \ t \in [0.6, 0.9], \\[4pt] \vdots \end{cases}$$

and

$$\sin_{0.3} 3t = \begin{cases} 3(t+0.3), \ t \in [-0.3, 0), \\[4pt] 3(t+0.3) - 3^3\frac{t^3}{3!}, \ t \in [0, 0.3), \\[4pt] 3(t+0.3) - 3^3\frac{t^3}{3!} + 3^5\frac{(t-0.3)^5}{5!}, \ t \in [0.3, 0.6), \\[4pt] 3(t+0.3) - 3^3\frac{t^3}{3!} + 3^5\frac{(t-0.3)^5}{5!} - 3^7\frac{(t-0.6)^7}{7!}, \ t \in [0.6, 0.9]. \\[4pt] \vdots \end{cases}$$

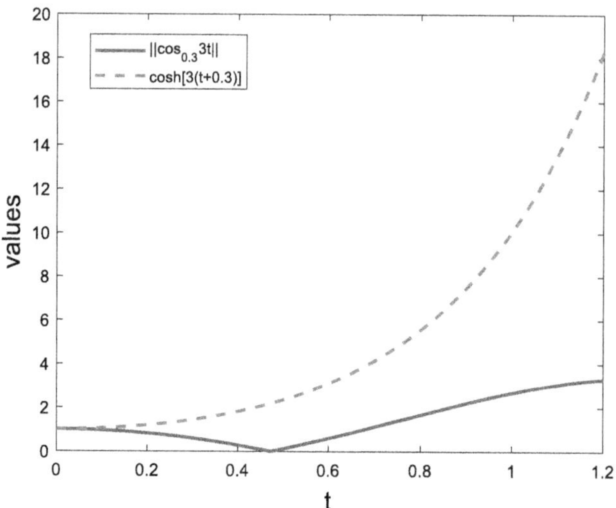

FIGURE 3.1 $\| \cos_{0.3} 3t \|$ and $\cosh[3(t + 0.3)]$.

It follows from Fig. 3.1 and Fig. 3.2 that the inequalities in Lemma 3.2 and Lemma 3.3 hold.

The images of the delayed cosine $\cos_{0.3} 3t$ and the delayed sine $\sin_{0.3} 3t$ are shown in Fig. 3.3 and Fig. 3.4, respectively. Obviously, we can see that the delayed cosine and delayed sine do have some similar properties to the classical cosine and sine functions, such as a wave shape, monotonicity, and periodicity.

However, the differences between the delayed cosine and delayed sine and the cosine and sine are that the delayed cosine and delayed sine appear in the interval segment $[-\tau, 0]$ and with increasing variables t, the upper and lower bounds of the delayed cosine and delayed sine are also increasing. When the delay $\tau = 0$, by (3.3) and (3.4), the delayed cosine and delayed sine coincide with the cosine and sine functions.

3.1.1.1 *Finite time stability results for linear systems*

In this section, we present some sufficient conditions for finite time stability results for the desired system (3.1) by using three possible formulas of solutions, which do enrich the design methods for practical problems.

Theorem 3.1. *System (3.1) is finite time stable with respect to* $\{0, J, \delta, \epsilon, \tau\}$ *if*

$$\cosh(\|\Omega\|T) < \frac{\epsilon - \delta(1 + \tau)\|\Omega^{-1}\| \sinh[\|\Omega\|(T + \tau)]}{\delta}, \tag{3.10}$$

where δ and ϵ are defined in Definition 3.1.

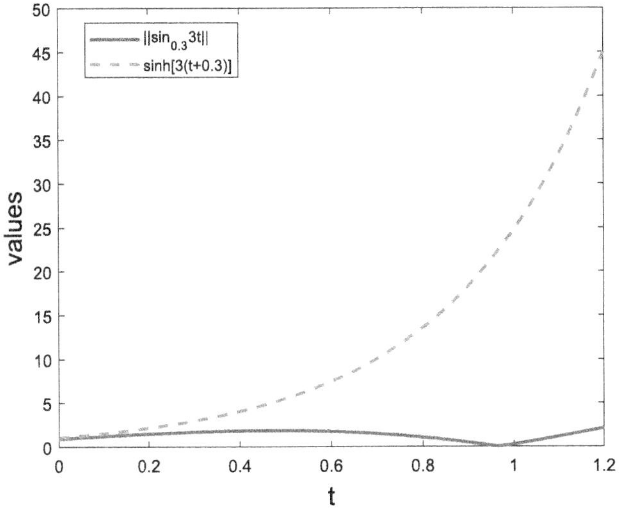

FIGURE 3.2 $\| \sin_{0.3} 3t \|$ and $\sinh[3(t + 0.3)]$.

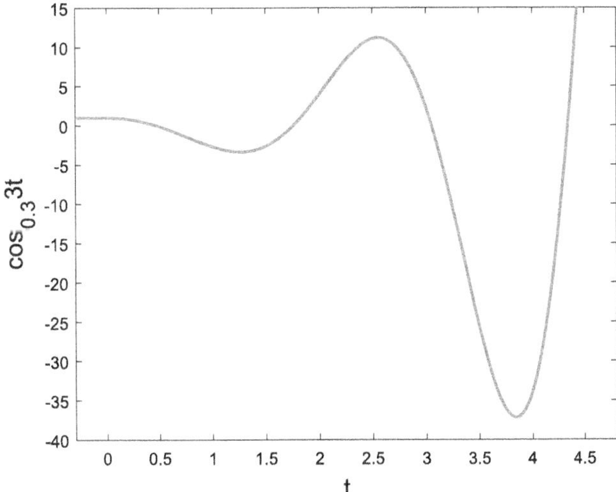

FIGURE 3.3 The delayed cosine function $\cos_{0.3} 3t$.

Proof. By using formula (3.2), via some fundamental computations one can get

$$\|x(t)\| = \left\| \cos_\tau \Omega t \varphi(-\tau) + \Omega^{-1} \sin_\tau \Omega t \dot{\varphi}(-\tau) \right.$$

$$\left. + \Omega^{-1} \int_{-\tau}^{0} \sin_\tau \Omega (t - \tau - s) \ddot{\varphi}(s) ds \right\|$$

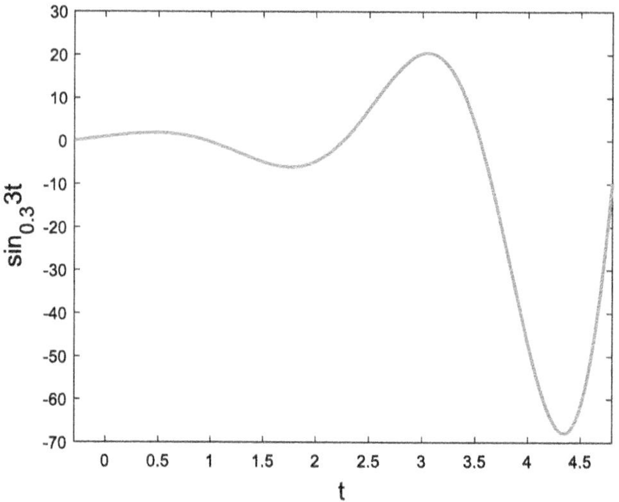

FIGURE 3.4 The delayed sine function $\sin_{0.3} 3t$.

$$\leq \| \cos_\tau \Omega t \| \|\varphi(-\tau)\| + \|\Omega^{-1}\| \| \sin_\tau \Omega t \| \|\dot{\varphi}(-\tau)\|$$
$$+ \left\| \Omega^{-1} \int_{-\tau}^{0} \sin_\tau \Omega(t - \tau - s)\ddot{\varphi}(s)ds \right\|$$
$$\leq \| \cos_\tau \Omega t \| \|\varphi(-\tau)\| + \|\Omega^{-1}\| \| \sin_\tau \Omega t \| \|\dot{\varphi}(-\tau)\|$$
$$+ \|\Omega^{-1}\| \int_{-\tau}^{0} \| \sin_\tau \Omega(t - \tau - s)\| \|\ddot{\varphi}(s)\|ds. \tag{3.11}$$

From (3.8) and (3.11), we have

$$\|x(t)\| \leq \delta \| \cos_\tau \Omega t \| + \delta \|\Omega^{-1}\| \| \sin_\tau \Omega t \|$$
$$+ \delta \|\Omega^{-1}\| \int_{-\tau}^{0} \| \sin_\tau \Omega(t - \tau - s)\|ds.$$

Next, according to Lemmas 3.2 and 3.3, we obtain

$$\|x(t)\| \leq \delta \cosh(\|\Omega\|t) + \delta \|\Omega^{-1}\| \sinh[\|\Omega\|(t + \tau)]$$
$$+ \delta\tau \|\Omega^{-1}\| \sinh[\|\Omega\|(t + \tau)], \tag{3.12}$$

where we use the fact that

$$\| \sin_\tau \Omega(t - \tau - s)\| \leq \sinh[\|\Omega\|(t - s)]$$
$$\leq \sinh[\|\Omega\|(t + \tau)], \quad -\tau \leq s \leq 0, \; t \in J. \tag{3.13}$$

Since $\sinh t$ and $\cosh t$ are both monotonically increasing functions when $t \geq 0$, $\|x(t)\| < \epsilon$ for $\forall t \in J$ can be obtained by combining (3.10) and (3.12).

Thus, system (3.1) is finite time stable by Definition 3.1. $\qquad\square$

Next, we use a new alternative representation of solution (3.6) to derive the following result.

Theorem 3.2. *System (3.1) is finite time stable with respect to* $\{0, J, \delta, \epsilon, \tau\}$ *if*

$$\cosh(\|\Omega\|T) < \frac{\epsilon - \delta\|\Omega^{-1}\|\sinh(\|\Omega\|T) - \delta\tau\mu}{\delta}, \tag{3.14}$$

where $\mu := \max\{\cosh(\|\Omega\|\tau), \cosh(\|\Omega\|T)\}$.

Proof. Similar to Theorem 3.1, we estimate the norm $\|\cdot\|$ of solution formula (3.6),

$$\|x(t)\| \leq \|\cos_\tau \Omega t\|\|\varphi(-\tau)\| + \|\Omega^{-1}\|\|\sin_\tau \Omega(t-\tau)\|\|\dot\varphi(0)\|$$

$$+ \left\|\int_{-\tau}^{0} \cos_\tau \Omega(t-\tau-s)\dot\varphi(s)ds\right\|$$

$$\leq \|\cos_\tau \Omega t\|\|\varphi(-\tau)\| + \|\Omega^{-1}\|\|\sin_\tau \Omega(t-\tau)\|\|\dot\varphi(0)\|$$

$$+ \int_{-\tau}^{0} \|\cos_\tau \Omega(t-\tau-s)\|\|\dot\varphi(s)\|ds. \tag{3.15}$$

By (3.8), inequality (3.15) implies

$$\|x(t)\| \leq \delta\|\cos_\tau \Omega t\| + \delta\|\Omega^{-1}\|\|\sin_\tau \Omega(t-\tau)\|$$

$$+ \delta\int_{-\tau}^{0} \|\cos_\tau \Omega(t-\tau-s)\|ds.$$

Then, according to Lemmas 3.2 and 3.3, one can get

$$\|x(t)\| \leq \delta\cosh(\|\Omega\|t) + \delta\|\Omega^{-1}\|\sinh(\|\Omega\|t) + \mu, \tag{3.16}$$

where we use the fact that

$$\|\cos_\tau \Omega(t-\tau-s)\| \leq \cosh[\|\Omega\|(t-\tau-s)] \leq \mu, \quad -\tau \leq s \leq 0, \ t \in J.$$

Linking (3.14) and (3.16), we obtain $\|x(t)\| < \epsilon, \forall\, t \in J$. Thus, system (3.1) is finite time stable. $\qquad\square$

Finally, we adopt another representation of solution (3.7) to derive another new result.

Theorem 3.3. *System (3.1) is finite time stable with respect to* $\{0, J, \delta, \epsilon, \tau\}$ *if*

$$\theta < \frac{\epsilon - \delta(\|\Omega^{-1}\| + \tau\|\Omega\|)\sinh(\|\Omega\|T)}{\delta}, \tag{3.17}$$

where $\theta := \max\{\cosh(\|\Omega\|\tau), \cosh(\|\Omega\|T - \tau)\}$.

Proof. By Lemmas 3.2 and 3.3 and taking the norm on both sides of (3.7) via (3.8), we have

$$\|x(t)\| \leq \| \cos_\tau \Omega(t - \tau)\| \|\varphi(0)\| + \|\Omega^{-1}\| \| \sin_\tau \Omega(t - \tau)\| \|\dot\varphi(0)\|$$

$$+ \|\Omega\| \left\| \int_{-\tau}^{0} \sin_\tau \Omega(t - 2\tau - s)\varphi(s)ds \right\|$$

$$\leq \| \cos_\tau \Omega(t - \tau)\| \|\varphi(0)\| + \|\Omega^{-1}\| \| \sin_\tau \Omega(t - \tau)\| \|\dot\varphi(0)\|$$

$$+ \|\Omega\| \|\varphi(s)\| \int_{-\tau}^{0} \| \sin_\tau \Omega(t - 2\tau - s)\|ds. \tag{3.18}$$

Note that $\sin_\tau \Omega t = \Theta$ if $t \in (-\infty, -\tau)$. For $-\tau \leq s \leq 0$, we get $\| \sin_\tau \Omega(t - 2\tau - s)\| = 0$ when $t - 2\tau - s < -\tau$. By Lemma 3.3, we get $\| \sin_\tau \Omega(t - 2\tau - s)\| \leq \sinh[\|\Omega\|(t - \tau - s)] \leq \sinh(\|\Omega\|t)$ when $t - 2\tau - s \geq -\tau$. In the end, we can obtain

$$\| \sin_\tau \Omega(t - 2\tau - s)\| \leq \sinh(\|\Omega\|t), \quad -\tau \leq s \leq 0, \ t \in J. \tag{3.19}$$

From (3.8), (3.18), (3.19), and Lemmas 3.2 and 3.3, we can get

$$\|x(t)\| \leq \delta \cosh[\|\Omega\|(t - \tau)] + \delta\|\Omega^{-1}\| \sinh(\|\Omega\|t) + \delta\tau\|\Omega\| \sinh(\|\Omega\|t)$$

$$\leq \delta\theta + \delta(\|\Omega^{-1}\| + \tau\|\Omega\|) \sinh(\|\Omega\|t). \tag{3.20}$$

Combining (3.17) with (3.20), we can finally obtain $\|x(t)\| < \epsilon, \ \forall \ t \in J$. Thus, system (3.1) is finite time stable. $\qquad\square$

Remark 3.4. By the results in Theorems 3.1–3.3, we can analyze that when $\alpha < \beta$ and $\alpha < \rho$, the result of Theorem 3.1 is optimal. When $\beta < \alpha$ and $\beta < \rho$, the result of Theorem 3.2 is optimal. When $\rho < \alpha$ and $\rho < \beta$, the result of Theorem 3.3 is optimal. We have

$$\alpha := \delta \cosh(\|\Omega\|T)$$

$$+ \delta\|\Omega^{-1}\| \sinh[\|\Omega\|(T + \tau)] + \delta\tau\|\Omega^{-1}\| \sinh[\|\Omega\|(T + \tau)],$$

$$\beta := \delta \cosh(\|\Omega\|T) + \delta\|\Omega^{-1}\| \sinh(\|\Omega\|T) + \delta\tau\mu,$$

$$\rho := \delta\theta + \delta(\|\Omega^{-1}\| + \tau\|\Omega\|) \sinh(\|\Omega\|T).$$

Remark 3.5. We have studied the finite time stability of system (3.1) in Theorems 3.1–3.3. Now we analyze the stability of solution (3.2) to system (3.1) when $t \to \infty$. In fact,

$$\|x(t)\| \leq \| \cos_\tau \Omega t\| \|\varphi(-\tau)\| + \|\Omega^{-1}\| \| \sin_\tau \Omega t\dot\varphi(-\tau)\|$$

$$+ \|\Omega^{-1}\| \int_{-\tau}^{0} \| \sin_\tau \Omega(t - \tau - s)\ddot\varphi(s)\|ds,$$

which implies that it is impossible to guarantee that $\|x(t)\| \to 0$ when $t \to \infty$ since the first term $\|\cos_\tau \Omega t\|\|\varphi(-\tau)\| \le \cosh(\|\Omega\|t)\|\varphi(-\tau)\| = \frac{e^{\|\Omega\|t}+e^{-\|\Omega\|t}}{2}\|\varphi(-\tau)\| \to \infty$ when $t \to \infty$. We can even assume the strong condition $\|\Omega^{-1}\| \le e^{-\nu(t+\tau)}$, $\nu > \|\Omega\|$ to guarantee the second and third terms tend to zero due to $\|\sin_\tau \Omega t\| \le \sinh[\|\Omega\|(t+\tau)] \le e^{\|\Omega\|(t+\tau)}$.

3.1.1.2 Extension to delay systems with nonlinear term

In this section, we consider the following delay differential equations with nonlinear term:

$$\begin{cases} \ddot{x}(t) + \Omega^2 x(t-\tau) = f(x(t)), \ \tau > 0, \ t \in J, \\ x(t) \equiv \varphi(t), \ \dot{x}(t) \equiv \dot{\varphi}(t), \ -\tau \le t \le 0, \end{cases} \tag{3.21}$$

where $f \in C(\mathbb{R}^n, \mathbb{R}^n)$.

Definition 3.2. (see [70, Definition 2]) System (3.21) satisfying initial conditions $x(t) \equiv \varphi(t)$ and $\dot{x}(t) \equiv \dot{\varphi}(t)$ for $-\tau \le t \le 0$ is finite time stable with respect to $\{0, J, \delta, \epsilon, \tau\}$ if and only if

$$\gamma^2 < \delta \tag{3.22}$$

implies

$$\|x(t)\|^2 < \epsilon, \ \forall\, t \in J,$$

where $\gamma = \max\{\|\varphi\|, \|\dot{\varphi}\|, \|\ddot{\varphi}\|\}$ denotes the initial time of observation of the system. In addition, δ, ϵ are real positive numbers.

Now we are ready to state our main result in the section.

Theorem 3.4. *Suppose that $f \in C(\mathbb{R}^n, \mathbb{R}^n)$ and there exists $P > 0$ such that $\|f(x)\| \le P\|x\|$ for all $x \in \mathbb{R}^n$. System (3.21) is finite time stable with respect to $\{0, J, \delta, \epsilon, \tau\}$ provided that*

$$e^{P\|\Omega^{-1}\|\|\Omega\|^{-1}[\cosh(\|\Omega\|t)-1]} < \frac{\sqrt{\epsilon}}{a}, \ \forall\, t \in J, \tag{3.23}$$

where

$$a = \sqrt{\delta}\cosh(\|\Omega\|T) + \sqrt{\delta}(\tau+1)\|\Omega^{-1}\|\sinh[\|\Omega\|(T+\tau)]. \tag{3.24}$$

Proof. By [5, Theorem 2, formula (14)], the solution of (3.21) has the form

$$x(t) = \cos_\tau \Omega t \varphi(-\tau) + \Omega^{-1}\sin_\tau \Omega t \dot{\varphi}(-\tau)$$

$$+ \Omega^{-1}\int_{-\tau}^{0} \sin_\tau \Omega(t-\tau-s)\ddot{\varphi}(s)ds$$

$$+ \Omega^{-1}\int_{0}^{t} \sin_\tau \Omega(t-\tau-s)f(x(s))ds, \tag{3.25}$$

where the matrix Ω is nonsingular. Taking the norm for (3.25), we obtain

$$\|x(t)\| \leq \|\cos_\tau \Omega t\| \|\varphi(-\tau)\| + \|\Omega^{-1}\| \|\sin_\tau \Omega t\| \|\dot{\varphi}(-\tau)\|$$

$$+ \|\Omega^{-1}\| \int_{-\tau}^{0} \|\sin_\tau \Omega(t - \tau - s)\| \|\ddot{\varphi}(s)\| ds$$

$$+ \|\Omega^{-1}\| \int_{0}^{t} \|\sin_\tau \Omega(t - \tau - s)\| \|f(x(s))\| ds. \tag{3.26}$$

From (3.22) and (3.26), we have

$$\|x(t)\| \leq \sqrt{\delta} \|\cos_\tau \Omega t\| + \sqrt{\delta} \|\Omega^{-1}\| \|\sin_\tau \Omega t\|$$

$$+ \sqrt{\delta} \|\Omega^{-1}\| \int_{-\tau}^{0} \|\cos_\tau \Omega(t - \tau - s)\| ds$$

$$+ \|\Omega^{-1}\| \int_{0}^{t} \|\sin_\tau \Omega(t - \tau - s)\| \|f(x(s))\| ds.$$

Next, according to Lemmas 3.2 and 3.3 and (3.13), we obtain

$$\|x(t)\| \leq \sqrt{\delta} \cosh(\|\Omega\| t) + \sqrt{\delta} \|\Omega^{-1}\| \sinh[\|\Omega\|(t + \tau)]$$

$$+ \sqrt{\delta} \tau \|\Omega^{-1}\| \sinh[\|\Omega\|(t + \tau)]$$

$$+ \int_{0}^{t} \|\Omega^{-1}\| \|f(x(s))\| \sinh[\|\Omega\|(t - s)] ds. \tag{3.27}$$

Note that $\|f(x)\| \leq P\|x\|$ for all $x \in \mathbb{R}^n$, then inequality (3.27) becomes

$$\|x(t)\| \leq a + \int_{0}^{t} k(t, s)\|x(s)\| ds,$$

where a is defined in (3.24) and $k(t, s) = P\|\Omega^{-1}\| \sinh[\|\Omega\|(t - s)]$.
Calculating the partial derivative $k_t(t, s)$ via (3.9), we obtain

$$k_t(t, s) = P\|\Omega^{-1}\| \|\Omega\| \cosh[\|\Omega\|(t - s)], \ 0 \leq s \leq t.$$

Note that $\|x(t)\|$, $k(t, s)$, and its partial derivative $k_t(t, s)$ are all nonnegative continuous functions and a is a constant. By the Gronwall inequality (see [78]), we get

$$\|x(t)\| \leq a e^{\int_0^t k(t,s)ds}, \ t \in J. \tag{3.28}$$

Next,

$$\int_{0}^{t} k(t, s)ds = \int_{0}^{t} P\|\Omega^{-1}\| \sinh[\|\Omega\|(t - s)] ds$$

$$= P\|\Omega^{-1}\| \|\Omega\|^{-1} [\cosh(\|\Omega\| t) - 1]. \tag{3.29}$$

Combining (3.29) with (3.28), by (3.23) we obtain

$$\|x(t)\| \le ae^{P\|\Omega^{-1}\|\|\Omega\|^{-1}[\cosh(\|\Omega\|t)-1]} < \sqrt{\epsilon},$$

which implies that $\|x(t)\|^2 < \epsilon, t \in J$. The proof is finished. $\qquad\square$

3.1.1.3 Numerical examples and discussion

Example 3.2. In this part, we consider the finite time stability of the following second order differential equations:

$$\begin{cases} \ddot{x}(t) + \Omega^2 x(t - 0.5) = 0, \ x \in \mathbb{R}^2, \ t \in J := [0, 1], \\ \varphi(t) = (0.1t^2, 0.2t)^\top, \ \dot{\varphi}(t) = (0.2t, 0.2)^\top, \\ \ddot{\varphi}(t) = (0.2, 0)^\top, \ -0.5 \le t \le 0, \end{cases} \tag{3.30}$$

where $\tau = 0.5, T = 1, n = 2$,

$$\Omega = \begin{pmatrix} 2 & 0 \\ 1 & 2 \end{pmatrix}, \ \Omega^{-1} = \begin{pmatrix} 0.5 & 0 \\ -0.25 & 0.5 \end{pmatrix}.$$

By (3.5), we get the solution of system (3.30) as follows:

$$\begin{aligned} x(t) = {}& \cos_{0.5} \Omega t \varphi(-0.5) + \Omega^{-1} \sin_{0.5} \Omega t \dot{\varphi}(-0.5) \\ & + \Omega^{-1} \int_{-0.5}^{t-0.5} \sin_{0.5} \Omega(t - 0.5 - s)\ddot{\varphi}(s)ds \\ & + \Omega^{-1} \int_{t-0.5}^{0} \sin_{0.5} \Omega(t - 0.5 - s)\ddot{\varphi}(s)ds, \end{aligned} \tag{3.31}$$

where $0 \le t \le 0.5$ and

$$\begin{aligned} x(t) = {}& \cos_{0.5} \Omega t \varphi(-0.5) + \Omega^{-1} \sin_{0.5} \Omega t \dot{\varphi}(-0.5) \\ & + \Omega^{-1} \int_{-0.5}^{t-1} \sin_{0.5} \Omega(t - 0.5 - s)\ddot{\varphi}(s)ds \\ & + \Omega^{-1} \int_{t-1}^{0} \sin_{0.5} \Omega(t - 0.5 - s)\ddot{\varphi}(s)ds, \end{aligned} \tag{3.32}$$

where $0.5 \le t \le 1$.

Next, we get

$$\cos_{0.5} \Omega t = \begin{pmatrix} \cos_{0.5} 2t & 0 \\ \cos_{0.5} t & \cos_{0.5} 2t \end{pmatrix}, \ \sin_{0.5} \Omega t = \begin{pmatrix} \sin_{0.5} 2t & 0 \\ \sin_{0.5} t & \sin_{0.5} 2t \end{pmatrix}.$$

By (3.3) and (3.4), we obtain

$$
\cos_{0.5} t =
\begin{cases}
1, \ t \in [-0.5, 0), \\[4pt]
1 - \frac{t^2}{2}, \ t \in [0, 0.5), \\[4pt]
1 - \frac{t^2}{2} + \frac{(t-0.5)^4}{4!}, \ t \in [0.5, 1), \\[4pt]
\vdots
\end{cases}
$$

$$
\sin_{0.5} t =
\begin{cases}
(t + 0.5), \ t \in [-0.5, 0), \\[4pt]
(t + 0.5) - \frac{t^3}{3!}, \ t \in [0, 0.5), \\[4pt]
(t + 0.5) - \frac{t^3}{3!} + \frac{(t-0.5)^5}{5!}, \ t \in [0.5, 1), \\[4pt]
\vdots
\end{cases}
$$

and

$$
\cos_{0.5} 2t =
\begin{cases}
1, \ t \in [-0.5, 0), \\[4pt]
1 - 2^2 \frac{t^2}{2}, \ t \in [0, 0.5), \\[4pt]
1 - 2^2 \frac{t^2}{2} + 2^4 \frac{(t-0.5)^4}{4!}, \ t \in [0.5, 1), \\[4pt]
\vdots
\end{cases}
$$

$$
\sin_{0.5} 2t =
\begin{cases}
2(t + 0.5), \ t \in [-0.5, 0), \\[4pt]
2(t + 0.5) - 2^3 \frac{t^3}{3!}, \ t \in [0, 0.5), \\[4pt]
2(t + 0.5) - 2^3 \frac{t^3}{3!} + 2^5 \frac{(t-0.5)^5}{5!}, \ t \in [0.5, 1). \\[4pt]
\vdots
\end{cases}
$$

When $0 \le t \le 0.5$, by (3.31) we get

$$
\begin{aligned}
x(t) = {}& \begin{pmatrix} \cos_{0.5} 2t & 0 \\ \cos_{0.5} t & \cos_{0.5} 2t \end{pmatrix} \begin{pmatrix} 0.025 \\ -0.1 \end{pmatrix} \\[6pt]
& + \begin{pmatrix} 0.5 & 0 \\ -0.25 & 0.5 \end{pmatrix} \begin{pmatrix} \sin_{0.5} 2t & 0 \\ \sin_{0.5} t & \sin_{0.5} 2t \end{pmatrix} \begin{pmatrix} -0.1 \\ 0.2 \end{pmatrix} \\[6pt]
& + \begin{pmatrix} 0.5 & 0 \\ -0.25 & 0.5 \end{pmatrix} \begin{pmatrix} \int_{-0.5}^{t-0.5} 0.2 \sin_{0.5} 2(t - 0.5 - s)ds \\ \int_{-0.5}^{t-0.5} 0.2 \sin_{0.5}(t - 0.5 - s)ds \end{pmatrix} \\[6pt]
& + \begin{pmatrix} 0.5 & 0 \\ -0.25 & 0.5 \end{pmatrix} \begin{pmatrix} \int_{t-0.5}^{0} 0.2 \sin_{0.5} 2(t - 0.5 - s)ds \\ \int_{t-0.5}^{0} 0.2 \sin_{0.5}(t - 0.5 - s)ds \end{pmatrix} \\[6pt]
= {}& \begin{pmatrix} x_1(t) \\ x_2(t) \end{pmatrix}.
\end{aligned}
$$

Through basic calculations one can obtain

$$x_1(t) = 0.025\cos_{0.5} 2t - 0.05\sin_{0.5} 2t + 0.5 \int_{-0.5}^{t-0.5} 0.2\sin_{0.5} 2(t - 0.5 - s)ds$$

$$+ 0.5 \int_{t-0.5}^{0} 0.2\sin_{0.5} 2(t - 0.5 - s)ds$$

and

$$x_2(t) = 0.025\cos_{0.5} t - 0.1\cos_{0.5} 2t + 0.125\sin_{0.5} 2t - 0.05\sin_{0.5} t$$

$$- 0.25 \int_{-0.5}^{t-0.5} 0.2\sin_{0.5} 2(t - 0.5 - s)ds$$

$$+ 0.5 \int_{-0.5}^{t-0.5} 0.2\sin_{0.5}(t - 0.5 - s)ds$$

$$- 0.25 \int_{t-0.5}^{0} 0.2\sin_{0.5} 2(t - 0.5 - s)ds$$

$$+ 0.5 \int_{t-0.5}^{0} 0.2\sin_{0.5}(t - 0.5 - s)ds.$$

When $0.5 \le t \le 1$, by (3.32), we similarly get

$$x(t) = \begin{pmatrix} \cos_{0.5} 2t & 0 \\ \cos_{0.5} t & \cos_{0.5} 2t \end{pmatrix} \begin{pmatrix} 0.025 \\ -0.1 \end{pmatrix}$$

$$+ \begin{pmatrix} 0.5 & 0 \\ -0.25 & 0.5 \end{pmatrix} \begin{pmatrix} \sin_{0.5} 2t & 0 \\ \sin_{0.5} t & \sin_{0.5} 2t \end{pmatrix} \begin{pmatrix} -0.1 \\ 0.2 \end{pmatrix}$$

$$+ \begin{pmatrix} 0.5 & 0 \\ -0.25 & 0.5 \end{pmatrix} \begin{pmatrix} \int_{-0.5}^{t-1} 0.2\sin_{0.5} 2(t - 0.5 - s)ds \\ \int_{-0.5}^{t-1} 0.2\sin_{0.5}(t - 0.5 - s)ds \end{pmatrix}$$

$$+ \begin{pmatrix} 0.5 & 0 \\ -0.25 & 0.5 \end{pmatrix} \begin{pmatrix} \int_{t-1}^{0} 0.2\sin_{0.5} 2(t - 0.5 - s)ds \\ \int_{t-1}^{0} 0.2\sin_{0.5}(t - 0.5 - s)ds \end{pmatrix}$$

$$= \begin{pmatrix} x_1(t) \\ x_2(t) \end{pmatrix}.$$

Then one can obtain

$$x_1(t) = 0.025\cos_{0.5} 2t - 0.05\sin_{0.5} 2t + 0.5 \int_{-0.5}^{t-1} 0.2\sin_{0.5} 2(t - 0.5 - s)ds$$

$$+ 0.5 \int_{t-1}^{0} 0.2\sin_{0.5} 2(t - 0.5 - s)ds$$

and

$$x_2(t) = 0.025\cos_{0.5} t - 0.1\cos_{0.5} 2t + 0.125\sin_{0.5} 2t - 0.05\sin_{0.5} t$$

$$- 0.25\int_{-0.5}^{t-1} 0.2\sin_{0.5} 2(t - 0.5 - s)ds$$

$$+ 0.5\int_{-0.5}^{t-1} 0.2\sin_{0.5}(t - 0.5 - s)ds$$

$$- 0.25\int_{t-1}^{0} 0.2\sin_{0.5} 2(t - 0.5 - s)ds$$

$$+ 0.5\int_{t-1}^{0} 0.2\sin_{0.5}(t - 0.5 - s)ds.$$

By calculating we obtain $\gamma = \max\{\|\varphi\|, \|\dot\varphi\|, \|\ddot\varphi\|\} = 0.3$, $\|\Omega\| = 3$, $\|\Omega^{-1}\| = 0.75$. Then we set $\delta = 0.31 > 0.3 = \gamma$.

By Theorems 3.1–3.3, we can calculate that $\|x(t)\| \le 18.8158$, $\|x(t)\| \le 7.0106$, and $\|x(t)\| \le 7.7167$; we just take $\epsilon = 18.82$, 7.02, 7.72, respectively.

3.1.1.4   Conclusion

We obtain a norm estimation of the delayed matrix sine and cosine of polynomial degrees, which are used to establish sufficient conditions to guarantee our finite time stability results. Meanwhile, we extend our study to the same issue of a delay differential equation with nonlinearity by virtue of the Gronwall inequality approach. The results in this part are motivated from [90].

3.2   Controllability

3.2.1   Controllability of delay oscillating systems

Note that Khusainov et al. [5] studied the following Cauchy problem for a second order linear differential equation with pure delay:

$$\begin{cases} \ddot{x}(t) + \Omega^2 x(t - \tau) = f(t), \ t \ge 0, \ \tau > 0, \\ x(t) = \varphi(t), \ \dot{x}(t) = \dot\varphi(t), \ -\tau \le t \le 0, \end{cases} \tag{3.33}$$

where $f : [0, \infty) \to \mathbb{R}^n$, Ω is an $n \times n$ nonsingular matrix, τ is the time delay, and φ is an arbitrary twice continuously differentiable vector function. A solution of (3.33) has an explicit representation of the form [5, Theorem 2]

$$x(t) = (\cos_\tau \Omega t)\varphi(-\tau) + \Omega^{-1}(\sin_\tau \Omega t)\dot\varphi(-\tau)$$

$$+ \Omega^{-1}\int_{-\tau}^{0} \sin_\tau \Omega(t - \tau - s)\ddot\varphi(s)ds$$

$$+ \Omega^{-1}\int_{0}^{t} \sin_\tau \Omega(t - \tau - s)f(s)ds. \tag{3.34}$$

Diblík et al. [91] studied a control problem for a system governed by the following delay oscillating equations:

$$\begin{cases} \ddot{x}(t) + \Omega^2 x(t - \tau) = bu(t), \ t \in [0, t_1], \ \tau > 0, \ t_1 > 0, \\ x(t) = \varphi(t), \ \dot{x}(t) = \dot{\varphi}(t), \ t \in [-\tau, 0], \end{cases} \tag{3.35}$$

where $b \in \mathbb{R}^n$ and $u : [0, \infty) \to \mathbb{R}$ and they give sufficient and necessary conditions of relative controllability [91, Theorem 3.8] for (3.35) from the point of view of the rank criteria

$$\text{rank} \ (b, \Omega^2 b, \Omega^4 b, \cdots, \Omega^{2(n-1)} b) = n \tag{3.36}$$

provided by $t_1 > (n-1)\tau$. In addition, an explicit dependence of the control function related to $\sin_\tau \Omega$ and $\cos_\tau \Omega$ for (3.36) was given in [91, Theorem 3.9]:

$$u^*(t) = b^\top (\Omega^{-1} \sin_\tau \Omega(t_1 - \tau - t))^\top C_1^0 + b^\top (\cos_\tau \Omega(t_1 - \tau - t))^\top C_2^0,$$

where $C_1^0 = (c_1^0, \cdots, c_n^0)^\top$ and $C_2^0 = (c_{n+1}^0, \cdots, c_{2n}^0)^\top$ are the solutions of the algebraic equation in [91, (3.45)].

In this part, we use a different approach to that in [91] to study controllability of a system governed by the following Cauchy problem:

$$\begin{cases} \ddot{x}(t) + \Omega^2 x(t - \tau) = f(t, x(t)) + Bu(t), \ \tau > 0, \ t \in [0, t_1], \ t_1 > 0, \\ x(t) = \varphi(t), \ \dot{x}(t) = \dot{\varphi}(t), \ -\tau \leq t \leq 0, \end{cases}$$
$$\tag{3.37}$$

where $f : J \times \mathbb{R}^n \to \mathbb{R}^n$, B is an $n \times m$ matrix, and we have an input $u : [0, t_1] \to \mathbb{R}^m$

From (3.34), a solution of system (3.37) can be formulated as

$$x(t) = (\cos_\tau \Omega t)\varphi(-\tau) + \Omega^{-1}(\sin_\tau \Omega t)\dot{\varphi}(-\tau)$$
$$+ \Omega^{-1} \int_{-\tau}^0 \sin_\tau \Omega(t - \tau - s)\ddot{\varphi}(s)ds$$
$$+ \Omega^{-1} \int_0^t \sin_\tau \Omega(t - \tau - s)f(s, x(s))ds$$
$$+ \Omega^{-1} \int_0^t \sin_\tau \Omega(t - \tau - s)Bu(s)ds. \tag{3.38}$$

Definition 3.3. System (3.37) is controllable if there exists a control function $u^* : [0, t_1] \to \mathbb{R}^m$ such that

$$\ddot{x}(t) + \Omega^2 x(t - \tau) = f(t, x(t)) + Bu^*(t)$$

has a solution $x = x^* : [-\tau, t_1] \to \mathbb{R}^n$ satisfying

$$x^*(t) = \varphi(t), \ \dot{x}^*(t) = \dot{\varphi}(t), \ -\tau \leq t \leq 0,$$

$$x^*(t_1) = x_1, \ \dot{x}^*(t_1) = x_1',$$

where $x_1, x_1' \in \mathbb{R}^n$ are any finite terminal conditions and t_1 is an arbitrary given terminal point.

3.2.1.1 Controllability of linear delay systems

In this section, we study the controllability of a system governed by a second order linear delay differential equation:

$$\begin{cases} \ddot{x}(t) + \Omega^2 x(t - \tau) = Bu(t), \ t \in [0, t_1], \ \tau > 0, \\ x(t) = \varphi(t), \ \dot{x}(t) = \dot{\varphi}(t), \ t \in [-\tau, 0]. \end{cases} \tag{3.39}$$

We introduce a delay Grammian matrix as follows:

$$W_\tau[0, t_1] = \Omega^{-1} \int_0^{t_1} \sin_\tau \Omega(t_1 - \tau - s) BB^\top \sin_\tau \Omega^\top (t_1 - \tau - s) ds. \tag{3.40}$$

We give a new sufficient and necessary condition to guarantee (3.39) is controllable.

Theorem 3.5. *System (3.39) is controllable if and only if $W_\tau[0, t_1]$ defined in (3.40) is nonsingular.*

Proof. First, we establish the sufficiency. Since $W_\tau[0, t_1]$ is nonsingular, its inverse $W_\tau^{-1}[0, t_1]$ is well defined. Thus, for any finite terminal conditions $x_1, x_1' \in \mathbb{R}^n$, one can construct the corresponding control input $u(t)$ as

$$u(t) = B^\top \sin_\tau \Omega^\top (t_1 - \tau - t) W_\tau^{-1}[0, t_1] \zeta, \tag{3.41}$$

where

$$\zeta = x_1 - (\cos_\tau \Omega t_1)\varphi(-\tau) - \Omega^{-1}(\sin_\tau \Omega t_1)\dot{\varphi}(-\tau)$$
$$- \Omega^{-1} \int_{-\tau}^0 \sin_\tau \Omega(t_1 - \tau - s)\ddot{\varphi}(s) ds. \tag{3.42}$$

From (3.38), the solution $x(t_1)$ of system (3.39) can be formulated as

$$x(t_1) = (\cos_\tau \Omega t_1)\varphi(-\tau) + \Omega^{-1}(\sin_\tau \Omega t_1)\dot{\varphi}(-\tau)$$
$$+ \Omega^{-1} \int_{-\tau}^0 \sin_\tau \Omega(t_1 - \tau - s)\ddot{\varphi}(s) ds$$
$$+ \Omega^{-1} \int_0^{t_1} \sin_\tau \Omega(t_1 - \tau - s) Bu(s) ds. \tag{3.43}$$

Putting (3.41) into (3.43), we obtain

$$x(t_1) = (\cos_\tau \Omega t_1)\varphi(-\tau) + \Omega^{-1}(\sin_\tau \Omega t_1)\dot{\varphi}(-\tau)$$

$$+ \Omega^{-1} \int_{-\tau}^{0} \sin_\tau \Omega(t_1 - \tau - s)\ddot{\varphi}(s)ds + \Omega^{-1} \int_{0}^{t_1} \sin_\tau \Omega(t_1 - \tau - s)B$$

$$\times B^{\top} \sin_\tau \Omega^{\top}(t_1 - \tau - s)ds\, W_\tau^{-1}[0, t_1]\zeta. \tag{3.44}$$

Now (3.40), (3.42), and (3.44) give

$$x(t_1) = (\cos_\tau \Omega t_1)\varphi(-\tau) + \Omega^{-1}(\sin_\tau \Omega t_1)\dot{\varphi}(-\tau)$$

$$+ \Omega^{-1} \int_{-\tau}^{0} \sin_\tau \Omega(t_1 - \tau - s)\ddot{\varphi}(s)ds + \zeta = x_1,$$

and we use Lemmas 3.1 and 3.3 to obtain

$$\dot{x}(t_1) = \frac{d}{dt}\{(\cos_\tau \Omega t_1)\varphi(-\tau) + \Omega^{-1}(\sin_\tau \Omega t_1)\dot{\varphi}(-\tau)$$

$$+ \Omega^{-1} \int_{-\tau}^{0} \sin_\tau \Omega(t_1 - \tau - s)\ddot{\varphi}(s)ds + \zeta\}$$

$$= x_1'.$$

Next, we check the initial conditions $x(t) = \varphi(t)$, $\dot{x}(t) = \dot{\varphi}(t)$ hold when $-\tau \leq t \leq 0$. From (3.3) and (3.4), the following relations hold:

$$\cos_\tau \Omega t = I, \quad \sin_\tau \Omega t = \Omega(t + \tau), \quad -\tau \leq t \leq 0,$$

$$\sin_\tau \Omega(t - \tau - s) = \begin{cases} \Theta, & t < s \leq 0, \\ \Omega(t - s), & -\tau \leq s \leq t. \end{cases}$$

Linking (3.38) and the above relations, the solution of (3.39) can be expressed by

$$x(t) = \varphi(-\tau) + (t + \tau)\dot{\varphi}(-\tau) + \Omega^{-1} \int_{-\tau}^{t} \sin_\tau \Omega(t - \tau - s)\ddot{\varphi}(s)ds. \tag{3.45}$$

Integrating by parts and using Lemma 3.3 yields

$$\int_{-\tau}^{t} \sin_\tau \Omega(t - \tau - s)\ddot{\varphi}(s)ds$$

$$= \int_{-\tau}^{t} \sin_\tau \Omega(t - \tau - s)d\dot{\varphi}(s)$$

$$= \sin_\tau \Omega(t - \tau - s)\dot{\varphi}(s)|_{-\tau}^{t} - \int_{-\tau}^{t} \dot{\varphi}(s)d \sin_\tau \Omega(t - \tau - s)$$

$$= -(t + \tau)\Omega\dot{\varphi}(-\tau) + \Omega\varphi(t) - \Omega\varphi(-\tau). \tag{3.46}$$

Putting (3.46) into (3.45), we get

$$x(t) = \varphi(-\tau) + (t + \tau)\dot{\varphi}(-\tau) + \Omega^{-1}[-\Omega(t + \tau)\dot{\varphi}(-\tau) + \Omega\varphi(t) - \Omega\varphi(-\tau)]$$

$$= \varphi(t).$$

Now $\dot{x}(t) = \dot{\varphi}(t)$ holds. Thus, (3.39) is controllable according to Definition 3.3.

Next we establish the necessity. Assume the delay Grammian matrix $W_\tau[0, t_1]$ is singular. Then $W_\tau[0, t_1][\Omega^{-1}]^\top$ is singular too. Thus, there exists at least one nonzero state $\bar{x} \in \mathbb{R}^n$ such that

$$\bar{x}^\top W_\tau[0, t_1][\Omega^{-1}]^\top \bar{x} = 0.$$

It follows from (3.40) that

$$\begin{aligned}
0 &= \bar{x}^\top W_\tau[0, t_1][\Omega^{-1}]^\top \bar{x} \\
&= \int_0^{t_1} \bar{x}^\top \Omega^{-1} \sin_\tau \Omega(t_1 - \tau - s) B B^\top \sin_\tau \Omega^\top(t_1 - \tau - s)[\Omega^{-1}]^\top \bar{x} ds \\
&= \int_0^{t_1} \left[\bar{x}^\top \Omega^{-1} \sin_\tau \Omega(t_1 - \tau - s) B\right] \left[\bar{x}^\top \Omega^{-1} \sin_\tau \Omega(t_1 - \tau - s) B\right]^\top ds \\
&= \int_0^{t_1} \left\| \bar{x}^\top \Omega^{-1} \sin_\tau \Omega(t_1 - \tau - s) B \right\|^2 ds.
\end{aligned}$$

This implies that

$$\bar{x}^\top \Omega^{-1} \sin_\tau \Omega(t_1 - \tau - s) B = \mathbf{0}^\top, \quad \forall s \in J. \tag{3.47}$$

Since (3.39) is controllable, it can be driven from any continuously differentiable initial vector functions $\varphi, \dot{\varphi} : [-\tau, 0] \to \mathbb{R}^n$ to an arbitrary state $x(t_1) \in \mathbb{R}^n$. Hence there exists a control $u_0(t)$ that drives the initial state to zero. This means that

$$\begin{aligned}
x(t_1) = {} &\cos_\tau \Omega t_1 \varphi(-\tau) + \Omega^{-1} \sin_\tau \Omega t_1 \dot{\varphi}(-\tau) \\
&+ \Omega^{-1} \int_{-\tau}^0 \sin_\tau \Omega(t_1 - \tau - s) \ddot{\varphi}(s) ds \\
&+ \Omega^{-1} \int_0^{t_1} \sin_\tau \Omega(t_1 - \tau - s) B u_0(s) ds = \mathbf{0}. \tag{3.48}
\end{aligned}$$

Moreover, there exists a control $\tilde{u}(t)$ that drives the initial state to the state $\bar{x}$, so

$$\begin{aligned}
x(t_1) = {} &\cos_\tau \Omega t_1 \varphi(-\tau) + \Omega^{-1} \sin_\tau \Omega t_1 \dot{\varphi}(-\tau) \\
&+ \Omega^{-1} \int_{-\tau}^0 \sin_\tau \Omega(t_1 - \tau - s) \ddot{\varphi}(s) ds \\
&+ \Omega^{-1} \int_0^{t_1} \sin_\tau \Omega(t_1 - \tau - s) B \tilde{u}(s) ds = \bar{x}. \tag{3.49}
\end{aligned}$$

Combining (3.48) and (3.49) gives

$$\bar{x} = \Omega^{-1} \int_0^{t_1} \sin_\tau \Omega(t_1 - \tau - s) B[\tilde{u}(s) - u_0(s)] ds.$$

Multiplying both sides of the equality by $\bar{x}^\top$, we get

$$\bar{x}^\top \bar{x} = \int_0^{t_1} \bar{x}^\top \Omega^{-1} \sin_\tau \Omega(t_1 - \tau - s) B[\tilde{u}(s) - u_0(s)] ds.$$

Note that with (3.47), we obtain $\bar{x}^\top \bar{x} = 0$. That is, $\bar{x} = \mathbf{0}$, which conflicts with $\bar{x}$ being nonzero. Thus, the delay Grammian matrix $W_\tau[0, t_1]$ is nonsingular. $\qquad\square$

3.2.1.2 Controllability of nonlinear systems

In this section, we apply a fixed point method to establish a sufficient condition of controllability for (3.37).

We assume the following:

(H_1) $f : J \times \mathbb{R}^n \to \mathbb{R}^n$ is continuous (here $J = [0, t_1]$), and there exist $L_f \in L^q(J, \mathbb{R}^+)$ and $q > 1$ such that

$$\|f(t, x_1) - f(t, x_2)\| \le L_f(t)\|x_1 - x_2\|,$$

and set $M_f = \sup_{t \in J} \|f(t, 0)\|$.

(H_2) Consider the operator $W : L^2(J, \mathbb{R}^m) \to \mathbb{R}^n$ given by

$$W = \Omega^{-1} \int_0^{t_1} \sin_\tau \Omega(t_1 - \tau - s) Bu(s) ds.$$

Suppose that W^{-1} exists and there exists a constant $M_1 > 0$ such that

$$\|W^{-1}\|_{L_b(\mathbb{R}^n, L^2(J, \mathbb{R}^m)/\ker W)} \le M_1.$$

Next, consider a control function u_x of the form

$$
\begin{aligned}
u_x(t) = W^{-1}\Big[& x_1 - (\cos_\tau \Omega t_1)\varphi(-\tau) - \Omega^{-1}(\sin_\tau \Omega t_1)\dot{\varphi}(-\tau) \\
& - \Omega^{-1} \int_{-\tau}^0 \sin_\tau \Omega(t_1 - \tau - s)\ddot{\varphi}(s) ds \\
& - \Omega^{-1} \int_0^{t_1} \sin_\tau \Omega(t_1 - \tau - s) f(s, x(s)) ds \Big](t), \ t \in J. \qquad (3.50)
\end{aligned}
$$

We define an operator $\Gamma : C([-\tau, t_1], \mathbb{R}^n) \to C([-\tau, t_1], \mathbb{R}^n)$ as follows:

$$
\begin{aligned}
(\Gamma x)(t) = {} & (\cos_\tau \Omega t)\varphi(-\tau) + \Omega^{-1}(\sin_\tau \Omega t)\dot{\varphi}(-\tau) \\
& + \Omega^{-1} \int_{-\tau}^0 \sin_\tau \Omega(t - \tau - s)\ddot{\varphi}(s) ds
\end{aligned}
$$

$$+ \Omega^{-1} \int_0^t \sin_\tau \Omega(t - \tau - s) f(s, x(s)) ds$$

$$+ \Omega^{-1} \int_0^t \sin_\tau \Omega(t - \tau - s) B u_x(s) ds. \tag{3.51}$$

For each positive number ξ, let $O_\xi = \{x \in C([-\tau, t_1], \mathbb{R}^n) : \|x\| = \sup_{t \in [-\tau, t_1]} \|x(t)\| \leq \xi\}$. Now O_ξ is a bounded, closed, and convex set of $C([-\tau, t_1], \mathbb{R}^n)$.

Now we use Krasnoselskii's fixed point theorem to prove our result. We first prove that the operator Γ has a fixed point x, which is a solution of (3.37). Then we check $(\Gamma x)(t) = \varphi(t)$, $\frac{d}{dt}(\Gamma x)(t) = \dot{\varphi}(t)$ when $-\tau \leq t \leq 0$ and $(\Gamma x)(t_1) = x_1$, $\frac{d}{dt}(\Gamma x)(t_1) = x_1'$ via the control u_x defined in (3.50), and this means system (3.37) is controllable.

Theorem 3.6. *Suppose (H_1) and (H_2) are satisfied. Then (3.37) is controllable if*

$$M_2 \left[1 + \frac{\cosh(\|\Omega\| t_1) - 1}{\|\Omega\|} \|\Omega^{-1}\| \|B\| M_1 \right] < 1, \tag{3.52}$$

where $M_2 = \|\Omega^{-1}\| \left[\frac{1}{2^p \|\Omega\| p} (e^{\|\Omega\| p t_1} - 1) \right]^{\frac{1}{p}} \|L_f\|_{L^q(J, \mathbb{R}^+)}$, $\frac{1}{p} + \frac{1}{q} = 1$, p, $q > 1$.

Proof. We divide our proof into three steps to verify the conditions required in Krasnoselskii's fixed point theorem.

Step 1. We show $\Gamma(O_\xi) \subseteq O_\xi$ for some positive number ξ.

Consider any positive number ξ and let $x^\xi \in O_\xi$.

Let $t \in [0, t_1]$. From (H_1) and the Hölder inequality, we obtain

$$\int_0^t \sinh\left[\|\Omega\|(t - s)\right] L_f(s) ds$$

$$\leq \left(\int_0^t \left(\sinh[\|\Omega\|(t - s)]\right)^p ds\right)^{\frac{1}{p}} \left(\int_0^t L_f^q(s) ds\right)^{\frac{1}{q}}$$

$$\leq \left(\int_0^t \frac{e^{\|\Omega\| p(t-s)}}{2^p} ds\right)^{\frac{1}{p}} \|L_f\|_{L^q(J, \mathbb{R}^+)}$$

$$= \left[\frac{1}{2^p \|\Omega\| p} (e^{\|\Omega\| p t} - 1)\right]^{\frac{1}{p}} \|L_f\|_{L^q(J, \mathbb{R}^+)}, \tag{3.53}$$

where we use the fact that $\sinh t = \frac{e^t - e^{-t}}{2} \leq \frac{e^t}{2}$, for $\forall\, t \in \mathbb{R}$. Next,

$$\int_0^t \sinh\left[\|\Omega\|(t - s)\right] \|f(s, 0)\| ds \leq M_f \int_0^t \sinh\left[\|\Omega\|(t - s)\right] ds$$

$$\leq \frac{M_f}{\|\Omega\|}\left[\cosh(\|\Omega\|t) - 1\right]. \qquad (3.54)$$

From (3.50), (H_1), (H_2), (3.53), (3.54), and Lemmas 3.2 and 3.3, we obtain (here $\|\varphi\| = \max\limits_{s\in[-\tau,0]}\|\varphi(s)\|$, $\|\dot\varphi\| = \max\limits_{s\in[-\tau,0]}\|\dot\varphi(s)\|$, and $\|\ddot\varphi\| = \max\limits_{s\in[-\tau,0]}\|\ddot\varphi(s)\|$)

$$\|u_x(t)\| \leq \|W^{-1}\|_{L(\mathbb{R}^n, L^2(J,\mathbb{R}^m))/\ker W}\Bigg(\|x_1\| + \|\cos_\tau \Omega t\|\|\varphi(-\tau)\|$$

$$+ \|\Omega^{-1}\|\|\sin_\tau \Omega t\|\|\dot\varphi(-\tau)\|$$

$$+ \|\Omega^{-1}\|\int_{-\tau}^0 \|\sin_\tau \Omega(t-\tau-s)\|\|\ddot\varphi(s)\|ds$$

$$+ \|\Omega^{-1}\|\int_0^t \|\sin_\tau \Omega(t-\tau-s)\|\|f(s,x(s))\|ds\Bigg)$$

$$\leq M_1\|x_1\| + M_1\cosh(\|\Omega\|t)\|\varphi\| + M_1\|\Omega^{-1}\|\sinh\left[\|\Omega\|(t+\tau)\right]\|\dot\varphi\|$$

$$+ M_1\|\Omega^{-1}\|\|\ddot\varphi\|\int_{-\tau}^0 \sinh\left[\|\Omega\|(t-s)\right]ds$$

$$+ M_1\|\Omega^{-1}\|\int_0^t \sinh\left[\|\Omega\|(t-s)\right]L_f(s)\|x(s)\|ds$$

$$+ M_1\|\Omega^{-1}\|\int_0^t \sinh\left[\|\Omega\|(t-s)\right]\|f(s,0)\|ds$$

$$\leq M_1\|x_1\| + M_1\cosh(\|\Omega\|t)\|\varphi\| + M_1\|\Omega^{-1}\|\sinh\left[\|\Omega\|(t+\tau)\right]\|\dot\varphi\|$$

$$+ \frac{M_1\|\Omega^{-1}\|\|\ddot\varphi\|}{\|\Omega\|}\left(\cosh[\|\Omega\|(t+\tau)] - \cosh(\|\Omega\|t)\right)$$

$$+ M_1\|\Omega^{-1}\|\left[\frac{1}{2^p\|\Omega\|p}(e^{\|\Omega\|pt} - 1)\right]^{\frac{1}{p}}\|L_f\|_{L^q(J,\mathbb{R}^+)}\|x\|$$

$$+ M_1\|\Omega^{-1}\|\frac{M_f}{\|\Omega\|}\left[\cosh(\|\Omega\|t) - 1\right]$$

$$\leq M_1\|x_1\| + M_1\vartheta(t) + M_1 M_2\xi$$

$$\leq M_1\|x_1\| + M_1\vartheta(t_1) + M_1 M_2\xi,$$

where

$$\vartheta(t) = \cosh(\|\Omega\|t)\|\varphi\| + \|\Omega^{-1}\|\sinh\left[\|\Omega\|(t+\tau)\right]\|\dot\varphi\| + \|\Omega^{-1}\|\frac{M_f}{\|\Omega\|}$$

$$\times \left[\cosh(\|\Omega\|t) - 1\right] + \frac{\|\Omega^{-1}\|\|\ddot\varphi\|}{\|\Omega\|}\left(\cosh[\|\Omega\|(t+\tau)] - \cosh(\|\Omega\|t)\right)$$

(here we used the fact that $\frac{d}{dt}\vartheta(t) > 0$, $\forall\, t \in J$).

Now

$$\|(\Gamma x^{\xi})(t)\| \leq \|\cos_{\tau} \Omega t\| \|\varphi(-\tau)\| + \|\Omega^{-1}\| \|\sin_{\tau} \Omega t\| \|\dot{\varphi}(-\tau)\|$$

$$+ \|\Omega^{-1}\| \int_{-\tau}^{0} \|\sin_{\tau} \Omega(t-\tau-s)\| \|\ddot{\varphi}(s)\| ds$$

$$+ \|\Omega^{-1}\| \int_{0}^{t} \|\sin_{\tau} \Omega(t-\tau-s)\| \|f(s,x(s))\| ds$$

$$+ \|\Omega^{-1}\| \int_{0}^{t} \|\sin_{\tau} \Omega(t-\tau-s)\| \|B\| \|u_x(s)\| ds$$

$$\leq \cosh(\|\Omega\| t) \|\varphi\| + \|\Omega^{-1}\| \sinh\left[\|\Omega\|(t+\tau)\right] \|\dot{\varphi}\|$$

$$+ \frac{\|\Omega^{-1}\| \|\ddot{\varphi}\|}{\|\Omega\|} \left(\cosh[\|\Omega\|(t+\tau)] - \cosh(\|\Omega\| t) \right)$$

$$+ \|\Omega^{-1}\| \int_{0}^{t} \sinh\left[\|\Omega\|(t-s)\right] L_f(s) \|x(s)\| ds$$

$$+ \|\Omega^{-1}\| \int_{0}^{t} \sinh\left[\|\Omega\|(t-s)\right] \|f(s,0)\| ds + \|\Omega^{-1}\|$$

$$\times \int_{0}^{t} \sinh\left[\|\Omega\|(t-s)\right] \|B\| \left(M_1 \|x_1\| + M_1 \vartheta(t_1) + M_1 M_2 \xi \right) ds$$

$$\leq \vartheta(t_1) + M_2 \xi + \frac{\cosh(\|\Omega\| t) - 1}{\|\Omega\|} \|\Omega^{-1}\| \|B\| M_1 \|x_1\|$$

$$+ \frac{\cosh(\|\Omega\| t) - 1}{\|\Omega\|} \|\Omega^{-1}\| \|B\| M_1 \vartheta(t_1)$$

$$+ \frac{\cosh(\|\Omega\| t) - 1}{\|\Omega\|} \|\Omega^{-1}\| \|B\| M_1 M_2 \xi$$

$$\leq \vartheta(t_1) \left[1 + \frac{\cosh(\|\Omega\| t_1) - 1}{\|\Omega\|} \|\Omega^{-1}\| \|B\| M_1 \right]$$

$$+ \frac{\cosh(\|\Omega\| t_1) - 1}{\|\Omega\|} \|\Omega^{-1}\| \|B\| M_1 \|x_1\|$$

$$+ M_2 \left[1 + \frac{\cosh(\|\Omega\| t_1) - 1}{\|\Omega\|} \|\Omega^{-1}\| \|B\| M_1 \right] \xi.$$

Thus for some ξ sufficiently large (which we take for the rest of the proof), from (3.52) we have $\Gamma(x^{\xi}) \in O_{\xi}$, and as a result, $\Gamma(O_{\xi}) \subseteq O_{\xi}$.

Now we write the operator Γ defined in (3.51) as $\Gamma_1 + \Gamma_2$, where

$$(\Gamma_1 x)(t) = (\cos_{\tau} \Omega t)\varphi(-\tau) + \Omega^{-1}(\sin_{\tau} \Omega t)\dot{\varphi}(-\tau)$$

$$+ \Omega^{-1} \int_{-\tau}^{0} \sin_{\tau} \Omega(t-\tau-s)\ddot{\varphi}(s)ds$$

$$+ \Omega^{-1} \int_0^t \sin_\tau \Omega(t - \tau - s) B u_x(s) ds, \tag{3.55}$$

$$(\Gamma_2 x)(t) = \Omega^{-1} \int_0^t \sin_\tau \Omega(t - \tau - s) f(s, x(s)) ds. \tag{3.56}$$

Step 2. We show $\Gamma_1 : O_\xi \to C([-\tau, t_1], \mathbb{R}^n)$ is a contraction.
Let $t \in [0, t_1]$. From (3.50), (3.53), (H_1), and (H_2), for $\forall\, x, y \in O_\xi$, we have

$$\|u_x(t) - u_y(t)\| \le M_1 \|\Omega^{-1}\| \int_0^t \|\sin_\tau \Omega(t - \tau - s)\| L_f(s) \|x(s) - y(s)\| ds$$

$$\le M_1 \|\Omega^{-1}\| \|x - y\| \int_0^t \sinh\left[\|\Omega\|(t - s)\right] L_f(s) ds$$

$$\le M_1 M_2 \|x - y\|.$$

Then from (3.55), we have

$$\|(\Gamma_1 x)(t) - (\Gamma_1 y)(t)\|$$

$$\le \|\Omega^{-1}\| \int_0^t \|\sin_\tau \Omega(t - \tau - s)\| \|B\| \|u_x(s) - u_y(s)\| ds$$

$$\le \|\Omega^{-1}\| \|B\| M_1 M_2 \|x - y\| \int_0^t \sinh\left[\|\Omega\|(t - s)\right] ds$$

$$\le \varpi \|x - y\|,$$

where $\varpi =: \frac{\cosh(\|\Omega\|t_1)-1}{\|\Omega\|} \|\Omega^{-1}\| \|B\| M_1 M_2$. From (3.52), $\varpi < 1$ implies Γ_1 is a contraction.

Step 3. We show that $\Gamma_2 : O_\xi \to C([-\tau, t_1], \mathbb{R}^n)$ is a continuous compact operator.

Let $x_n \in O_\xi$ with $x_n \to x$ in O_ξ, let $F_n(\cdot) = f(\cdot, x_n(\cdot))$ and $F(\cdot) = f(\cdot, x(\cdot))$, and note

$$\sinh\left[\|\Omega\|(\cdot - s)\right] F_n(s) \to \sinh\left[\|\Omega\|(\cdot - s)\right] F(s), \quad a.e.\ s \in J = [0, t_1].$$

From (H_1), we get

$$\sinh\left[\|\Omega\|(\cdot - s)\right] \|F_n(s) - F(s)\| \le 2\xi \sinh\left[\|\Omega\|(\cdot - s)\right] L_f(s) \in L^1(J, \mathbb{R}^+).$$

Then using (3.56) and Lebesgue's dominated convergence theorem, we obtain

$$\|(\Gamma_2 x_n)(t) - (\Gamma_2 x)(t)\|$$

$$\le \|\Omega^{-1}\| \int_0^t \sinh\left[\|\Omega\|(t - s)\right] \|F_n(s) - F(s)\| ds \to 0 \text{ as } n \to \infty.$$

Thus $\Gamma_2 : O_\xi \to C([-\tau, t_1], \mathbb{R}^n)$ is continuous.

Next we show $\Gamma_2(O_\xi) \subset C([\tau, t_1], \mathbb{R}^n)$ is equicontinuous.

For $x \in O_\xi$ and $0 < t \leq t + h \leq t_1$, from (3.56), we have

$$(\Gamma_2 x)(t + h) - (\Gamma_2 x)(t) = \Omega^{-1} \int_0^{t+h} \sin_\tau \Omega(t + h - \tau - s) F(s) ds$$

$$- \Omega^{-1} \int_0^t \sin_\tau \Omega(t - \tau - s) F(s) ds$$

$$= \mathbf{K}_1 + \mathbf{K}_2,$$

where

$$\mathbf{K}_1 = \Omega^{-1} \int_t^{t+h} \sin_\tau \Omega(t + h - \tau - s) F(s) ds$$

and

$$\mathbf{K}_2 = \Omega^{-1} \int_0^t \left[\sin_\tau \Omega(t + h - \tau - s) - \sin_\tau \Omega(t - \tau - s) \right] F(s) ds.$$

Thus

$$\|(\Gamma_2 x)(t + h) - (\Gamma_2 x)(t)\| \leq \|\mathbf{K}_1\| + \|\mathbf{K}_2\|. \tag{3.57}$$

Now, we check $\|\mathbf{K}_i\| \to 0$ as $h \to 0$, $i = 1, 2$. For $\mathbf{K}_1$ (similar to (3.53)) we obtain

$$\int_t^{t+h} \sinh\left[\|\Omega\|(t + h - s) \right] L_f(s) ds$$

$$\leq \left[\frac{1}{2^{p_1} \|\Omega\| p_1} (e^{\|\Omega\| p_1 h} - 1) \right]^{\frac{1}{p_1}} \|L_f\|_{L^{q_1}(J, \mathbb{R}^+)}, \tag{3.58}$$

where $\frac{1}{p_1} + \frac{1}{q_1} = 1$, $p_1, q_1 > 1$. Then using (H_1), (3.58), and Lemmas 3.2 and 3.3, we get

$$\|\mathbf{K}_1\| \leq \|\Omega^{-1}\| \int_t^{t+h} \sinh\left[\|\Omega\|(t + h - s) \right] \|F(s)\| ds$$

$$\leq \|\Omega^{-1}\| \int_t^{t+h} \sinh\left[\|\Omega\|(t + h - s) \right]$$

$$\times (\|f(s, x(s)) - f(s, 0)\| + \|f(s, 0)\|) ds$$

$$\leq \|\Omega^{-1}\| \int_t^{t+h} \sinh\left[\|\Omega\|(t + h - s) \right] L_f(s) \|x(s)\| ds$$

$$+ M_f \|\Omega^{-1}\| \int_t^{t+h} \sinh\left[\|\Omega\|(t + h - s) \right] ds$$

$$\leq \xi \|\Omega^{-1}\| \left[\frac{1}{2^{p_1}\|\Omega\|p_1} (e^{\|\Omega\|p_1 h} - 1) \right]^{\frac{1}{p_1}} \|L_f\|_{L^{q_1}(J,\mathbb{R}^+)}$$

$$+ M_f \|\Omega^{-1}\| \frac{\cosh(\|\Omega\|h) - 1}{\|\Omega\|} \longrightarrow 0 \quad \text{as} \quad h \to 0.$$

For $\mathbf{K}_2$, from the Hölder inequality, we have

$$\int_0^t \|\sin_\tau \Omega(t+h-\tau-s) - \sin_\tau \Omega(t-\tau-s)\| L_f(s) ds$$

$$\leq \left(\int_0^t \|\sin_\tau \Omega(t+h-\tau-s) - \sin_\tau \Omega(t-\tau-s)\|^{p_2} ds \right)^{\frac{1}{p_2}} \|L_f\|_{L^{q_2}(J,\mathbb{R}^+)},$$

where $\frac{1}{p_2} + \frac{1}{q_2} = 1$, $p_2, q_2 > 1$. Then we get

$$\|\mathbf{K}_2\| \leq \|\Omega^{-1}\| \int_0^t \|\sin_\tau \Omega(t+h-\tau-s) - \sin_\tau \Omega(t-\tau-s)\| \|F(s)\| ds$$

$$\leq \xi \|\Omega^{-1}\| \int_0^t \|\sin_\tau \Omega(t+h-\tau-s) - \sin_\tau \Omega(t-\tau-s)\| L_f(s) ds$$

$$+ M_f \|\Omega^{-1}\| \int_0^t \|\sin_\tau \Omega(t+h-\tau-s) - \sin_\tau \Omega(t-\tau-s)\| ds$$

$$\leq \xi \|\Omega^{-1}\| \left(\int_0^{t_1} \|\sin_\tau \Omega(t+h-\tau-s) - \sin_\tau \Omega(t-\tau-s)\|^{p_2} ds \right)^{\frac{1}{p_2}}$$

$$\times \|L_f\|_{L^{q_2}(J,\mathbb{R}^+)}$$

$$+ M_f \|\Omega^{-1}\| \int_0^{t_1} \|\sin_\tau \Omega(t+h-\tau-s) - \sin_\tau \Omega(t-\tau-s)\| ds.$$

From (3.4), we know that the delayed matrix function $\sin_\tau \Omega t$ is uniformly continuous for $\forall\, t \in J$, and thus, we get $\|\sin_\tau \Omega(t+h-\tau-s) - \sin_\tau \Omega(t-\tau-s)\| \to 0$ as $h \to 0$. Finally, we get $\|\mathbf{K}_2\| \to 0$. Now $\|\mathbf{K}_1\| \to 0$ and $\|\mathbf{K}_2\| \to 0$ with (3.57) yield

$$\|(\Gamma_2 x)(t+h) - (\Gamma_2 x)(t)\| \to 0 \quad \text{as} \quad h \to 0,$$

for all $x \in O_\xi$. The other cases are treated similarly. From the Arzela–Ascoli theorem we know $\Gamma_2 : O_\xi \to C([\tau, t_1], \mathbb{R}^n)$ is compact.

From Krasnoselskii's fixed point theorem (see [82]), Γ has a fixed point x on O_ξ. From the definition of operator Γ, x is also the solution of system (3.37). Note $x(t_1) = x_1$ via the control function $u_x(t)$. Also $\dot{x}(t_1) = x_1'$. Finally, we get the initial conditions $x(t) = \varphi(t)$, $\dot{x}(t) = \dot{\varphi}(t)$ when $-\tau \leq t \leq 0$ using the same procedure as in the proof of (3.39) in Theorem 3.5. Thus, system (3.37) is controllable. $\qquad\square$

3.2.1.3 Numerical examples and discussion

Example 3.3. Consider the controllability of the following linear delay differential controlled system:

$$\begin{cases} \ddot{x}(t) + \Omega^2 x(t - 0.6) = Bu(t), \ t \in [0, 1.2], \\ x(t) \equiv \varphi(t), \ \dot{x}(t) \equiv \dot{\varphi}(t), \ t \in [-0.6, 0], \end{cases} \tag{3.59}$$

where

$$\Omega = \begin{pmatrix} 1 & 2 \\ 0 & 1 \end{pmatrix}, \ B = \begin{pmatrix} 1 \\ 1 \end{pmatrix}, \ \varphi(t) = \begin{pmatrix} 3t \\ 2t \end{pmatrix}, \ \dot{\varphi}(t) = \begin{pmatrix} 3 \\ 2 \end{pmatrix}.$$

Noting that B is an $n \times m$ matrix, with an input $u : [0, t_1] \to \mathbb{R}^m$, we can see $n = 2$, $m = 1$, $\tau = 0.6$, $t_1 = 1.2$. Constructing the corresponding delay Grammian matrix of system (3.59) via (3.40), we obtain

$$W_{0.6}[0, 1.2] = \Omega^{-1} \int_0^{1.2} \sin_{0.6} \Omega(0.6 - s) BB^\top \sin_{0.6} \Omega^\top (0.6 - s) ds$$
$$:= E_1 + E_2,$$

where

$$E_1 = \Omega^{-1} \int_0^{0.6} \sin_{0.6} \Omega(0.6 - s) BB^\top \sin_{0.6} \Omega^\top (0.6 - s) ds,$$
$$(0.6 - s) \in (0, 0.6),$$
$$E_2 = \Omega^{-1} \int_{0.6}^{1.2} \sin_{0.6} \Omega(0.6 - s) BB^\top \sin_{0.6} \Omega^\top (0.6 - s) ds,$$
$$(0.6 - s) \in (-0.6, 0),$$

and

$$\cos_{0.6} \Omega t = \begin{cases} \Theta, \ t \in (-\infty, -0.6), \\ I, \ t \in [-0.6, 0), \\ I - \Omega^2 \frac{t^2}{2!}, \ t \in [0, 0.6), \\ I - \Omega^2 \frac{t^2}{2!} + \Omega^4 \frac{(t-0.6)^4}{4!}, \ t \in [0.6, 1.2), \\ \vdots \end{cases}$$

$$\sin_{0.6} \Omega t = \begin{cases} \Theta, \ t \in (-\infty, -0.6), \\ \Omega(t + 0.6), \ t \in [-0.6, 0), \\ \Omega(t + 0.6) - \Omega^3 \frac{t^3}{3!}, \ t \in [0, 0.6), \\ \Omega(t + 0.6) - \Omega^3 \frac{t^3}{3!} + \Omega^5 \frac{(t-0.6)^5}{5!}, \ t \in [0.6, 1.2). \\ \vdots \end{cases} \tag{3.60}$$

Next, we can calculate that

$$E_1 = \begin{pmatrix} \frac{21681}{15625} & \frac{102717}{218750} \\ \frac{227259}{156250} & \frac{269307}{546875} \end{pmatrix}, \quad E_2 = \begin{pmatrix} \frac{27}{125} & \frac{9}{125} \\ \frac{27}{125} & \frac{9}{125} \end{pmatrix}.$$

Then, we get

$$W_{0.6}[0, 1.2] = \begin{pmatrix} \frac{25056}{15625} & \frac{118467}{218750} \\ \frac{261009}{156250} & \frac{308682}{546875} \end{pmatrix},$$

$$W_{0.6}^{-1}[0, 1.2] = \begin{pmatrix} \frac{428725000}{364257} & \frac{-411343750}{364257} \\ \frac{-422931250}{121419} & \frac{406000000}{121419} \end{pmatrix}.$$

Thus, system (3.59) is controllable by Theorem 3.5. In addition, for any finite terminal conditions $x(t_1) = x_1 = (x_{11}, x_{12})^\top$, $\dot{x}(t_1) = x_1' = (x_{11}', x_{12}')^\top$, it follows (see (3.41)) that one can construct the corresponding control input $u(t) \in \mathbb{R}$ as

$$u(t) = B^\top \sin_{0.6} \Omega^\top (0.6 - t) W_{0.6}^{-1}[0, 1.2]\zeta, \tag{3.61}$$

where

$$\zeta = x_1 - (\cos_{0.6} \Omega 1.2)\varphi(-0.6) - \Omega^{-1}(\sin_{0.6} \Omega 1.2)\dot{\varphi}(-0.6)$$

$$= \begin{pmatrix} x_{11} - \frac{36132604990374860091}{70368744177766400000} \\ x_{12} - \frac{9439319638987191039}{3518437208883200000} \end{pmatrix}.$$

From (3.38) and (3.61), the solution of system (3.59) has the following form:

$$x(t) = (\cos_{0.6} \Omega t)\varphi(-0.6) + \Omega^{-1}(\sin_{0.6} \Omega t)\dot{\varphi}(-0.6)$$

$$+ \Omega^{-1} \int_0^t \sin_{0.6} \Omega(t - 0.6 - s) B$$

$$\times B^\top \sin_{0.6} \Omega^\top (0.6 - s)ds \; W_{0.6}^{-1}[0, 1.2]\zeta. \tag{3.62}$$

Now we consider the integral term
$\int_0^t \sin_{0.6} \Omega(t - 0.6 - s)BB^\top \sin_{0.6} \Omega^\top (0.6 - s)ds$ in (3.62).

For $0 < t < 0.6$, we can obtain $-0.6 < t - 0.6 - s < t - 0.6 < 0$ and $0 < 0.6 - t < 0.6 - s < 0.6$, so the solution (3.62) can be expressed as the following form via (3.60):

$$x(t) = \left[I - \Omega^2 \frac{t^2}{2}\right] \varphi(-0.6) + \Omega^{-1} \left[\Omega(t + 0.6) - \Omega^3 \frac{t^3}{6}\right] \dot{\varphi}(-0.6)$$

$$+ \Omega^{-1} \int_0^t [\Omega(t - s)] BB^\top \left[\Omega^\top (1.2 - s) - (\Omega^\top)^3 \frac{(0.6 - s)^3}{6}\right] ds$$

$$\times W_{0.6}^{-1}[0, 1.2]\zeta.$$

For $0.6 < t < 1.2$, we get $0 < t - 0.6 - s < t - 0.6 < 0.6$ when $0 < s < t - 0.6$ and $-0.6 < t - 0.6 - s < 0$ when $t - 0.6 < s < t$. We can also obtain $0 < 0.6 - s < 0.6$ when $0 < s < 0.6$ and $-0.6 < 0.6 - t < 0.6 - s < 0$ when $0.6 < s < t$. Finally, (3.62) can be expressed as the following formula via (3.60):

$$
x(t) = \left[I - \Omega^2 \frac{t^2}{2} + \Omega^4 \frac{(t-0.6)^4}{24} \right] \varphi(-0.6)
$$

$$
+ \Omega^{-1} \left[\Omega(t+0.6) - \Omega^3 \frac{t^3}{6} + \Omega^5 \frac{(t-0.6)^5}{120} \right] \dot\varphi(-0.6)
$$

$$
+ \Omega^{-1} \int_0^{t-0.6} \left[\Omega(t-s) - \Omega^3 \frac{(t-0.6-s)^3}{6} \right] B
$$

$$
\times B^{\mathrm T} \left[\Omega^{\mathrm T}(1.2-s) - (\Omega^{\mathrm T})^3 \frac{(0.6-s)^3}{6} \right] ds\, W_{0.6}^{-1}[0,\,1.2]\zeta
$$

$$
+ \Omega^{-1} \int_{t-0.6}^{0.6} [\Omega(t-s)]\, B B^{\mathrm T} \left[\Omega^{\mathrm T}(1.2-s) \right.
$$

$$
\left. - (\Omega^{\mathrm T})^3 \frac{(0.6-s)^3}{6} \right] ds\, W_{0.6}^{-1}[0,\,1.2]\zeta
$$

$$
+ \Omega^{-1} \int_{0.6}^{t} [\Omega(t-s)]\, B B^{\mathrm T} \left[\Omega^{\mathrm T}(1.2-s) \right] ds\, W_{0.6}^{-1}[0,\,1.2]\zeta.
$$

Example 3.4. In this example, we consider the following nonlinear delay differential controlled system:

$$
\begin{cases} \ddot{x}(t) + \Omega^2 x(t-0.4) = f(t, x(t)) + Bu(t), \ t \in [0, 0.8], \\ x(t) \equiv \varphi(t), \ \dot{x}(t) \equiv \dot\varphi(t), \ t \in [-0.4, 0], \end{cases} \tag{3.63}
$$

where we set

$$
\Omega = \begin{pmatrix} 0 & 1 \\ -1 & 0 \end{pmatrix}, \ B = I_{2\times 2}, \varphi(t) = \begin{pmatrix} 5t+1 \\ 2t^2 \end{pmatrix}, \ \dot\varphi(t) = \begin{pmatrix} 5 \\ 4t \end{pmatrix},
$$

$$
f(t, x(t)) = \begin{pmatrix} 0.3(t-0.4)\sin[x_1(t)] \\ 0.3(t-0.4)\sin[x_2(t)] \end{pmatrix}.
$$

Now, we set $u(t) = \tilde{x}$, where $\tilde{x} = \sum_{n=1}^{2} \langle \tilde{x}, e_n \rangle e_n$, where e_n is the orthonormal basis of $\mathbb{R}^2$. From the definition of W in (H_2), we get

$$
W = \Omega^{-1} \int_0^{0.8} \sin_{0.4} \Omega(0.4 - s) B ds\, \tilde{x}
$$

$$
= \Omega^{-1} \int_0^{0.4} \sin_{0.4} \Omega(0.4 - s) ds\, \tilde{x} + \Omega^{-1} \int_{0.4}^{0.8} \sin_{0.4} \Omega(0.4 - s) ds\, \tilde{x}
$$

$$= \begin{pmatrix} \frac{452}{1875} & 0 \\ 0 & \frac{452}{1875} \end{pmatrix} \widetilde{x} + \begin{pmatrix} \frac{2}{25} & 0 \\ 0 & \frac{2}{25} \end{pmatrix} \widetilde{x}$$

$$= \begin{pmatrix} \frac{602}{1875} & 0 \\ 0 & \frac{602}{1875} \end{pmatrix} \widetilde{x}.$$

Define the inverse $W^{-1} : \mathbb{R}^2 \to L^2(J_1, \mathbb{R}^2)$ by

$$(W^{-1}\widetilde{x})(t) := \begin{pmatrix} \frac{1875}{602} & 0 \\ 0 & \frac{1875}{602} \end{pmatrix} \widetilde{x},$$

where $J_1 = [0, 0.8]$.

Then, we get

$$\|(W^{-1}\widetilde{x})(t)\| \leq \left\| \begin{pmatrix} \frac{1875}{602} & 0 \\ 0 & \frac{1875}{602} \end{pmatrix} \right\| \|\widetilde{x}\| = 3.1146\|\widetilde{x}\|,$$

and thus, we obtain $\|W^{-1}\| \leq 3.1146 := M_1$. Hence, W satisfies assumption (H_2).

Next, note that $|\sin a - \sin b| \leq |a - b|$, $\forall\, a, b \in \mathbb{R}$. We have

$$\|f(t, x) - f(t, y)\|$$

$$= |0.3(t - 0.4)|\sqrt{(\sin[x_1(t)] - \sin[y_1(t)])^2 + (\sin[x_2(t)] - \sin[y_2(t)])^2}$$

$$\leq |0.3(t - 0.4)|\sqrt{[x_1(t) - y_1(t)]^2 + [x_2(t) - y_2(t)]^2}$$

$$- |0.3(t - 0.4)| \|x - y\|, \quad \forall\, t \in J_1, x(t), y(t) \in \mathbb{R}^2.$$

We can set $L_f = |0.3(t - 0.4)| \in L^q(J_1, \mathbb{R}^+)$ in (H_1). When we choose $p = q = 2$, we get

$$\|L_f\|_{L^2(J_1, \mathbb{R}^+)} = \left(\int_0^{0.8} [0.3(s - 0.4)]^2 ds \right)^{\frac{1}{2}} = 0.0620.$$

Then, we obtain

$$M_2 = \|\Omega^{-1}\| \left[\frac{1}{2^3 \|\Omega\|} (e^{1.6\|\Omega\|} - 1) \right]^{\frac{1}{2}} \|L_f\|_{L^2(J, \mathbb{R}^+)} = 0.0436.$$

Finally, we calculate that

$$M_2 \left[1 + \frac{\cosh(0.8\|\Omega\|) - 1}{\|\Omega\|} \|\Omega^{-1}\| \|B\| M_1 \right] = 0.0894 < 1,$$

which implies that condition (3.52) holds.

Now all the conditions required in Theorem 3.6 are satisfied, thus, system (3.63) is controllable.

3.2.1.4 Conclusion

We give sufficient and necessary conditions for the controllability for the linear second order delay differential system from the point of view of the delay Grammian matrix. In addition, we construct a specific control function for the controllability problem of transferring an initial function to a prescribed point in the phase space. Then, we construct a specific control function involving a nonlinear term and apply the fixed point theorem to establish a sufficient condition of controllability for the nonlinear system by using properties of the delayed matrix sine and the delayed matrix cosine. The results in this part are motivated from [16].

3.3 Iterative learning control

3.3.1 Iterative learning control for an oscillating system

In this part, we discuss the ILC problem via a new approach, that is, the delayed matrix sine and cosine of polynomial degrees methods, for an oscillating system with pure delay of the following form:

$$\begin{cases} \ddot{x}_k(t) + \Omega^2 x_k(t - \tau) = u_k(t), \ t \in [0, T], \\ x_k(t) = \varphi(t), \ -\tau \leq t < 0, \ \tau > 0, \\ y_k(t) = C x_k(t) + D u_k(t), \end{cases} \tag{3.64}$$

where $\varphi \in C^2([-\tau, 0], \mathbb{R}^n)$ and C, D are two $m \times n$ matrices. The index k denotes the k-th learning iteration and the variables $x_k(t), u_k(t) \in \mathbb{R}^n$ and $y_k(t) \in \mathbb{R}^m$ denote the state, input, and output, respectively.

Let y_d be a desired trajectory, let the output error be

$$e_k(t) = y_d(t) - y_k(t), \tag{3.65}$$

and define

$$\Delta u_k(t) = u_{k+1}(t) - u_k(t).$$

For system (3.64), we consider the following set of ILC updating laws:

(i) open-loop P-type ILC updating law,

$$u_{k+1}(t) = u_k(t) + K_1 e_k(t); \tag{3.66}$$

(ii) closed-loop P-type ILC updating law,

$$u_{k+1}(t) = u_k(t) + K_2 e_{k+1}(t); \tag{3.67}$$

(iii) open-closed-loop P-type ILC updating law,

$$u_{k+1}(t) = u_k(t) + K_{\rho_1} e_k(t) + K_{\rho_2} e_{k+1}(t); \tag{3.68}$$

(iv) open-loop D-type ILC updating law,

$$u_{k+1}(t) = u_k(t) + K_3 \dot{e}_k(t); \tag{3.69}$$

(v) closed-loop D-type ILC updating law,

$$u_{k+1}(t) = u_k(t) + K_4 \dot{e}_{k+1}(t); \tag{3.70}$$

(vi) open-closed-loop D-type ILC updating law,

$$u_{k+1}(t) = u_k(t) + K_{\rho_3} \dot{e}_k(t) + K_{\rho_4} \dot{e}_{k+1}(t). \tag{3.71}$$

Let $f : J \to \mathbb{R}^n$ be a continuous function. Consider the following delayed system:

$$\begin{cases} \ddot{x}(t) + \Omega^2 x(t - \tau) = f(t), \ t \geq 0, \ \tau > 0, \\ x(t) = \varphi(t), \qquad\qquad -\tau \leq t < 0. \end{cases} \tag{3.72}$$

For a given nonsingular matrix Ω, a solution of system (3.72) is derived by [5, Theorem 1]

$$x(t) = \varphi(-\tau) \cos_\tau \Omega t + \Omega^{-1} \dot{\varphi}(-\tau) \sin_\tau \Omega t$$
$$+ \Omega^{-1} \int_{-\tau}^0 \sin_\tau \Omega(t - \tau - s) \ddot{\varphi}(s) ds + \Omega^{-1} \int_0^t \sin_\tau \Omega(t - \tau - s) f(s) ds.$$

Lemma 3.4. *The solution $x(\cdot)$ of system (3.72) can be expressed by the following formula:*

$$x(t) = \varphi(-\tau) \cos_\tau \Omega t + \Omega^{-1} \dot{\varphi}(-\tau) \sin_\tau \Omega t$$
$$+ \int_{-\tau}^0 \sum_{i=0}^{\eta-1} (-1)^i \Omega^{2i} \frac{(t - i\tau - s)^{2i+1}}{(2i + 1)!} \ddot{\varphi}(s) ds$$
$$+ \int_{-\tau}^{t-\eta\tau} (-1)^\eta \Omega^{2\eta} \frac{(t - \eta\tau - s)^{2\eta+1}}{(2\eta + 1)!} \ddot{\varphi}(s) ds$$
$$+ \sum_{l=0}^{\eta-1} \int_0^{t-l\tau} (-1)^l \Omega^{2l} \frac{(t - l\tau - s)^{2l+1}}{(2l + 1)!} f(s) ds,$$

where $(\eta - 1)\tau \leq t < \eta\tau$, $\eta = 1, 2, \ldots, N$.

Proof. To achieve our aim, we only need to show that

$$J_1 := \int_{-\tau}^{0} \sin_\tau \Omega(t - \tau - s)\ddot{\varphi}(s)ds$$

$$= \int_{-\tau}^{0} \sum_{i=0}^{\eta-1} (-1)^i \Omega^{2i+1} \frac{(t - i\tau - s)^{2i+1}}{(2i+1)!} \ddot{\varphi}(s)ds$$

$$+ \int_{-\tau}^{t-\eta\tau} (-1)^\eta \Omega^{2\eta+1} \frac{(t - \eta\tau - s)^{2\eta+1}}{(2\eta+1)!} \ddot{\varphi}(s)ds$$

and

$$J_2 := \int_{0}^{t} \sin_\tau \Omega(t - \tau - s)f(s)ds$$

$$= \sum_{l=0}^{\eta-1} \int_{0}^{t-l\tau} (-1)^l \Omega^{2l+1} \frac{(t - l\tau - s)^{2l+1}}{(2l+1)!} f(s)ds.$$

We divide the proof into two parts.

(i) For $-\tau < s < 0$, we obtain $t - \tau < t - \tau - s < t$, and then $(\eta - 2)\tau < t - \tau - s < \eta\tau$. When $-\tau < s < t - \eta\tau$, we obtain $(\eta - 1)\tau < t - \tau - s < t$, and then $(\eta - 1)\tau < t - \tau - s < \eta\tau$. When $t - \eta\tau < s < 0$, we obtain $t - \tau < t - \tau - s < (\eta - 1)\tau$, and then $(\eta - 2)\tau < t - \tau - s < (\eta - 1)\tau$. Thus,

$$J_1 = \int_{-\tau}^{t-\eta\tau} \sin_\tau \Omega(t - \tau - s)\ddot{\varphi}(s)ds + \int_{t-\eta\tau}^{0} \sin_\tau \Omega(t - \tau - s)\ddot{\varphi}(s)ds$$

$$= \int_{-\tau}^{t-\eta\tau} \sum_{i=0}^{\eta} (-1)^i \Omega^{2i+1} \frac{(t - i\tau - s)^{2i+1}}{(2i+1)!} \ddot{\varphi}(s)ds$$

$$+ \int_{t-\eta\tau}^{0} \sum_{i=0}^{\eta-1} (-1)^i \Omega^{2i+1} \frac{(t - i\tau - s)^{2i+1}}{(2i+1)!} \ddot{\varphi}(s)ds$$

$$= \int_{-\tau}^{0} \sum_{i=0}^{\eta-1} (-1)^i \Omega^{2i+1} \frac{(t - i\tau - s)^{2i+1}}{(2i+1)!} \ddot{\varphi}(s)ds$$

$$+ \int_{-\tau}^{t-\eta\tau} (-1)^\eta \Omega^{2\eta+1} \frac{(t - \eta\tau - s)^{2\eta+1}}{(2\eta+1)!} \ddot{\varphi}(s)ds.$$

(ii) For $0 < s < t$, we obtain $-\tau < t - \tau - s < t - \tau$, and then $-\tau < t - \tau - s < (\eta-1)\tau$. When $0 < s < t - (\eta - 1)\tau$, we obtain $(\eta-2)\tau < t - \tau - s < t - \tau$, and then $(\eta - 2)\tau < t - \tau - s < (\eta - 1)\tau$. When $t - (j + 1)\tau < s < t - j\tau$, we have $(j - 1)\tau < t - \tau - s < j\tau$, $j = 0, 1, \ldots, \eta - 2$. Thus,

$$J_2 = \int_{0}^{t-(\eta-1)\tau} \sin_\tau \Omega(t - \tau - s)f(s)ds$$

$$+ \sum_{j=0}^{\eta-2} \int_{t-(j+1)\tau}^{t-j\tau} \sin_\tau \Omega(t - \tau - s) f(s) ds$$

$$= \int_0^{t-(\eta-1)\tau} \sum_{l=0}^{\eta-1} (-1)^l \Omega^{2l+1} \frac{(t - l\tau - s)^{2l+1}}{(2l+1)!} f(s) ds$$

$$+ \sum_{j=0}^{\eta-2} \int_{t-(j+1)\tau}^{t-j\tau} \sum_{l=0}^{j} (-1)^l \Omega^{2l+1} \frac{(t - l\tau - s)^{2l+1}}{(2l+1)!} f(s) ds$$

$$= \sum_{l=0}^{\eta-1} \int_0^{t-(\eta-1)\tau} (-1)^l \Omega^{2l+1} \frac{(t - l\tau - s)^{2l+1}}{(2l+1)!} f(s) ds$$

$$+ \sum_{l=0}^{\eta-2} \int_{t-(\eta-1)\tau}^{t-l\tau} (-1)^l \Omega^{2l+1} \frac{(t - l\tau - s)^{2l+1}}{(2l+1)!} f(s) ds$$

$$= \sum_{l=0}^{\eta-2} \int_0^{t-l\tau} (-1)^l \Omega^{2l+1} \frac{(t - l\tau - s)^{2l+1}}{(2l+1)!} f(s) ds$$

$$+ \int_0^{t-(\eta-1)\tau} (-1)^{\eta-1} \Omega^{2\eta-1} \frac{[t - (\eta-1)\tau - s]^{2\eta-1}}{(2\eta-1)!} f(s) ds$$

$$= \sum_{l=0}^{\eta-1} \int_0^{t-l\tau} (-1)^l \Omega^{2l+1} \frac{(t - l\tau - s)^{2l+1}}{(2l+1)!} f(s) ds.$$

The proof is finished. $\square$

Remark 3.6. By Lemma 3.4, the solution $x_k(\cdot)$ of (3.64) has the following form:

$$x_k(t) = \varphi(-\tau) \cos_\tau \Omega t + \Omega^{-1} \dot{\varphi}(-\tau) \sin_\tau \Omega t$$

$$+ \int_{-\tau}^{0} \sum_{i=0}^{\eta-1} (-1)^i \Omega^{2i} \frac{(t - i\tau - s)^{2i+1}}{(2i+1)!} \ddot{\varphi}(s) ds$$

$$+ \int_{-\tau}^{t-\eta\tau} (-1)^\eta \Omega^{2\eta} \frac{(t - \eta\tau - s)^{2p+1}}{(2\eta+1)!} \ddot{\varphi}(s) ds$$

$$+ \sum_{l=0}^{\eta-1} \int_0^{t-l\tau} (-1)^l \Omega^{2l} \frac{(t - l\tau - s)^{2l+1}}{(2l+1)!} u_k(s) ds, \tag{3.73}$$

where $(\eta - 1)\tau \le t < \eta\tau$, $\eta = 1, 2, \ldots, n$.

Remark 3.7. By using (3.73) and [5, Lemmas 1, 2], one can get the following derivative form of the solutions of system (3.72):

$$\dot{x}(t) = -\Omega \varphi(-\tau) \sin_\tau \Omega(t - \tau) + \dot{\varphi}(-\tau) \cos_\tau \Omega t$$

$$
+ \int_{-\tau}^{0} \sum_{i=0}^{\eta-1} (-1)^i \Omega^{2i} \frac{(t - i\tau - s)^{2i}}{(2i)!} \ddot{\varphi}(s)\,ds
$$

$$
+ \int_{-\tau}^{t-\eta\tau} (-1)^{\eta} \Omega^{2\eta} \frac{(t - \eta\tau - s)^{2\eta}}{(2\eta)!} \ddot{\varphi}(s)\,ds
$$

$$
+ \sum_{l=0}^{\eta-1} \int_{0}^{t-l\tau} (-1)^l \Omega^{2l} \frac{(t - l\tau - s)^{2l}}{(2l)!} f(s)\,ds. \tag{3.74}
$$

Proof. Using the fact that

$$
\frac{d}{dt} \cos_{\tau} \Omega t = -\Omega \sin_{\tau} \Omega(t - \tau), \qquad \frac{d}{dt} \sin_{\tau} \Omega t = \Omega \cos_{\tau} \Omega t,
$$

$$
\frac{d}{dt} \int_{-\tau}^{0} \sum_{i=0}^{\eta-1} (-1)^i \Omega^{2i} \frac{(t - i\tau - s)^{2i+1}}{(2i + 1)!} \ddot{\varphi}(s)\,ds
$$

$$
= \int_{-\tau}^{0} \sum_{i=0}^{\eta-1} (-1)^i \Omega^{2i} \frac{(t - i\tau - s)^{2i}}{(2i)!} \ddot{\varphi}(s)\,ds,
$$

$$
\frac{d}{dt} \int_{-\tau}^{t-\eta\tau} (-1)^{\eta} \Omega^{2\eta} \frac{(t - \eta\tau - s)^{2\eta+1}}{(2\eta + 1)!} \ddot{\varphi}(s)\,ds
$$

$$
= \int_{-\tau}^{t-\eta\tau} (-1)^{\eta} \Omega^{2\eta} \frac{(t - \eta\tau - s)^{2p}}{(2\eta)!} \ddot{\varphi}(s)\,ds,
$$

$$
\frac{d}{dt} \int_{0}^{t-l\tau} (-1)^l \Omega^{2l} \frac{(t - l\tau - s)^{2l+1}}{(2l + 1)!} f(s)\,ds
$$

$$
= \int_{0}^{t-l\tau} (-1)^l \Omega^{2l} \frac{(t - l\tau - s)^{2l}}{(2l)!} f(s)\,ds,
$$

one can get (3.74) from (3.73). $\qquad\square$

3.3.1.1 Convergence analysis of P-type ILC

In this section, we give convergence results of P-type.

Theorem 3.7. *For the given system (3.64) and the open-loop P-type ILC law (3.66),* $\lim_{k \to \infty} y_k(t) = y_d(t)$ *uniformly on* $[0, T]$ *in the* λ*-norm sense if the condition* $\|I - DK_1\| < 1$ *is met.*

Proof. Consider $(\eta - 1)\tau \le t < \eta\tau, \eta = 0, 1, \ldots, n$. By (3.65), (3.66), and (3.73) we have

$$
e_{k+1}(t) = e_k(t) + y_k(t) - y_{k+1}(t)
$$

$$
= e_k(t)(I - DK_1) - C \sum_{l=0}^{\eta-1} \int_{0}^{t-l\tau} (-1)^l \Omega^{2l} \frac{(t - l\tau - s)^{2l+1}}{(2l + 1)!} \Delta u_k(t)\,ds.
$$

Taking the norm $\| \cdot \|$ on $\mathbb{R}^n$ via fundamental computations, we have

$$\|e_{k+1}(t)\| \leq \|I - DK_1\| \|e_k(t)\|$$

$$+ \|C\| \sum_{l=0}^{\eta-1} \int_0^{t-l\tau} \|\Omega^{2l}\| \frac{(t-l\tau-s)^{2l+1}}{(2l+1)!} \|\Delta u_k(t)\| ds$$

$$\leq \|I - DK_1\| \|e_k(t)\|$$

$$+ \|C\| \|\Omega^{2l}\| \|K_1\| \|e_k\|_\lambda \sum_{l=0}^{\eta-1} \int_0^{t-l\tau} \frac{(t-l\tau-s)^{2l+1}}{(2l+1)!} e^{\lambda s} ds.$$

$$(3.75)$$

Next, we show that

$$\int_0^{t-l\tau} \frac{(t-l\tau-s)^{2l+1}}{(2l+1)!} e^{\lambda s} ds = -\sum_{j=1}^{2l+2} \frac{1}{\lambda^j} \frac{(t-l\tau)^{2l-j+2}}{2l-j+2} + \frac{e^{\lambda(t-l\tau)}}{\lambda^{2l+2}}. \quad (3.76)$$

In fact, integration by parts gives

$$\int_0^{t-l\tau} \frac{(t-l\tau-s)^{2l+1}}{(2l+1)!} e^{\lambda s} ds = -\frac{1}{\lambda} \frac{(t-l\tau)^{2l+1}}{(2l+1)!} + \frac{1}{\lambda} \int_0^{t-l\tau} \frac{(t-l\tau-s)^{2l}}{(2l)!} e^{\lambda s} ds$$

$$\vdots$$

$$= -\frac{1}{\lambda} \frac{(t-l\tau)^{2l+1}}{(2l+1)!} - \frac{1}{\lambda^2} \frac{(t-l\tau)^{2l}}{(2l)!} - \cdots$$

$$- \frac{1}{\lambda^{2l+1}} \frac{(t-l\tau)^1}{1!} + \frac{1}{\lambda^{2l+1}} \int_0^{t-l\tau} e^{\lambda s} ds$$

$$= -\sum_{j=1}^{2l+2} \frac{1}{\lambda^j} \frac{(t-l\tau)^{2l-j+2}}{2l-j+2} + \frac{e^{\lambda(t-l\tau)}}{\lambda^{2l+2}}.$$

Combining (3.76) with (3.75), we obtain

$$\|e_{k+1}(t)\| \leq \|I - DK_1\| \|e_k(t)\| + \|C\| \|\Omega^{2l}\| \|K_1\| \|e_k\|_\lambda \theta_\lambda,$$

where

$$\theta_\lambda := \sum_{l=0}^{\eta-1} \left[-\sum_{j=1}^{2l+2} \frac{1}{\lambda^j} \frac{(t-l\tau)^{2l-j+2}}{2l-j+2} + \frac{e^{\lambda(t-l\tau)}}{\lambda^{2l+2}} \right].$$

Taking the λ-norm, we arrive at

$$\|e_{k+1}\|_\lambda \leq \|I - DK_1\| \|e_k\|_\lambda$$

$$+ \|C\| \|\Omega^{2l}\| \|K_1\| \|e_k\|_\lambda \sum_{l=0}^{\eta-1} \left[-\sum_{j=1}^{2l+2} \frac{1}{\lambda^j} \frac{(t-l\tau)^{2l-j+2}}{2l-j+2} + \frac{e^{-\lambda l\tau}}{\lambda^{2l+2}} \right].$$

Note that η is a finite number and $\sum_{l=0}^{p-1} \left[-\sum_{j=1}^{2l+2} \frac{1}{\lambda^j} \frac{(t-l\tau)^{2l-j+2}}{2l-j+2} + \frac{e^{-\lambda l\tau}}{\lambda^{2l+2}} \right]$ tends to zero when λ is sufficiently large. By the condition $\|I - DK_1\| < 1$, one can choose a λ sufficiently large such that

$$\|e_{k+1}\|_\lambda < \|e_k\|_\lambda,$$

which implies $\lim_{k\to\infty} \|e_k\|_\lambda = 0$ and $\lim_{k\to\infty} y_k(t) = y_d(t)$ uniformly on $[0, T]$ in the λ-norm sense. $\qquad\square$

Next, we give the second convergence result of closed-loop P-type.

Theorem 3.8. *For the given system (3.64) and the closed-loop P-type ILC law (3.67), we have* $\lim_{k\to\infty} y_k(t) = y_d(t)$ *uniformly on* $[0, T]$ *in the λ-norm sense if the condition* $\|I + DK_2\| > 1$ *is met.*

Proof. Without loss of generality, we consider $(\eta - 1)\tau \leq t < \eta\tau$, $\eta = 0, 1, \ldots, n$. By (3.65), (3.67), and (3.73) we have

$$e_{k+1}(t) = e_k(t) + y_k(t) - y_{k+1}(t)$$

$$= e_k(t) - DK_2 e_{k+1}(t)$$

$$- C\sum_{l=0}^{\eta-1} \int_0^{t-l\tau} (-1)^l \Omega^{2l} \frac{(t-l\tau-s)^{2l+1}}{(2l+1)!} \Delta u_k(t) ds.$$

Then

$$(I + DK_2)e_{k+1}(t) = e_k(t) - C\sum_{l=0}^{\eta-1} \int_0^{t-l\tau} (-1)^l \Omega^{2l} \frac{(t-l\tau-s)^{2l+1}}{(2l+1)!} \Delta u_k(t) ds.$$

Taking the norm $\|\cdot\|$ on $\mathbb{R}^n$, one has

$$\|I + DK_2\| \|e_{k+1}(t)\|$$

$$\leq \|e_k(t)\| + \|C\| \sum_{l=0}^{\eta-1} \int_0^{t-l\tau} \|\Omega^{2l}\| \frac{(t-l\tau-s)^{2l+1}}{(2l+1)!} \|\Delta u_k(t)\| ds$$

$$\leq \|e_k(t)\| + \|C\| \|\Omega^{2l}\| \|K_2\| \|e_{k+1}\|_\lambda \sum_{l=0}^{\eta-1} \int_0^{t-l\tau} \frac{(t-l\tau-s)^{2l+1}}{(2l+1)!} e^{\lambda s} ds.$$

According to (3.76), we obtain

$$\|I + DK_2\|\|e_{k+1}(t)\| \leq \|e_k(t)\| + \|C\|\|\Omega^{2l}\|\|K_2\|\|e_{k+1}\|_\lambda \theta_\lambda.$$

Taking the λ-norm, we arrive at

$$\|e_{k+1}\|_\lambda \leq \frac{1}{\|I + DK_2\|}\|e_k\|_\lambda + \frac{1}{\|I + DK_2\|}\|C\|\|\Omega^{2l}\|\|K_2\|\|e_{k+1}\|_\lambda$$
$$\times \sum_{l=0}^{\eta-1}\left[-\sum_{j=1}^{2l+2}\frac{1}{\lambda^j}\frac{(t - l\tau)^{2l-j+2}}{2l - j + 2} + \frac{e^{-\lambda l\tau}}{\lambda^{2l+2}}\right].$$

Thus, for some λ sufficiently large and $\|I + DK_2\| > 1$ we obtain $\|e_{k+1}\|_\lambda < \|e_k\|_\lambda$, which implies $\lim_{k\to\infty}\|e_k\|_\lambda = 0$. The proof is completed. $\square$

To end this section, we conclude with the following result of open-closed-loop P-type. Since the proof is similar to that of the above two theorems, we omit it here.

Theorem 3.9. *For the given system (3.64) and the open-closed-loop P-type ILC law (3.68), we have $\lim_{k\to\infty} y_k(t) = y_d(t)$ uniformly on $[0, T]$ in the λ-norm sense if the condition $\frac{\|I - DK_{\rho_1}\|}{\|I + DK_{\rho_2}\|} < 1$ is met.*

3.3.1.2 Convergence analysis of D-type ILC

In this section, we give the convergence results of D-type.

Theorem 3.10. *For the given system (3.64) associated with*

$$y_k(t) = Cx_k(t) + D\int_0^t u_k(t)ds$$

and the open-loop D-type ILC law (3.69), we have $\lim_{k\to\infty} y_k(t) = y_d(t)$ uniformly on $[0, T]$ in the λ-norm sense if the condition $\|I - DK_3\| < 1$ is met.

Proof. Consider $(\eta - 1)\tau \leq t < \eta\tau, \eta = 0, 1, \ldots, n$. By (3.65), (3.69), and (3.74) we have

$$\dot{e}_{k+1}(t) - \dot{e}_k(t) = \dot{y}_k(t) - \dot{y}_{k+1}(t) = C[\dot{x}_k(t) - \dot{x}_{k+1}(t)] - D\Delta u_k(t)$$
$$= C\sum_{l=0}^{\eta-1}\int_0^{t-l\tau}(-1)^{l+1}\Omega^{2l}\frac{(t - l\tau - s)^{2l}}{(2l)!}\Delta u_k(t)ds - DK_3\dot{e}_k(t).$$

Then

$$\dot{e}_{k+1}(t) = (I - DK_3)\dot{e}_k(t) + C\sum_{l=0}^{\eta-1}\int_0^{t-l\tau}(-1)^{l+1}\Omega^{2l}\frac{(t - l\tau - s)^{2l}}{(2l)!}\Delta u_k(t)ds.$$

Taking the norm $\| \cdot \|$ on $\mathbb{R}^n$

$\|\dot{e}_{k+1}(t)\|$

$$\leq \|I - DK_3\|\|\dot{e}_k(t)\| + \|C\| \sum_{l=0}^{\eta-1} \int_0^{t-l\tau} \|\Omega^{2l}\| \frac{(t - l\tau - s)^{2l}}{(2l)!} \|\Delta u_k(t)\| ds$$

$$\leq \|I - DK_3\|\|\dot{e}_k(t)\| + \|C\|\|\Omega^{2l}\|\|K_3\|\|\dot{e}_k\|_\lambda$$

$$\times \sum_{l=0}^{\eta-1} \int_0^{t-l\tau} \frac{(t - l\tau - s)^{2l}}{(2l)!} e^{\lambda s} ds.$$

Similar to (3.76), we have

$$\int_0^{t-l\tau} \frac{(t - l\tau - s)^{2l}}{(2l)!} e^{\lambda s} ds = - \sum_{j=1}^{2l+1} \frac{1}{\lambda^j} \frac{(t - l\tau)^{2l-j+1}}{2l - j + 1} + \frac{e^{\lambda(t - l\tau)}}{\lambda^{2l+1}}. \qquad (3.77)$$

Then

$$\|\dot{e}_{k+1}(t)\|e^{-\lambda t} \leq \|I - DK_3\|\|\dot{e}_k(t)\|e^{-\lambda t} + \|C\|\|\Omega^{2l}\|\|K_3\|\|\dot{e}_k\|_\lambda$$

$$\times \sum_{l=0}^{\eta-1} \left[-e^{-\lambda t} \sum_{j=1}^{2l+1} \frac{1}{\lambda^j} \frac{(t - l\tau)^{2l-j+1}}{2l - j + 1} + \frac{e^{-\lambda l\tau}}{\lambda^{2l+1}} \right].$$

Taking the λ-norm, we arrive at

$$\|\dot{e}_{k+1}\|_\lambda \leq \|I - DK_3\|\|\dot{e}_k\|_\lambda$$

$$+ \|C\|\|\Omega^{2l}\|\|K_3\|\|\dot{e}_k\|_\lambda \sum_{l=0}^{\eta-1} \left[-\sum_{j=1}^{2l+1} \frac{1}{\lambda^j} \frac{(t - l\tau)^{2l-j+1}}{2l - j + 1} + \frac{e^{-\lambda l\tau}}{\lambda^{2l+1}} \right].$$

So for some λ sufficiently large and $\|I - DK_3\| < 1$, we have $\|\dot{e}_{k+1}\|_\lambda < \|\dot{e}_k\|_\lambda$. This yields $\lim_{k \to \infty} \|\dot{e}_k\|_\lambda = 0$. Due to the fact that $e_k(0) = 0$, we get $\|e_k\|_\lambda \leq \frac{1}{\lambda}\|\dot{e}_k\|_\lambda$; consequently, we find $\lim_{k \to \infty} \|e_k\|_\lambda = 0$. The proof is completed. $\qquad \square$

Next, we give the second convergence result of closed-loop D-type.

Theorem 3.11. *For the given system (3.64) associated with $y_k(t) = Cx_k(t) + D \int_0^t u_k(t)ds$ and the closed-loop D-type ILC law (3.70), we have $\lim_{k \to \infty} y_k(t) = y_d(t)$ uniformly on $[0, T]$ in the λ-norm sense if the condition $\|I + DK_4\| > 1$ is met.*

Proof. Consider $(\eta - 1)\tau \leq t < \eta\tau, \eta = 0, 1, \ldots, n$. By (3.65), (3.70), and (3.74) we have

$$(I + DK_4)\dot{e}_{k+1}(t) = \dot{e}_k(t) + C \sum_{l=0}^{\eta-1} \int_0^{t-l\tau} (-1)^{l+1} \Omega^{2l} \frac{(t - l\tau - s)^{2l}}{(2l)!} \Delta u_k(t) ds.$$

Taking the norm $\| \cdot \|$ on $\mathbb{R}^n$, we have

$$\|\dot{e}_{k+1}(t)\| \leq \frac{1}{\|I + DK_4\|} \|\dot{e}_k(t)\|$$

$$+ \frac{1}{\|I + DK_4\|} \|C\| \|\Omega^{2l}\| \|K_4\| \|\dot{e}_k\|_\lambda \sum_{l=0}^{\eta-1} \int_0^{t-l\tau} \frac{(t - l\tau - s)^{2l}}{(2l)!} e^{\lambda s} ds.$$

Taking the λ-norm via (3.77), we have

$$\|\dot{e}_{k+1}\|_\lambda \leq \frac{1}{\|I + DK_4\|} \|\dot{e}_k\|_\lambda + \frac{1}{\|I + DK_4\|} \|C\| \|\Omega^{2l}\| \|K_4\| \|\dot{e}_k\|_\lambda$$

$$\times \sum_{l=0}^{\eta-1} \left[-\sum_{j=1}^{2l+1} \frac{1}{\lambda^j} \frac{(t - l\tau)^{2l-j+1}}{2l - j + 1} + \frac{e^{-\lambda l\tau}}{\lambda^{2l+1}} \right].$$

Thus, for some λ sufficiently large and $\|I + DK_4\| > 1$, we have $\|\dot{e}_{k+1}\|_\lambda < \|\dot{e}_k\|_\lambda$, which implies $\lim_{k\to\infty} \|\dot{e}_k\|_\lambda = 0$. We get $\lim_{k\to\infty} \|e_k\|_\lambda = 0$. The proof is completed. $\square$

To end this section, we conclude with the following result of open-closed-loop D-type. Since the proof is similar to that of the above two theorems, we omit it here.

Theorem 3.12. *For the given system (3.64) associated with* $y_k(t) = Cx_k(t) + D \int_0^t u_k(t) ds$ *and the closed-loop D-type ILC law (3.71), we have* $\lim_{k\to\infty} y_k(t) = y_d(t)$ *uniformly on* $[0, T]$ *in the λ-norm sense if the condition* $\frac{\|I - DK_{\rho3}\|}{\|I + DK_{\rho4}\|} < 1$ *is met.*

3.3.1.3 Numerical examples and discussion

In this section, several numerical examples are presented to demonstrate the validity of the designed method.

Example 3.5. Consider second order differential equations with pure delay of the form

$$\begin{cases} \ddot{x}_k(t) + x_k(t - 0.5) = u_k(t), \ x_k(t), \ u_k(t) \in \mathbb{R}^2, \ t \in [0, 1], \\ x_k(t) = (2t, 3t)^\top, \ t \in [-0.5, 0), \\ y_k(t) = (1, 2)x_k(t) + (0.5, 0.6)u_k(t) \end{cases} \tag{3.78}$$

and choose an ILC updating law (3.66) as follows:

$$u_{k+1}(t) = u_k(t) + (1, 0.5)^\top e_k(t).$$

The desired continuous trajectory is given as

$$y_d(t) = 4\sin(3\pi t), \ t \in [0, 1].$$

Obviously, we can see $T = 1$, $\tau = 0.5$, $\varphi(t) = (2t, 3t)^\top$, $n = 2$, $m = 1$, $\Omega = I$, $C = (1, 2)$, $D = (0.5, 0.6)$, $K_1 = (1, 0.5)^\top$, and $u_k(t) = (u_{1k}(t), u_{2k}(t))^\top$. Clearly,

$$\cos_{0.5} t = \begin{cases} -\frac{t^2}{2} + 1, \ t \in [0, 0.5), \\ \frac{(t-0.5)^4}{24} - \frac{t^2}{2} + 1, \ t \in [0.5, 1], \end{cases} \tag{3.79}$$

and

$$\sin_{0.5} t = \begin{cases} -\frac{t^3}{6} + t + \frac{1}{2}, \ t \in [0, 0.5), \\ \frac{(t-0.5)^5}{120} - \frac{t^3}{6} + t + \frac{1}{2}, \ t \in [0.5, 1]. \end{cases} \tag{3.80}$$

Combining (3.79), (3.80), and $\varphi(t)$ with (3.73), we can get the state of the k-th iteration:

$$x_k(t) = \begin{pmatrix} -\frac{t^3}{6} + \frac{t^2}{2} + t - \frac{1}{2} + \int_0^t (t-s)u_{1k}(s)ds \\ -\frac{t^3}{6} + \frac{3}{4}t^2 + t - 1 + \int_0^t (t-s)u_{2k}(s)ds \end{pmatrix}, \ t \in [0, 0.5),$$

$$x_k(t) = \begin{pmatrix} \frac{(t-0.5)^5}{120} - \frac{(t-0.5)^4}{24} - \frac{t^3}{6} + \frac{t^2}{2} + t - \frac{1}{2} \\ + \int_0^t (t-s)u_{1k}(s)ds - \int_0^{t-0.5} \frac{(t-s-0.5)^3}{6} u_{1k}(s)ds \\ \frac{(t-0.5)^5}{120} - \frac{(t-0.5)^4}{16} - \frac{t^3}{6} + \frac{3}{4}t^2 + t - 1 \\ + \int_0^t (t-s)u_{2k}(s)ds - \int_0^{t-0.5} \frac{(t-s-0.5)^3}{6} u_{2k}(s)ds \end{pmatrix}, \ t \in [0.5, 1].$$

Moreover, $\|I - DK_1\| = 0.2$. Now all the conditions of Theorem 3.7 are required, y_k uniformly converges to y_d for all $t \in [0, 1]$.

The continuous curve of the upper figure of Fig. 3.5 shows the output y_k of the first 10 iterations of system (3.78) and the reference trajectory y_d. The lower figure of Fig. 3.5 shows the L^2-norm of the tracking error in each iteration.

Clearly, the output of the system $y_k(\cdot)$ can fast track on our desired trajectory $y_d(\cdot)$, and the 10th iteration error is 3.98×10^{-5}, which is relatively small.

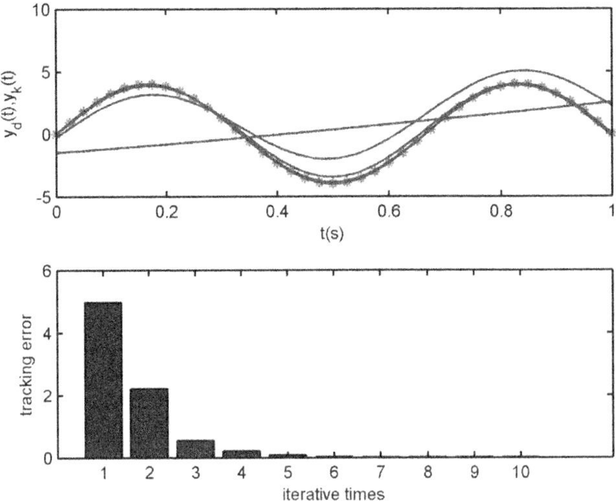

FIGURE 3.5   The tracking performance of system (3.78) of Example 3.5 and the L^2-norm of the tracking error.

Example 3.6. Conditions for this example are the same as in Example 3.5, except we changed the desired trajectory to a discontinuous trajectory:

$$y_d(t) = \begin{cases} 2\sin(3\pi t), \ t \in [0, 0.4), \\ 2\cos(3\pi t) - 1, \ t \in [0.4, 1]. \end{cases}$$

The image of Fig. 3.6 shows the output y_k of the first 10 iterations of system (3.78), the referenced trajectory y_d, and the L^2-norm of the tracking error in each iteration. The 10th iteration error is 2.71×10^{-5}, which is small too.

Example 3.7. In this example, we simulate different situations for $\varphi(\cdot)$. Again, the conditions in this example are the same as in Example 3.5, except in this example we changed $\varphi(t)$ to a constant vector $(3, 1)^\top$. Simulation results are shown in Fig. 3.7.

Due to the change of $\varphi(\cdot)$, we can get the state $x_k(t)$ from Example 3.5 as follows:

$$x_k(t) = \begin{pmatrix} -\frac{3}{2}t^2 + 3 + \int_0^t (t - s)u_{1k}(s)ds \\ -\frac{t^2}{2} + 1 + \int_0^t (t - s)u_{2k}(s)ds \end{pmatrix}, \ t \in [0, 0.5),$$

$$x_k(t)$$

$$= \begin{pmatrix} \frac{(t-0.5)^4}{8} - \frac{3}{2}t^2 + 3 + \int_0^t (t - s)u_{1k}(s)ds - \int_0^{t-0.5} \frac{(t-s-0.5)^3}{6} u_{1k}(s)ds \\ \frac{(t-0.5)^4}{24} - \frac{t^2}{2} + 1 + \int_0^t (t - s)u_{2k}(s)ds - \int_0^{t-0.5} \frac{(t-s-0.5)^3}{6} u_{2k}(s)ds \end{pmatrix},$$

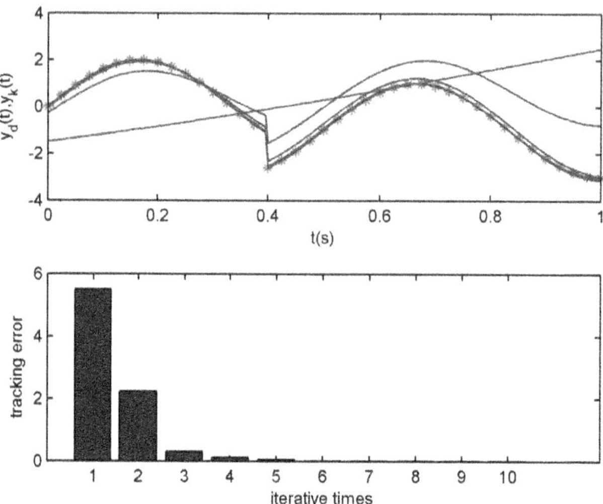

FIGURE 3.6 The tracking performance of Example 3.6 and the L^2-norm of the tracking error.

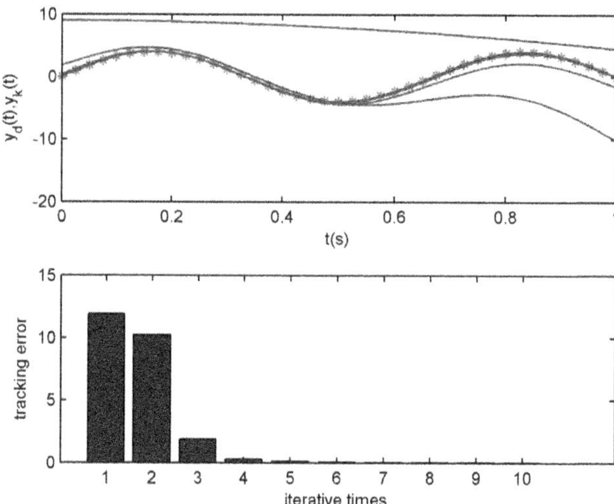

FIGURE 3.7 The tracking performance of Example 3.7 and the L^2-norm of the tracking error.

$t \in [0.5, 1]$.

Fig. 3.7 shows the output y_k of the first 10 iterations, the referenced trajectory y_d, and the L^2-norm of the tracking error in each iteration. The 10th iteration error is 6.82×10^{-5}.

The simulation results show that for different continuous functions $\varphi(\cdot)$, our method is also effective.

3.3.1.4   Conclusion

We give P-type and D-type convergence results, and several examples are given to demonstrate the applicability of our main results. The results in this part are motivated from [92].

Chapter 4

Impulsive delay systems

4.1 Asymptotical stability

4.1.1 Basic estimation and Gronwall-type inequalities

Lemma 4.1. *(see [7, Lemma 12]) If* $\|B\| \leq \alpha e^{\alpha\tau}$, $\alpha \in \mathbb{R}^+$, *then* $\|e_\tau^{B(t-\tau)}\| \leq e^{\alpha t}$, $t \in \mathbb{R}$.

Lemma 4.2. *For all* $t \geq 0$, *we have* $\|e_\tau^{Bt}\| \leq e^{\|B\|t}$.

Proof. From (2.4), without loss of generality, for $(k-1)\tau \leq t < k\tau$, $k = 1, 2, \cdots$, we have

$$\|e_\tau^{Bt}\| = \left\| \sum_{n=0}^{k} B^n \frac{(t-(n-1)\tau)^n}{n!} \right\| \leq \sum_{n=0}^{k} \|B\|^n \frac{t^n}{n!}$$

$$\leq \sum_{n=0}^{\infty} \|B\|^n \frac{t^n}{n!} = e^{\|B\|t}.$$

The result is proved. $\qquad\square$

Lemma 4.3. *(see [93]) Let* $y(\cdot), b(\cdot) \in C([t_0, T], \mathbb{R}^+)$ *and* $k \geq 0$, $M \geq 0$. *Further,* $\omega(\cdot) \in C([0, \infty), \mathbb{R}^+)$ *and* $\omega(\cdot)$ *is nondecreasing function. Then the inequality*

$$y(t) \leq k + M \int_{t_0}^{t} b(s)\omega(y(s))ds, \ t \in [t_0, T],$$

implies the inequality

$$y(t) \leq W^{-1}\left(W(k) + M \int_{t_0}^{t} b(s)ds \right), \ t \in [t_0, \tilde{t}], \ \tilde{t} \leq T,$$

where

$$W(\mu) = \int_{\mu_0}^{\mu} \frac{dz}{\omega(z)}, \ \mu_0 > 0, \ \mu > 0,$$

and $W^{-1}(\cdot)$ *means the inverse function of* $W(\cdot)$.

Stability and Controls Analysis for Delay Systems. https://doi.org/10.1016/B978-0-32-399792-8.00010-4
Copyright © 2023 Elsevier Inc. All rights reserved.

Denote $PC(\mathbb{R}^+, \mathbb{R}^n) := \{v : \mathbb{R}^+ \to \mathbb{R}^n : v \in C((t_i, t_{i+1}], \mathbb{R}^n)\}$, there exist $v(t_i^-)$ and $v(t_i^+)$ with $v(t_i^-) = v(t_i)$ for any $i = 1, 2, \cdots\}$ and $PC^1(\mathbb{R}^+, \mathbb{R}^n) := \{v : \mathbb{R}^+ \to \mathbb{R}^n : v' \in PC(\mathbb{R}^+, \mathbb{R}^n)\}$. Denote the space $PC(\Omega, \mathbb{R}^n)$ of vector-valued piecewise continuous functions from $\Omega \to \mathbb{R}^n$ endowed with the norm $\|v\|_{PC} = \sup_{t \in \Omega} \|v(t)\|$, where $\|\cdot\|$ is the norm on $\mathbb{R}^n$. In addition, $\|\psi\|_{PC} = \sup_{t \in [-\tau, 0]} \|\psi(t)\|$.

Let Y_1, Y_2 be two Banach spaces and let $L_b(Y_1, Y_2)$ denote the space of all bounded linear operators from Y_1 to Y_2. Now $L^p(\Omega, Y_2)$ denotes the Banach space of functions $y : \Omega \to Y_2$ which are Bochner integrable with norm $\|y\|_{L^p(\Omega, Y_2)}$ (here $1 < p < \infty$).

4.1.2 Linear impulsive delay differential systems

In this section, we introduce the impulsive delayed Cauchy matrix (which will be used to seek the formula of solutions) for the following linear impulsive delay differential systems:

$$
\begin{cases}
v'(t) = Av(t) + Bv(t - \tau), \ t \geq 0, \ \tau > 0, \ t \neq t_i, \\
\Delta v(t_i) = C_i v(t_i), \ i = 1, 2, \cdots, \\
v(t) = \psi(t), \ -\tau \leq t \leq 0,
\end{cases}
\tag{4.1}
$$

where A, B, C_i are constant $n \times n$ matrices, $AB = BA$, $AC_i = C_i A$, and $BC_i = C_i B$ for each $i = 1, 2, \cdots$, $\psi \in C_\tau^1 := C^1([-\tau, 0], \mathbb{R}^n)$, $v(t) \in \mathbb{R}^n$, time sequences $\{t_i\}_{k=1}^\infty$ satisfy $0 = t_0 < t_1 < \cdots < t_i < \cdots$, impulsive conditions $\Delta v(t_i) := v(t_i^+) - v(t_i^-)$, and $v(t_i^+) = \lim_{\epsilon \to 0^+} v(t_i + \epsilon)$ and $v(t_i^-) = v(t_i)$ represent the right and left limits of $v(t)$ at $t = t_i$ and $\lim_{i \to +\infty} t_i = \infty$, respectively.

4.1.2.1 Impulsive delayed Cauchy matrix and its properties

In what follows, we introduce the concept of impulsive delay matrix functions, an extension of delay matrix functions for linear delay differential equations, which help us to seek explicit formulas of solutions to impulsive delay differential equations.

By using (2.4), we define $Y(\cdot, \cdot) : \mathbb{R} \times \mathbb{R} \to \mathbb{R}^{n \times n}$ and

$$
Y(t, s) = e^{A(t-s)} X(t, s + \tau), \ t > s,
\tag{4.2}
$$

where

$$
X(t, s) = e_\tau^{\check{B}(t-s)} + \sum_{s-\tau < t_j \leq t} C_j e_\tau^{\check{B}(t-\tau-t_j)} X(t_j, s), \ \check{B} = e^{-A\tau} B.
\tag{4.3}
$$

Here, we call $Y(\cdot, \cdot)$ defined in (4.2) the impulsive delayed Cauchy matrix associated with (4.1).

Remark 4.1. Obviously, $Y(t,t) = X(t, t+\tau) = e_\tau^{\check{B}(t-t-\tau)} = I$ and $Y(t,s) = e^{A(t-s)} X(t, s+\tau) = e^{A(t-s)} e_\tau^{\check{B}(t-s-\tau)} = \Theta, t < s$.

Lemma 4.4. *The impulsive delayed Cauchy matrix $Y(\cdot,\cdot)$ is the fundamental matrix of (4.1).*

Proof. We divide our proof into two steps.

Step 1. We verify that $Y(\cdot,\cdot)$ satisfies differential equation $\frac{d}{dt} Y(t,s) = AY(t,s) + BY(t-\tau,s), t \in (t_i, t_{i+1}]$. In fact, having differentiated $Y(t,s)$ for $t \in (t_i, t_{i+1}]$ and $t > s$, by using Lemma 2.2, we obtain

$$
\begin{aligned}
\frac{d}{dt} & Y(t,s) \\
&= Ae^{A(t-s)} X(t, s+\tau) \\
&\quad + e^{A(t-s)} \left\{ \check{B} e_\tau^{\check{B}(t-2\tau-s)} + \check{B} \sum_{s < t_j \le t} C_j e_\tau^{\check{B}(t-2\tau-t_j)} X(t_j, s+\tau) \right\} \\
&= AY(t,s) + Be^{A(t-\tau-s)} \left\{ e_\tau^{\check{B}(t-2\tau-s)} + \sum_{s < t_j \le t-\tau} C_j e_\tau^{\check{B}(t-2\tau-t_j)} X(t_j, s+\tau) \right\} \\
&= AY(t,s) + BY(t-\tau, s).
\end{aligned}
$$

Step 2. We verify that $Y(t_i^+, s) - Y(t_i^-, s) = C_i Y(t_i, s)$. Note that $e_\tau^{Bt^+} = e_\tau^{Bt^-}$ for all $t > -\tau$, then

$$
\begin{aligned}
Y(t_i^+, s) &- e^{A(t_i^+ - s)} \left\{ e_\tau^{\check{B}(t_i^+ - \tau - s)} + \sum_{s < t_j \le t_i^+} C_j e_\tau^{\check{B}(t_i^+ - \tau - t_j)} X(t_j, s+\tau) \right\} \\
&= e^{A(t_i^- - s)} \left\{ e_\tau^{\check{B}(t_i^- - \tau - s)} + \sum_{s < t_j \le t_i^-} C_j e_\tau^{\check{B}(t_i^- - \tau - t_j)} X(t_j, s+\tau) \right. \\
&\quad \left. + C_i e_\tau^{\check{B}(t_i^+ - \tau - t_i)} X(t_i, s+\tau) \right\} \\
&= e^{A(t_i^- - s)} X(t_i^-, s+\tau) + C_i e^{A(t_i^- - s)} X(t_i, s+\tau) \\
&= Y(t_i^-, s) + C_i Y(t_i, s).
\end{aligned}
$$

This ends the proof. $\qquad\qquad\square$

To give the norm estimation of $Y(\cdot,\cdot)$, we first consider the norm estimation of $\check{B} = e^{-A\tau} B$.

For a given $\check{B}$, one can find an $\alpha \in \mathbb{R}^+$ such that

$$
\|\check{B}\| \le \alpha e^{\alpha\tau}. \tag{4.4}
$$

In fact, by [94, 2.28, p. 44] and [95, (3.7), p. 109], we have $\|\check{B}\| = \|e^{-A\tau}B\| \leq K(\rho(B) + \varepsilon)e^{(\alpha(-A)+\varepsilon)\tau}$. Choosing $\alpha \geq \max\{K(\rho(B) + \varepsilon), \alpha(-A) + \varepsilon\}$, $\varepsilon > 0$, we obtain (4.4). In addition, we can calculate the parameter α for a given matrix by using MATLAB® software.

Lemma 4.5. *For any $\varepsilon > 0$ there exists $K \geq 1$ such that*

$$\|X(t,s)\| \leq \left(\prod_{s-\tau < t_j \leq t} (\rho(C_j) + 1 + \varepsilon) \right) e^{\alpha(t+\tau-s)} \tag{4.5}$$

and

$$\|Y(t,s)\| \leq K \left(\prod_{s < t_j \leq t} (\rho(C_j) + 1 + \varepsilon) \right) e^{(\alpha(A)+\alpha+\varepsilon)(t-s)}. \tag{4.6}$$

Proof. Without loss of generality, we suppose that $t_m < s - \tau \leq t_{m+1}$ and $t_{m+n} < t \leq t_{m+n+1}$, $m, n = 0, 1, 2, \cdots$. Next, we apply mathematical induction to complete our proof.

(i) For $n = 0$, by Lemma 4.1 via (4.4),

$$\|X(t,s)\| \leq \|e_\tau^{\check{B}(t-s)}\| \leq e^{\alpha(t+\tau-s)}.$$

(ii) For $n = 1$, by Lemma 4.1 and [94, 2.28, p. 44] via (4.4), we have

$$\begin{aligned}
\|X(t,s)\| &\leq \|e_\tau^{\check{B}(t-s)} + C_{m+1}e_\tau^{\check{B}(t-\tau-t_{m+1})}X(t_{m+1},s)\| \\
&\leq e^{\alpha(t+\tau-s)} + (\rho(C_{m+1}) + \varepsilon)e^{\alpha(t-t_{m+1})}e^{\alpha(t_{m+1}+\tau-s)} \\
&\leq (\rho(C_{m+1}) + 1 + \varepsilon)e^{\alpha(t+\tau-s)}.
\end{aligned}$$

(iii) Let $i := i(0, t)$ be the number of impulsive points which belong to $(0, t)$. For $n = k$, suppose that

$$\begin{aligned}
\|X(t,s)\| &\leq \left(\prod_{j=m+1}^{m+k} (\rho(C_j) + 1 + \varepsilon) \right) e^{\alpha(t+\tau-s)} \\
&= \left(\prod_{s-\tau < t_j \leq t} (\rho(C_j) + 1 + \varepsilon) \right) e^{\alpha(t+\tau-s)}.
\end{aligned}$$

(iv) For $n = k + 1$, by Lemma 4.1 and [94, 2.28, p. 44] via (4.4) again,

$$\|X(t,s)\|$$

$$\leq \left\| e_\tau^{\check{B}(t-s)} + \sum_{s-\tau < t_j \leq t} C_j e_\tau^{\check{B}(t-\tau-t_j)}X(t_j,s) \right\|$$

$$\leq e^{\alpha(t+\tau-s)} + \sum_{j=m+1}^{m+k+1} (\rho(C_j) + \varepsilon) e^{\alpha(t-t_j)} \left(\prod_{r=m+1}^{j-1} (\rho(C_r) + 1 + \varepsilon) \right) e^{\alpha(t_j+\tau-s)}$$

$$\leq \left(1 + \sum_{j=m+1}^{m+k+1} (\rho(C_j) + \varepsilon) \prod_{r=m+1}^{j-1} (\rho(C_r) + 1 + \varepsilon) \right) e^{\alpha(t+\tau-s)}$$

$$= \left(\prod_{j=m+1}^{m+k+1} (\rho(C_j) + 1 + \varepsilon) \right) e^{\alpha(t+\tau-s)} = \left(\prod_{s-\tau<t_j\leq t} (\rho(C_j) + 1 + \varepsilon) \right) e^{\alpha(t+\tau-s)}.$$

Linking mathematical induction, one can obtain (4.5).

Finally, using (4.2) and (4.5) via [95, (3.7), p. 109], one can derive (4.6) immediately. The proof is finished. $\qquad\square$

Lemma 4.6. *For any $t > s$ we have*

$$\|X(t,s)\| \leq \left(\prod_{s-\tau<t_j\leq t} (1 + \|C_j\| e^{-\|\check{B}\|\tau}) \right) e^{\|\check{B}\|(t-s)} \tag{4.7}$$

and

$$\|Y(t,s)\| \leq \left(\prod_{s<t_j\leq t} (1 + \|C_j\| e^{-\|\check{B}\|\tau}) \right) e^{-\|\check{B}\|\tau} e^{a(t-s)}, \tag{4.8}$$

where $a = \|A\| + \|\check{B}\|$.

Proof. Without loss of generality, we suppose that $t_m < s - \tau \leq t_{m+1}$ and $t_{m+n} < t \leq t_{m+n+1}$, $m, n = 0, 1, 2, \cdots$. To complete our proof we use mathematical induction.

(i) For $n = 0$, from Lemma 4.2 via (4.3),

$$\|X(t,s)\| \leq \|e_\tau^{\check{B}(t-s)}\| \leq e^{\|\check{B}\|(t-s)}.$$

(ii) For $n = 1$, from Lemma 4.2 via (4.3), we have

$$\|X(t,s)\| \leq \|e_\tau^{\check{B}(t-s)} + C_{m+1} e_\tau^{\check{B}(t-\tau-t_{m+1})} X(t_{m+1},s)\|$$

$$\leq e^{\|\check{B}\|(t-s)} + \|C_{m+1}\| e^{\|\check{B}\|(t-\tau-t_{m+1})} e^{\|\check{B}\|(t_{m+1}-s)}$$

$$= (1 + \|C_{m+1}\| e^{-\|\check{B}\|\tau}) e^{\|\check{B}\|(t-s)}.$$

(iii) For $n = z$, suppose that

$$\|X(t,s)\| \leq \left(\prod_{j=m+1}^{m+z} (1 + \|C_j\| e^{-\|\check{B}\|\tau}) \right) e^{\|\check{B}\|(t-s)}$$

$$= \left(\prod_{s-\tau < t_j \le t} (1 + \|C_j\| e^{-\|\check{B}\|\tau}) \right) e^{\|\check{B}\|(t-s)}.$$

For $n = z + 1$, from Lemma 4.2 via (4.3) again,

$$\|X(t,s)\| \le \left(\prod_{j=m+1}^{m+z} (1 + \|C_j\| e^{-\|\check{B}\|\tau}) \right) e^{\|\check{B}\|(t-s)}$$

$$+ \|C_{m+z+1}\| e^{\|\check{B}\|(t-\tau-t_{m+z+1})} \left(\prod_{j=m+1}^{m+z} (1 + \|C_j\| e^{-\|\check{B}\|\tau}) \right)$$

$$\times e^{\|\check{B}\|(t_{m+z+1}-s)}$$

$$= \left(\prod_{j=m+1}^{m+z+1} (1 + \|C_j\| e^{-\|\check{B}\|\tau}) \right) e^{\|\check{B}\|(t-s)}$$

$$= \left(\prod_{s-\tau < t_j \le t} (1 + \|C_j\| e^{-\|\check{B}\|\tau}) \right) e^{\|\check{B}\|(t-s)}.$$

From the mathematical induction principle we obtain (4.7).

Finally, using (4.2) via (4.7) and $\|e^{At}\| \le e^{\|A\|t}$, one obtains (4.8) immediately. $\qquad\square$

4.1.2.2 Representation of solutions

In this section, we seek explicit formulas of solutions to (4.1) and linear impulsive nonhomogeneous delay differential equations

$$\begin{cases} v'(t) = Av(t) + Bv(t-\tau) + g(t), \ t \ge 0, \ \tau > 0, \ t \ne t_i, \ g \in C(\mathbb{R}^+, \mathbb{R}^n), \\ \Delta v(t_i) = C_i v(t_i), \ i = 1, 2, \cdots, \\ v(t) = \psi(t), \ -\tau \le t \le 0. \end{cases}$$

$$(4.9)$$

Definition 4.1. A function $v \in C^1([-\tau, 0], \mathbb{R}^n) \cup PC^1(\mathbb{R}^+, \mathbb{R}^n)$ is called the solution of (4.1) (or (4.9)) if v satisfies $v(t) = \psi(t)$ for $-\tau \le t \le 0$ and the first and second equations in (4.1) (or (4.9)).

We derive explicit formulas of solutions to linear impulsive homogeneous delay systems.

Theorem 4.1. *The solution of (4.1) has the form*

$$v(t) = Y(t, -\tau)\psi(-\tau) + \int_{-\tau}^{0} Y(t,s)[\psi'(s) - A\psi(s)]ds, \ t \ge -\tau. \quad (4.10)$$

Proof. Concerning on the solution of (4.1) and Lemma 4.4, one should search in the form

$$v(t) = Y(t, -\tau)\varsigma + \int_{-\tau}^{0} Y(t, s)\hbar(s)ds,$$

where ς is an unknown constant and $\hbar$ is an unknown continuously differentiable function. Moreover, it satisfies the initial condition $v(t) = \psi(t)$, $-\tau \le t \le 0$, i.e.,

$$v(t) = Y(t, -\tau)\varsigma + \int_{-\tau}^{0} Y(t, s)\hbar(s)ds := \psi(t), \quad -\tau \le t \le 0.$$

Letting $t = -\tau$, we have

$$Y(-\tau, s) = \begin{cases} \Theta, & -\tau < s \le 0, \\ I, & s = -\tau. \end{cases}$$

Thus $\varsigma = \psi(-\tau)$. Since $-\tau \le t \le 0$, one obtains

$$Y(t, s) = e^{A(t-s)}e_{\tau}^{\check{B}(t-\tau-s)} = \begin{cases} \Theta, & t < s \le 0, \\ e^{A(t-s)}, & -\tau \le s \le t. \end{cases}$$

Thus on the interval $-\tau \le t \le 0$, one can derive that

$$\begin{aligned} \psi(t) &= Y(t, -\tau)\psi(-\tau) + \int_{-\tau}^{0} Y(t, s)\hbar(s)ds \\ &= Y(t, -\tau)\psi(-\tau) + \int_{-\tau}^{t} Y(t, s)\hbar(s)ds + \int_{t}^{0} Y(t, s)\hbar(s)ds \\ &= e^{A(t+\tau)}\psi(-\tau) + \int_{-\tau}^{t} e^{A(t-s)}\hbar(s)ds. \end{aligned} \tag{4.11}$$

Having differentiated (4.11), we obtain

$$\psi'(t) = Ae^{A(t+\tau)}\psi(-\tau) + A\int_{-\tau}^{t} e^{A(t-s)}\hbar(s)ds + \hbar(t) = A\psi(t) + \hbar(t).$$

Therefore,

$$\hbar(t) = \psi'(t) - A\psi(t).$$

The desired result holds. $\quad\square$

Next, the solution of (4.9) can be written as a sum $v(t) = v_0(t) + \overline{v(t)}$, where $v_0(t)$ is a solution of (4.1) and $\overline{v(t)}$ is a solution of (4.9) satisfying the zero initial condition.

The following result shows us how to derive the formula of $\overline{v(\cdot)}$.

Theorem 4.2. *The solution $\overline{v(t)}$ of (4.9) satisfying the zero initial condition has the form*

$$\overline{v(t)} = \sum_{j=0}^{i-1} \int_{t_j}^{t_{j+1}} Y(t,s)g(s)ds + \int_{t_i}^{t} Y(t,s)g(s)ds, \ t \geq 0. \qquad (4.12)$$

Proof. By using the method of variation of constants, we will search $\overline{v(t)}$ in the following form:

$$\overline{v(t)} = \sum_{j=0}^{i-1} \int_{t_j}^{t_{j+1}} Y(t,s)c_j(s)ds + \int_{t_i}^{t} Y(t,s)c_i(s)ds, \qquad (4.13)$$

where $c_j(t)$, $j = 0, 1, \cdots, i(0,t)$, is an unknown vector function. We divide our proof into several steps:

(i) For any $0 < t \leq t_1$, we have $\overline{v(t)} = \int_0^t Y(t,s)c_0(s)ds$. Let us differentiate $\overline{v(t)}$ to obtain

$$\frac{d}{dt}\overline{v(t)} = A\overline{v(t)} + B\int_0^{t-\tau} Y(t-\tau,s)c_0(s)ds$$

$$+ B\int_{t-\tau}^{t} Y(t-\tau,s)c_0(s)ds + c_0(t)$$

$$= A\overline{v(t)} + B\overline{v(t-\tau)} + g(t).$$

From Remark 4.1, we know $Y(t-\tau,s) = \Theta$, $s > t - \tau$, then

$$c_0(t) = g(t).$$

(ii) For any $t_1 < t \leq t_2$, we have

$$\overline{v(t)} = \int_0^{t_1} Y(t,s)g(s)ds + \int_{t_1}^{t} Y(t,s)c_1(s)ds.$$

Let us differentiate $\overline{v(t)}$ again to obtain

$$\frac{d}{dt}\overline{v(t)} = \int_0^{t_1} [AY(t,s) + BY(t-\tau,s)]g(s)ds$$

$$+ \int_{t_1}^{t} [AY(t,s) + BY(t-\tau,s)]c_1(s)ds + c_1(t)$$

$$= A\overline{v(t)} + B\overline{v(t-\tau)} + g(t),$$

which implies that

$$c_1(t) = g(t).$$

(iii) Suppose that $c_{i-1}(t) = g(t)$ holds on subintervals $(t_{i-1}, t_i]$, $i = 2, 3 \cdots$. For any $t_i < t \leq t_{i+1}$, we have

$$\overline{v(t)} = \sum_{j=0}^{i-1} \int_{t_j}^{t_{j+1}} Y(t,s)g(s)ds + \int_{t_i}^{t} Y(t,s)c_i(s)ds.$$

Let us differentiate $\overline{v(t)}$ again to obtain

$$\frac{d}{dt}\overline{v(t)} = A\overline{v(t)} + B\Bigg\{ \sum_{j=0}^{i(0,t-\tau)-1} \int_{t_j}^{t_{j+1}} Y(t-\tau,s)g(s)ds$$

$$+ \int_{t_{i(0,t-\tau)}}^{t-\tau} Y(t-\tau,s)c_{i(0,t-\tau)}(s)ds \Bigg\} + c_i(t)$$

$$= A\overline{v(t)} + B\overline{v(t-\tau)} + g(t).$$

This yields $c_i(t) = g(t)$. According to the mathematical induction, one can obtain $c_i(t) = g(t)$, $i = 1, 2 \cdots$. Linking with (4.13), (4.12) is derived. $\qquad\square$

Corollary 4.1. *The solution of (4.9) has the form*

$$v(t) = Y(t,-\tau)\psi(-\tau) + \int_{-\tau}^{0} Y(t,s)[\psi'(s) - A\psi(s)]ds$$

$$+ \sum_{j=0}^{i-1} \int_{t_j}^{t_{j+1}} Y(t,s)g(s)ds + \int_{t_i}^{t} Y(t,s)g(s)ds.$$

4.1.2.3 Asymptotical stability results

In this section, we discuss asymptotical stability of the trivial solution of (4.1).

Definition 4.2. The trivial solution of (4.1) is called locally asymptotically stable if there exists $\delta > 0$ such that $\|\psi\|_1 := \max_{[-\tau,0]} \|\psi(t)\| + \max_{[-\tau,0]} \|\psi'(t)\| < \delta$, the following holds:

$$\lim_{t\to\infty} \|v(t)\| = 0.$$

We introduce the following conditions.

(H_1) Suppose that the distance between the impulsive points t_k and t_{k+1} satisfies

$$0 < \theta_1 \leq t_{k+1} - t_k \leq \theta_2, \ k = 0, 1, 2 \cdots,$$

and set

$$\theta = \begin{cases} \theta_1, & \alpha(A) + \alpha < 0, \\ \theta_2, & \alpha(A) + \alpha \geq 0, \end{cases}$$

for some α given in (4.4).

(H_2) Let $\eta_1 := \alpha(A) + \alpha + \frac{1}{\theta}\ln(\rho(C) + 1)$, where $\rho(C) := \max\{\rho(C_j) : j = 1, \cdots, i(0, t)\}$, and assume that $\eta_1 < 0$.

(H_3) There exists a positive constant p such that

$$\lim_{t \to \infty} \frac{i(0, t)}{t} = \tilde{p}.$$

(H_4) Suppose that $\eta_2 < 0$, where

$$\eta_2 := \alpha(A) + \alpha + \tilde{p}\ln(\rho(C) + 1). \tag{4.14}$$

Now, we are ready to state our first stability result for the trivial solution of (4.1).

Theorem 4.3. *If (H_1) and (H_2) are satisfied, then the trivial solution of (4.1) is locally asymptotically stable.*

Proof. Using (4.10), Lemma 4.5, and Theorem 4.1, we have

$$\|v(t)\| \le \|Y(t, -\tau)\|\|\psi(-\tau)\| + \int_{-\tau}^{0} \|Y(t, s)\|\|\psi'(s) - A\psi(s)\|ds$$

$$\le K\left(\prod_{j=1}^{i}(\rho(C_j) + 1 + \varepsilon)\right)e^{(\alpha(A)+\alpha+\varepsilon)(t+\tau)}\|\psi(-\tau)\|$$

$$+ \int_{-\tau}^{0} K\left(\prod_{j=1}^{i}(\rho(C_j) + 1 + \varepsilon)\right)e^{(\alpha(A)+\alpha+\varepsilon)(t-s)}\|\psi'(s) - A\psi(s)\|ds$$

$$\le T_1 e^{(\alpha(A)+\alpha+\varepsilon)t}(\rho(C) + 1 + \varepsilon)^{i(0,t)}$$

$$= T_1 e^{(\alpha(A)+\alpha+\varepsilon)(t-t_{i(0,t)})}e^{(\alpha(A)+\alpha+\varepsilon)t_{i(0,t)}}(\rho(C) + 1 + \varepsilon)^{i(0,t)}$$

$$\le T_1 e^{(\alpha(A)+\alpha+\varepsilon)\theta}e^{(\alpha(A)+\alpha+\varepsilon)i(0,t)\theta}(\rho(C) + 1 + \varepsilon)^{i(0,t)}$$

$$= T_1 e^{(\alpha(A)+\alpha+\varepsilon)\theta}[e^{(\alpha(A)+\alpha+\varepsilon)\theta}(\rho(C) + 1 + \varepsilon)]^{i(0,t)}, \tag{4.15}$$

where

$$T_1 := K(e^{(\alpha(A)+\alpha+\varepsilon)\tau}\|\psi(-\tau)\| + \int_{-\tau}^{0} e^{-(\alpha(A)+\alpha+\varepsilon)s}\|\psi'(s) - A\psi(s)\|ds) > 0.$$

It follows from condition (H_2) that one can choose $\varepsilon > 0$ such that

$$e^{(\alpha(A)+\alpha+\varepsilon)\theta}(\rho(C) + 1 + \varepsilon) \le e^{\frac{\theta\eta_1}{2}} < 1.$$

Then (4.15) gives

$$\|v(t)\| \le T_1 e^{(\alpha(A)+\alpha+\varepsilon)\theta}e^{\frac{\theta\eta_1}{2}i(0,t)} \to 0 \text{ for } t \to \infty\,(\text{implies } i(0, t) \to \infty).$$

The proof is completed. $\qquad\square$

Next, we give the second stability result of the trivial solution of (4.1).

Theorem 4.4. *If (H_3) and (H_4) are satisfied, then the trivial solution of (4.1) is locally asymptotically stable.*

Proof. The proof is similar to that of Theorem 4.3, so we only give the details of the main differences. Note that with (H_3), we have $i(0,t) < t(\tilde{p} + \varepsilon)$, $\varepsilon > 0$ arbitrarily small. Linking formula (4.15), one has

$$\|v(t)\| \le T_1 e^{(\alpha(A)+\alpha+\varepsilon)t}(\rho(C)+1+\varepsilon)^{i(0,t)} \le T_1 e^{[\alpha(A)+\alpha+\varepsilon+(\tilde{p}+\varepsilon)\ln(\rho(C)+1+\varepsilon)]t}.$$

By virtue of (H_4) we can find ε such that

$$\|v(t)\| \le T_1 e^{[\alpha(A)+\alpha+\varepsilon+(\tilde{p}+\varepsilon)\ln(\rho(C)+1+\varepsilon)]t} \le T_1 e^{\frac{\eta_2}{2}t} \to 0 \text{ as } t \to \infty.$$

The desired result holds. $\qquad\qquad\square$

To end this section, we extend the above results to the nonlinear case. Consider the following linear impulsive delay system with nonlinear term:

$$\begin{cases} v'(t) = Av(t) + Bv(t-\tau) + f(v(t)), \ t \ge 0, \ \tau > 0, \ t \ne t_i, \\ \Delta v(t_i) = C_i v(t_i), \ i = 1, 2, \cdots, \\ v(t) = \psi(t), \ -\tau \le t \le 0, \end{cases} \tag{4.16}$$

where $f \in C(\mathbb{R}^n, \mathbb{R}^n)$.

We impose the following additional assumptions:

(H_5) Suppose that there exists $l > 0$ such that $\|f(v)\| < l\|v\|$.

(H_6) Suppose that $Kl + \frac{\eta_2}{2} < 0$, where η_2 is defined in (4.14).

Theorem 4.5. *Assume that (H_3), (H_5), and (H_6) are satisfied. Then the trivial solution of (4.16) is locally asymptotically stable.*

Proof. By Corollary 4.1, any solution of (4.16) should have the form

$$v(t) = Y(t, -\tau)\psi(-\tau) + \int_{-\tau}^{0} Y(t, s)[\psi'(s) - A\psi(s)]ds$$

$$+ \sum_{j=0}^{i-1} \int_{t_j}^{t_{j+1}} Y(t, s)f(v(s))ds + \int_{t_i}^{t} Y(t, s)f(v(s))ds. \tag{4.17}$$

By the conditions of (H_3), (4.6) reduces to

$$\|Y(t, s)\| \le K(\rho(C)+1+\varepsilon)^{i(s,t)} e^{(\alpha(A)+\alpha+\varepsilon)(t-s)}$$

$$\le K e^{[\alpha(A)+\alpha+\varepsilon+(\tilde{p}+\varepsilon)\ln(\rho(C)+1+\varepsilon)](t-s)}$$

$$\le K e^{\frac{\eta_2}{2}(t-s)}. \tag{4.18}$$

Taking the norm on both sides of (4.17) via (4.18) and (H_5), one has

$$\|v(t)\| \leq K e^{\frac{\eta_2}{2}(t+\tau)}\|\psi(-\tau)\| + \int_{-\tau}^{0} K e^{\frac{\eta_2}{2}(t-s)}\|\psi'(s) - A\psi(s)\|ds$$

$$+ \sum_{j=0}^{i-1} \int_{t_j}^{t_{j+1}} K e^{\frac{\eta_2}{2}(t-s)}l\|v(s)\|ds + \int_{t_i}^{t} K e^{\frac{\eta_2}{2}(t-s)}l\|v(s)\|ds.$$

This yields

$$e^{-\frac{\eta_2}{2}t}\|v(t)\| \leq K e^{\frac{\eta_2}{2}\tau}\|\psi(-\tau)\| + \int_{-\tau}^{0} K e^{-\frac{\eta_2}{2}s}\|\psi'(s) - A\psi(s)\|ds$$

$$+ Kl \int_{0}^{t} e^{-\frac{\eta_2}{2}s}\|v(s)\|ds.$$

Let

$$T_2 = K e^{\frac{\eta_2}{2}\tau}\|\psi(-\tau)\| + \int_{-\tau}^{0} K e^{-\frac{\eta_2}{2}s}\|\psi'(s) - A\psi(s)\|ds > 0.$$

By using the well-known classical Gronwall inequality (see [96, Theorem 1]), we have

$$e^{-\frac{\eta_2}{2}t}\|v(t)\| \leq T_2 e^{Klt},$$

which yields

$$\|v(t)\| \leq T_2 e^{(Kl+\frac{\eta_2}{2})t} \to 0 \text{ as } t \to \infty$$

due to (H_6). The proof is finished. $\qquad\square$

4.1.2.4 Examples

In this section, we give two examples to illustrate our above theoretical results. Here, we use MATLAB software to compute some parameters and draw the figures for the examples.

Example 4.1. Consider (4.1) with $\vartheta = 0.2$ and

$$A = \begin{pmatrix} -3 & 0 \\ 0 & -2.5 \end{pmatrix}, \quad B = \begin{pmatrix} 1 & 0 \\ 0 & 0.8 \end{pmatrix}, \quad C_j = \begin{pmatrix} 2+\frac{1}{j} & 0 \\ 0 & 2 \end{pmatrix},$$

$$j = 1, 2, \cdots, \tag{4.19}$$

and

$$\psi(t) = \begin{pmatrix} 0.25 \\ 0.4 \end{pmatrix}, \quad i(0, t) = \left[\frac{t+1}{2}\right], \tag{4.20}$$

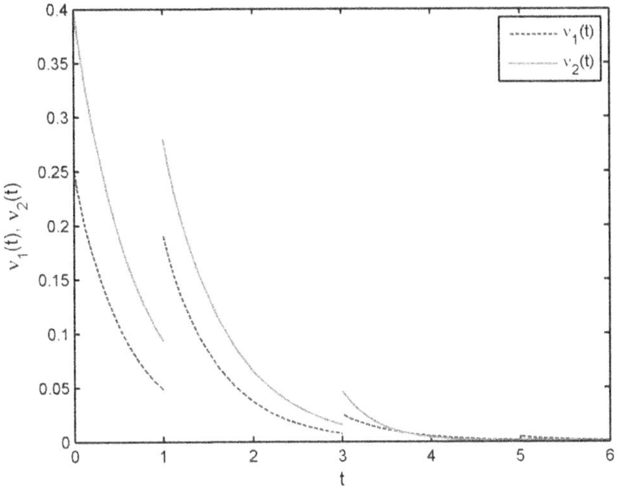

FIGURE 4.1 The state response $v(t)$ of (4.1), (4.19), and (4.20).

where $[x]$ is the biggest integer less than real x.

Obviously, $AB = BA$, $AC_j = C_j A$, $BC_j = C_j B$, $j = 1, 2, \cdots$, $\alpha(A) = -2.5$, and $\rho(C) = 3$. Next, $\|e^{At}\| = e^{-2.5t} \leq Ke^{(\alpha(A)+\varepsilon)t}$, where $K = 1$ and $\varepsilon > 0$. By computation, $\|\psi\|_1 = 0.4 = \|\psi(-\vartheta)\| < \delta := 0.41$, $\tilde{p} = \lim\limits_{t \to \infty} \frac{i(0,t)}{t} = \lim\limits_{t \to \infty} \frac{[\frac{t+1}{2}]}{t} = \frac{1}{2}$. Next, by using MATLAB software, we obtain

$$\|\check{B}\| = \|e^{-A\vartheta} B\| = \left\| \begin{pmatrix} 1.8221 & 0 \\ 0 & 1.3190 \end{pmatrix} \right\| \leq \alpha e^{0.2\alpha}, \text{ choose } \alpha = 1.3821,$$

$$T_1 \geq K \left(e^{(\alpha(A)+\alpha)\vartheta} \|\psi(-\vartheta)\| + \int_{-\vartheta}^{0} e^{-(\alpha(A)+\alpha)s} \|\psi'(s) - A\psi(s)\| ds \right)$$

$$= 0.5440 > 0,$$

and

$$\eta_2 = \alpha(A) + \alpha + p\ln(\rho(C) + 1) = -0.4248 < 0.$$

Now all the conditions of Theorem 4.4 are satisfied. Thus,

$$\|v(t)\| \leq T_1 e^{\frac{\eta_2}{2}t} = T_1 e^{-0.2124t} \to 0 \text{ as } t \to \infty,$$

that is, the trivial solution of (4.1), (4.19), and (4.20) is locally asymptotically stable (see Fig. 4.1).

Example 4.2. Consider (4.16) where $\vartheta = 0.2$,

$$
A = \begin{pmatrix} -3.3 & 0 \\ 0 & -3.3 \end{pmatrix}, \quad f(v(t)) = \begin{pmatrix} 0.5\sin v_1 \\ 0.5\sin v_2 \end{pmatrix},
$$

$$
B = \begin{pmatrix} 0.8 & 0.2 \\ 0 & 0.6 \end{pmatrix}, \quad C_j = \begin{pmatrix} 3 & 0.5 \\ 0 & 2.5 \end{pmatrix}, \ j = 1, 2, \cdots, \tag{4.21}
$$

and $\psi, i(0, t)$ are defined in (4.20).

By the calculation,

$$
AB = \begin{pmatrix} -2.64 & -0.66 \\ 0 & -1.98 \end{pmatrix} = BA, \quad AC_j = \begin{pmatrix} -9.9 & -1.65 \\ 0 & -8.25 \end{pmatrix} = C_j A,
$$

$$
BC_j = \begin{pmatrix} 2.4 & 0.9 \\ 0 & 1.5 \end{pmatrix} = C_j B, \quad j = 1, 2, \cdots.
$$

Obviously, $\alpha(A) = -3.3$ and $\|e^{At}\| = e^{-3.3t} \leq Ke^{(\alpha(A)+\varepsilon)t}$, where $K = 1$ and $\varepsilon > 0$. In addition, $\|f(v)\| < l\|v\|$, and we can choose $l = 0.5$. Next, by using MATLAB software, we obtain

$$
\|\check{B}\| = \|e^{-A\vartheta} B\| = \left\| \begin{pmatrix} 1.5478 & 03870 \\ 0 & 1.1609 \end{pmatrix} \right\| \leq \alpha e^{0.2\alpha}, \text{ and choose } \alpha = 1.4483.
$$

Moreover, $\eta_2 = \alpha(A) + \alpha + p\ln(\rho(C) + 1) = -1.1586$; thus,

$$
T_2 = Ke^{\frac{\eta_2}{2}\vartheta}\|\psi(-\vartheta)\| + \int_{-\vartheta}^{0} Ke^{-\frac{\eta_2}{2}s}\|\psi'(s) - A\psi(s)\|ds = 0.6228 > 0.
$$

Note that $Kl + \frac{\eta_2}{2} = -0.0793 < 0$. Now all the conditions of Theorem 4.5 are satisfied. Then,

$$
\|v(t)\| \leq T_2 e^{(Kl+\frac{\eta_2}{2})t} = 0.6228 e^{-0.0793t} \to 0 \text{ as } t \to \infty,
$$

that is, the trivial solution of (4.16), (4.20), and (4.21) is locally asymptotically stable (see Fig. 4.2).

4.1.2.5   Conclusion

We introduce the impulsive delayed matrix function and give its norm estimation. With the help of the impulsive delayed Cauchy matrix and the method of variation of constants, we obtain a representation of solutions to linear impulsive delay differential equations. Moreover, we derive some sufficient conditions to guarantee the trivial solution is locally asymptotically stable. The results in this part are motivated from [97].

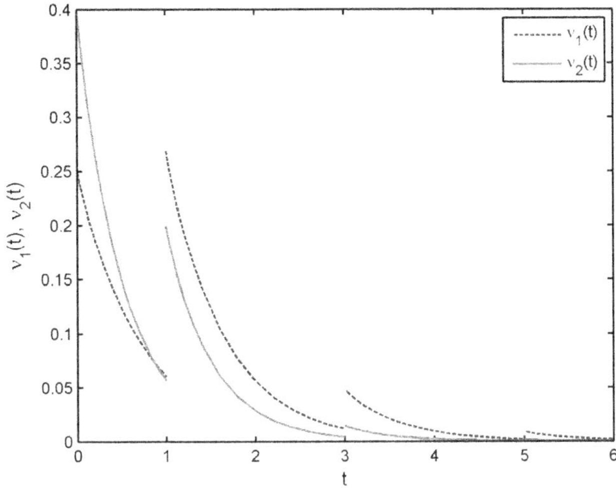

FIGURE 4.2 The state response $v(t)$ of (4.16), (4.20), and (4.21).

4.2 Finite time stability

4.2.1 Representation of solutions

Let $\mathscr{T}$ be the set of all impulsive points t_i. Consider the linear delay systems

$$\begin{cases} v'(t) = Av(t) + Bv(t-\tau), \ t \in \Omega := [0, T], \ t \notin \mathscr{T}, \ \tau > 0, \\ v(t) = \psi(t), \ -\tau \le t \le 0, \end{cases} \tag{4.22}$$

and the nonlinear delay system

$$\begin{cases} v'(t) = Av(t) + Bv(t-\tau) + F(t, v(t), v(t-\tau)), \ t \in \Omega, \ t \notin \mathscr{T}, \\ v(t) = \psi(t), \ -\tau \le t \le 0, \end{cases} \tag{4.23}$$

with the following linear impulsive conditions:

(I_1) $\Delta v(t_i) = C_i v(t_i), \ t_i \in \mathscr{T},$

(I_2) $\Delta v(t_i) = C_i v(t_i) + d_i, \ t_i \in \mathscr{T},$

where $A, B, C_i \in \mathbb{R}^{n \times n}$, $AB = BA$, $AC_i = C_i A$, $BC_i = C_i B$ for each $i \in \{1, 2, \cdots\}$ and $d_i \in \mathbb{R}^n$, $v \in C^1([-\tau, 0], \mathbb{R}^n) \cup PC^1(\Omega, \mathbb{R}^n)$, $\psi \in C_\tau^1$, $F \in C(\Omega \times \mathbb{R}^n \times \mathbb{R}^n, \mathbb{R}^n)$, and $v(t_i^{\pm}) = \lim_{\epsilon \to 0^+} v(t_i \pm \epsilon)$ and $v(t_i^-) = v(t_i)$ represent respectively the right and left limits of $v(t)$ at $t = t_i$. Set $\Omega \cap \mathscr{T} = \{t_1, t_2, \cdots, t_r\}$.

Any classical solution $v \in C^1([-\tau, 0], \mathbb{R}^n) \cup PC^1(\Omega, \mathbb{R}^n)$ of (4.22) with (I_1) has the form

$$v(t) = Y(t, -\tau)\psi(-\tau) + \int_{-\tau}^{0} Y(t, s)[\psi'(s) - A\psi(s)]ds. \tag{4.24}$$

By Corollary 4.1 any classical solution $v \in C^1([-\tau, 0], \mathbb{R}^n) \cup PC^1(\Omega, \mathbb{R}^n)$ of (4.23) with (I_1) has the form

$$
\begin{aligned}
v(t) = {} & Y(t, -\tau)\psi(-\tau) + \int_{-\tau}^{0} Y(t, s)[\psi'(s) - A\psi(s)]ds \\
& + \sum_{\gamma=0}^{i-1} \int_{t_\gamma}^{t_{\gamma+1}} Y(t, s)F(s, v(s), v(s - \tau))ds \\
& + \int_{t_i}^{t} Y(t, s)F(s, v(s), v(s - \tau))ds.
\end{aligned}
\tag{4.25}
$$

Define $H(\cdot, \cdot) : \mathbb{R} \times \mathbb{R} \to \mathbb{R}^{n \times n}$ given by

$$
\begin{aligned}
H(t, s) = {} & e^{A(t-s)} e_\tau^{\check{B}(t-\tau-s)} + \sum_{s < t_j \leq t} C_j e^{A(t-t_j)} e_\tau^{\check{B}(t-\tau-t_j)} H(t_j, s) \\
& + \sum_{s < t_j \leq t} e^{A(t-t_j)} e_\tau^{\check{B}(t-\tau-t_j)} d_j, \quad t > s.
\end{aligned}
\tag{4.26}
$$

Obviously, $H(t, t) = I$ and $H(t, s) = \Theta, t < s$.

Lemma 4.7. $H(\cdot, \cdot)$ *defined in (4.26) is the fundamental matrix of (4.22) with (I_2).*

Proof. We divide our proof into two steps.

Step 1. We verify that $H(\cdot, \cdot)$ satisfies the differential equation

$$
\frac{d}{dt} H(t, s) = AH(t, s) + BH(t - \tau, s), \quad t \in (t_i, t_{i+1}].
$$

In fact, having differentiated $H(t, s)$ for $t \in (t_i, t_{i+1}]$ and $t > s$, by using Lemma 2.2, we obtain

$$
\begin{aligned}
& \frac{d}{dt} H(t, s) \\
& = Ae^{A(t-s)} e_\tau^{\check{B}(t-\tau-s)} + \check{B} e^{A(t-s)} e_\tau^{\check{B}(t-2\tau-s)} \\
& \quad + \sum_{s < t_j \leq t} C_j A e^{A(t-t_j)} e_\tau^{\check{B}(t-\tau-t_j)} H(t_j, s) \\
& \quad + \sum_{s < t_j \leq t} C_j \check{B} e^{A(t-t_j)} e_\tau^{\check{B}(t-2\tau-t_j)} H(t_j, s) \\
& \quad + \sum_{s < t_j \leq t} A e^{A(t-t_j)} e_\tau^{\check{B}(t-\tau-t_j)} d_j + \sum_{s < t_j \leq t} \check{B} e^{A(t-t_j)} e_\tau^{\check{B}(t-2\tau-t_j)} d_j \\
& = A \Bigg[e^{A(t-s)} e_\tau^{\check{B}(t-\tau-s)} + \sum_{s < t_j \leq t} C_j e^{A(t-t_j)} e_\tau^{\check{B}(t-\tau-t_j)} H(t_j, s)
\end{aligned}
$$

$$+ \sum_{s<t_j\leq t} e^{A(t-t_j)} e_\tau^{\check{B}(t-\tau-t_j)} d_j \Bigg]$$

$$+ B\Bigg[e^{A(t-\tau-s)} e_\tau^{\check{B}(t-2\tau-s)} + \sum_{s<t_j\leq t-\tau} C_j e^{A(t-\tau-t_j)} e_\tau^{\check{B}(t-2\tau-t_j)} H(t_j,s)$$

$$+ \sum_{s<t_j\leq t-\tau} e^{A(t-\tau-t_j)} e_\tau^{\check{B}(t-2\tau-t_j)} d_j \Bigg]$$

$$= AH(t,s) + BH(t-\tau,s).$$

Step 2. We verify that

$$H(t_i^+,s) - H(t_i^-,s) = C_i H(t_i,s) + d_i.$$

Note that $e_\tau^{Bt^+} = e_\tau^{Bt^-}$ for all $t > -\tau$, then

$$H(t_i^+,s)$$

$$= e^{A(t_i^+-s)} e_\tau^{\check{B}(t_i^+-\tau-s)} + \sum_{s<t_j\leq t_i^+} C_j e^{A(t_i^+-t_j)} e_\tau^{\check{B}(t_i^+-\tau-t_j)} H(t_j,s)$$

$$+ \sum_{s<t_j\leq t_i^+} e^{A(t_i^+-t_j)} e_\tau^{\check{B}(t_i^+-\tau-t_j)} d_j$$

$$= e^{A(t_i^--s)} e_\tau^{\check{B}(t_i^--\tau-s)} + \sum_{s<t_j\leq t_i^-} C_j e^{A(t_i^--t_j)} e_\tau^{\check{B}(t_i^--\tau-t_j)} H(t_j,s)$$

$$+ C_i e^{A(t_i^+-t_i)} e_\tau^{\check{B}(t_i^+-\tau-t_i)} H(t_j,s)$$

$$+ \sum_{s<t_j\leq t_i^-} e^{A(t_i^--t_j)} e_\tau^{\check{B}(t_i^--\tau-t_j)} d_j + e^{A(t_i^+-t_i)} e_\tau^{\check{B}(t_i^+-\tau-t_i)} d_i$$

$$= H(t_i^-,s) + C_i H(t_i,s) + d_i.$$

This ends the proof. $\qquad\square$

Theorem 4.6. *The classical solution $v \in C^1([-\tau,0],\mathbb{R}^n) \cup PC^1(\Omega,\mathbb{R}^n)$ of (4.22) with (I_2) has the form*

$$v(t) = H(t,-\tau)\psi(-\tau) + \int_{-\tau}^0 H(t,s)[\psi'(s) - A\psi(s)]ds. \qquad (4.27)$$

Proof. Concerning the solution of (4.22) with (I_2), one should search the form

$$v(t) = H(t,-\tau)c + \int_{-\tau}^0 H(t,s)h(s)ds,$$

where c is an unknown constant and h is an unknown continuously differentiable function. Moreover, it satisfies the initial condition $v(t) = \psi(t)$, $-\tau \leq t \leq 0$, i.e.,

$$v(t) = H(t, -\tau)c + \int_{-\tau}^{0} H(t, s)h(s)ds = \psi(t), \quad -\tau \leq t \leq 0.$$

Letting $t = -\tau$, we have

$$H(-\tau, s) = \begin{cases} \Theta, & -\tau < s \leq 0, \\ I, & s = -\tau. \end{cases}$$

Thus, $c = \psi(-\tau)$. Since $-\tau \leq t \leq 0$, one obtains

$$H(t, s) = e^{A(t-s)}e_{\tau}^{\breve{B}(t-\tau-s)} = \begin{cases} \Theta, & t < s \leq 0, \\ e^{A(t-s)}, & -\tau \leq s \leq t. \end{cases}$$

Thus for $-\tau \leq t \leq 0$, one can derive

$$\psi(t) = H(t, -\tau)\psi(-\tau) + \int_{-\tau}^{0} H(t, s)h(s)ds$$

$$= e^{A(t+\tau)}\psi(-\tau) + \int_{-\tau}^{t} e^{A(t-s)}h(s)ds. \tag{4.28}$$

Having differentiated (4.28), we obtain

$$\psi'(t) = Ae^{A(t+\tau)}\psi(-\tau) + A\int_{-\tau}^{t} e^{A(t-s)}h(s)ds + h(t) = A\psi(t) + h(t).$$

Therefore,

$$h(t) = \psi'(t) - A\psi(t).$$

The desired result holds. $\qquad\qquad\qquad\qquad\qquad\qquad\qquad\qquad\square$

Lemma 4.8. *Let $g \in C(\mathbb{R}^+, \mathbb{R}^n)$. The classical solution $v \in C^1([-\tau, 0], \mathbb{R}^n) \cup PC^1(\Omega, \mathbb{R}^n)$ of*

$$\begin{cases} v'(t) = Av(t) + Bv(t - \tau) + g(t), \ t \in \Omega, \ t \notin \mathscr{T}, \\ v(t) = \psi(t), \quad -\tau \leq t \leq 0, \\ \Delta v(t_i) = C_i v(t_i) + d_i, \ t_i \in \mathscr{T}, \end{cases} \tag{4.29}$$

has the form

$$v(t) = H(t, -\tau)\psi(-\tau) + \int_{-\tau}^{0} H(t, s)[\psi'(s) - A\psi(s)]ds$$

$$+ \sum_{\gamma=0}^{i-1} \int_{t_\gamma}^{t_{\gamma+1}} H(t,s)g(s)ds + \int_{t_i}^{t} H(t,s)g(s)ds, \qquad (4.30)$$

where $H(\cdot,\cdot)$ is defined in (4.26).

Proof. Obviously, any solution of (4.29) can be written as a sum

$$v(t) = v_0(t) + \overline{v(t)},$$

where $v_0(t)$ is formula (4.27) and $\overline{v(t)}$ is a solution of (4.29) satisfying the zero initial condition. By using the method of variation of constants, one will search $\overline{v(t)}$ in the following form:

$$\overline{v(t)} = \sum_{\gamma=0}^{i-1} \int_{t_\gamma}^{t_{\gamma+1}} H(t,s)c_\gamma(s)ds + \int_{t_i}^{t} H(t,s)c_i(s)ds, \qquad (4.31)$$

where $c_\gamma(t)$, $\gamma = 0, 1, \cdots, i(0, t]$, are unknown vector functions.

We divide our proofs into several steps as follows:

(i) For any $0 < t \le t_1$, we have

$$\overline{v(t)} = \int_0^t H(t,s)c_0(s)ds.$$

Let us differentiate $\overline{v(t)}$ to obtain

$$\frac{d}{dt}\overline{v(t)} = A\overline{v(t)} + B \int_0^{t-\tau} H(t-\tau,s)c_0(s)ds$$

$$+ B \int_{t-\tau}^{t} H(t-\tau,s)c_0(s)ds + c_0(t)$$

$$= A\overline{v(t)} + B\overline{v(t-\tau)} + g(t).$$

Note that $H(t-\tau,s) = \Theta$, $s > t - \tau$, then

$$c_0(t) = g(t).$$

(ii) For any $t_1 < t \le t_2$, we have

$$\overline{v(t)} = \int_0^{t_1} H(t,s)g(s)ds + \int_{t_1}^{t} H(t,s)c_1(s)ds.$$

Let us differentiate $\overline{v(t)}$ again to obtain

$$\frac{d}{dt}\overline{v(t)} = \int_0^{t_1} [AH(t,s) + BH(t-\tau,s)]g(s)ds$$

$$+ \int_{t_1}^{t} [AH(t,s) + BH(t-\tau,s)]c_1(s)ds + c_1(t)$$

$$= A\overline{v(t)} + B\overline{v(t-\tau)} + g(t),$$

which implies that

$$c_1(t) = g(t).$$

(iii) Suppose that $c_{i-1}(t) = g(t)$ holds on subintervals $(t_{i-1}, t_i]$, $i = 2, 3, \cdots$.

For any $t_i < t \leq t_{i+1}$, we have

$$\overline{v(t)} = \sum_{\gamma=0}^{i-1} \int_{t_\gamma}^{t_{\gamma+1}} H(t,s)g(s)ds + \int_{t_i}^{t} H(t,s)c_i(s)ds.$$

Let us differentiate $\overline{v(t)}$ again to obtain

$$\frac{d}{dt}\overline{v(t)} = A\overline{v(t)} + B\left\{ \sum_{\gamma=0}^{i(0,t-\tau)-1} \int_{t_\gamma}^{t_{\gamma+1}} H(t-\tau,s)g(s)ds \right.$$

$$\left. + \int_{t_{i(0,t-\tau)}}^{t-\tau} H(t-\tau,s)c_{i(0,t-\tau)}(s)ds \right\} + c_i(t)$$

$$= A\overline{v(t)} + B\overline{v(t-\tau)} + g(t).$$

This yields $c_i(t) = g(t)$. According to mathematical induction, one can obtain

$$c_\gamma(t) = g(t), \ \gamma = 1, 2, \cdots, i(0,t].$$

Linking with (4.31), (4.30) is derived. $\qquad\qquad\square$

4.2.2 Finite time stability results

Lemma 4.9. *For any $t > s$ we have*

$$\|H(t,s)\| \leq \left(\prod_{s<t_j\leq t} (1 + \|C_j\|e^{-\|\check{B}\|\tau}) \right) e^{-\|\check{B}\|\tau} e^{a(t-s)}$$

$$+ \sum_{s<t_j\leq t} \left(\prod_{t_j<t_k\leq t} (1 + \|C_k\|e^{-\|\check{B}\|\tau}) \right) e^{-\|\check{B}\|\tau} e^{a(t-t_j)} \|d_j\|.$$

Proof. The proof is similar to that of Lemma 4.6, so we omit it here. $\qquad\square$

Definition 4.3. (see [70, Definition 2])System (4.22) (or (4.23)) with impulsive condition (I_1) (or (I_2)) is finite time stable with respect to $\{0, \Omega, \eta, \delta, \tau\}$ if and only if $\|\psi\|_C^2 < \eta$ implies

$$\|v(t)\|^2 < \delta, \ \forall\, t \in \Omega,$$

where $\eta, \delta \in \mathbb{R}^+$ and $\delta > \eta$.

Let us assume that the nonlinear term F satisfies the following conditions:

(H_7) $F(t, v(t), v(t - \tau)) = f(t, v(t))$, $\|f(t, v(t))\| \le P\|v(t)\|^\sigma$, $P > 0$, $\sigma > 0$.

(H_8) $F(t, v(t), v(t - \tau)) = g(t, v(t - \tau))$, $\|g(t, v(t - \tau))\| \le L\|v(t - \tau)\|^m$, $L > 0$, $m > 0$.

4.2.2.1 Impulsive condition (I_1) holds

Theorem 4.7. *The solution of (4.22) with (I_1) is finite time stable with respect to $\{0, \Omega, \eta, \delta, \tau\}$ if the inequality*

$$e^{a(t+\tau)} < \frac{\sqrt{\delta}}{N\left(\sqrt{\eta} + \int_{-\tau}^{0} \|\psi'(s) - A\psi(s)\|ds \right)}, \quad \forall\, t \in \Omega, \tag{4.32}$$

holds, where $N = \left(\prod_{j=1}^{r}(1 + \|C_j\|e^{-\|\check{B}\|\tau}) \right)e^{-\|\check{B}\|\tau}$ and η, δ are defined in Definition 4.3.

Proof. Taking the norm for (4.24) via Lemma 4.6, we obtain

$$\|v(t)\|$$

$$\le \|Y(t, -\tau)\|\|\psi(-\tau)\| + \int_{-\tau}^{0} \|Y(t, s)\|\|\psi'(s) - A\psi(s)\|ds$$

$$\le \left(\prod_{j=1}^{i(0,t]}(1 + \|C_j\|e^{-\|\check{B}\|\tau}) \right)e^{-\|\check{B}\|\tau}e^{a(t+\tau)}\|\psi(-\tau)\|$$

$$+ \int_{-\tau}^{0} \left(\prod_{j=1}^{i(0,t]}(1 + \|C_j\|e^{-\|\check{B}\|\tau}) \right)e^{-\|\check{B}\|\tau}e^{a(t-s)}\|\psi'(s) - A\psi(s)\|ds$$

$$\le \left(\prod_{j=1}^{r}(1 + \|C_j\|e^{-\|\check{B}\|\tau}) \right)e^{-\|\check{B}\|\tau}e^{a(t+\tau)}\left(\sqrt{\eta} + \int_{-\tau}^{0} \|\psi'(s) - A\psi(s)\|ds \right).$$

It is easy to see that $\|v(t)\|^2 < \delta, \forall\, t \in \Omega$ via (4.32). The proof is completed. $\square$

Theorem 4.8. *The solution of (4.22) with (I_1) is finite time stable with respect to $\{0, \Omega, \eta, \delta, \tau\}$ provided that*

$$e^{at} < \frac{\sqrt{\delta}}{\overline{M}\sqrt{\eta}}, \quad \forall\, t \in \Omega, \tag{4.33}$$

where $\overline{M} = N\left(1 + \frac{\|B\|(1-e^{-a\tau})}{a}\right)$, where N is defined in Theorem 4.7.

Proof. We note the following two facts:

$$\frac{d}{ds}X(t,s) = -\check{B}X(t,s+\tau),$$

$$\frac{d}{ds}Y(t,s) = -Ae^{A(t-s)}X(t,s+\tau) - \check{B}e^{A(t-s)}X(t,s+2\tau)$$

$$= -AY(t,s) - BY(t,s+\tau).$$

By using integration by parts, (4.24) can be changed to the following formula:

$$v(t) = Y(t,-\tau)\psi(-\tau) + \int_{-\tau}^{0} Y(t,s)[\psi'(s) - A\psi(s)]ds$$

$$= Y(t,-\tau)\psi(-\tau) - A\int_{-\tau}^{0} Y(t,s)\psi(s)ds + \int_{-\tau}^{0} Y(t,s)d(\psi(s))$$

$$= Y(t,0)\psi(0) + B\int_{-\tau}^{0} Y(t,s+\tau)\psi(s)ds. \tag{4.34}$$

Taking the norm for (4.34) via Lemma 4.6,

$$\|v(t)\| \leq \|Y(t,0)\|\|\psi(0)\| + \|B\|\int_{-\tau}^{0} \|Y(t,s+\tau)\|\|\psi(s)\|ds$$

$$\leq \left(\prod_{j=1}^{i(0,t]}(1+\|C_j\|e^{-\|\check{B}\|\tau})\right)e^{-\|\check{B}\|\tau}e^{at}\|\psi(0)\|$$

$$+ \|B\|\int_{-\tau}^{0}\left(\prod_{s+\tau<t_j\leq t}(1+\|C_j\|e^{-\|\check{B}\|\tau})\right)e^{-\|\check{B}\|\tau}e^{a(t-\tau-s)}\|\psi(s)\|ds$$

$$\leq \left(\prod_{j=1}^{i(0,t]}(1+\|C_j\|e^{-\|\check{B}\|\tau})\right)e^{-\|\check{B}\|\tau}e^{at}\sqrt{\eta}\left(1+\|B\|\int_{-\tau}^{0}e^{-a(\tau+s)}ds\right)$$

$$\leq Ne^{at}\sqrt{\eta}\left(1+\frac{\|B\|(1-e^{-a\tau})}{a}\right).$$

By (4.33), we have $\|v(t)\|^2 < \delta, \forall t \in \Omega$. The proof is finished. $\qquad\square$

Theorem 4.9. *The solution of (4.23) with (I_1) is finite time stable with respect to $\{0, \Omega, \eta, \delta, \tau\}$ if F satisfies (H_7) with $\sigma = 1$ and*

$$e^{(NP+a)t} < \frac{\sqrt{\delta}}{\overline{M}\sqrt{\eta}}, \quad \forall t \in \Omega, \tag{4.35}$$

where $\overline{M}$, N are defined in Theorem 4.8 and Theorem 4.7, respectively.

Proof. Combining (4.34) with (4.25) and using the norm on both sides, we obtain

$$\|v(t)\| \leq \|Y(t,0)\|\|\psi(0)\| + \|B\| \int_{-\tau}^{0} \|Y(t,s+\tau)\|\|\psi(s)\|ds$$

$$+ \sum_{\gamma=0}^{i-1} \int_{t_\gamma}^{t_{\gamma+1}} \|Y(t,s)\|\|F(s,v(s),v(s-\tau))\|ds$$

$$+ \int_{t_i}^{t} \|Y(t,s)\|\|F(s,v(s),v(s-\tau))\|ds$$

$$\leq \overline{M}\sqrt{\eta}e^{at} + \int_{0}^{t} \|Y(t,s)\|\|F(s,v(s),v(s-\tau))\|ds$$

$$\leq \overline{M}\sqrt{\eta}e^{at} + \int_{0}^{t} \left(\prod_{s<t_j\leq t} (1+\|C_j\|e^{-\|\check{B}\|\tau}) \right)$$

$$\times e^{-\|\check{B}\|\tau} e^{a(t-s)} \|F(s,v(s),v(s-\tau))\|ds$$

$$\leq \overline{M}\sqrt{\eta}e^{at} + N \int_{0}^{t} e^{a(t-s)} \|F(s,v(s),v(s-\tau))\|ds. \tag{4.36}$$

For F satisfies (H_7) and $\sigma = 1$, (4.36) yields

$$e^{-at}\|v(t)\| \leq \overline{M}\sqrt{\eta} + NP \int_{0}^{t} e^{-as}\|v(s)\|ds.$$

By using the well-known classical Gronwall inequality, we have

$$e^{-at}\|v(t)\| \leq \overline{M}\sqrt{\eta}e^{NPt},$$

that is,

$$\|v(t)\| \leq \overline{M}\sqrt{\eta}e^{(NP+a)t}.$$

By (4.35), we have $\|v(t)\|^2 < \delta$, $\forall\, t \in \Omega$. The proof is completed. $\square$

Theorem 4.10. *The solution of (4.23) with (I_1) is finite time stable with respect to $\{0, \Omega, \eta, \delta, \tau\}$ if F satisfies (H_7), $0 < \sigma < 1$, and*

$$e^{at} < \left[(\overline{M}\sqrt{\eta})^{1-\sigma} - NP\frac{e^{a(\sigma-1)t}-1}{a} \right]^{\frac{1}{\sigma-1}} \sqrt{\delta}, \ \forall\, t \in \Omega. \tag{4.37}$$

Proof. Note that (H_7) and $0 < \sigma < 1$. Inequality (4.36) becomes

$$e^{-at}\|v(t)\| \leq \overline{M}\sqrt{\eta} + NP \int_{0}^{t} e^{a(\sigma-1)s} e^{-a\sigma s} \|v(s)\|^{\sigma} ds.$$

By applying [98, Lemma 2], one can get

$$e^{-at}\|v(t)\| \le \left[(\overline{M}\sqrt{\eta})^{1-\sigma} + (1-\sigma)NP \int_0^t e^{a(\sigma-1)s}ds \right]^{\frac{1}{1-\sigma}}$$

$$= \left[(\overline{M}\sqrt{\eta})^{1-\sigma} - NP \frac{e^{a(\sigma-1)t}-1}{a} \right]^{\frac{1}{1-\sigma}}.$$

Thus,

$$\|v(t)\| \le e^{at} \left[(\overline{M}\sqrt{\eta})^{1-\sigma} - NP \frac{e^{a(\sigma-1)t}-1}{a} \right]^{\frac{1}{1-\sigma}}.$$

By (4.37), the proof is completed. $\qquad\square$

Theorem 4.11. *If F satisfies (H_7) and $\sigma > 1$, the solution of (4.23) with (I_1) is finite time stable with respect to $\{0, \Omega, \eta, \delta, \tau\}$ provided that*

$$(\overline{M}\sqrt{\eta})^{1-\sigma} - NP \frac{e^{a(\sigma-1)t}-1}{a} > 0 \tag{4.38}$$

and

$$e^{at} < \left[(\overline{M}\sqrt{\eta})^{1-\sigma} - NP \frac{e^{a(\sigma-1)t}-1}{a} \right]^{\frac{1}{\sigma-1}} \sqrt{\delta}, \ \forall \, t \in \Omega. \tag{4.39}$$

Proof. For $\sigma > 1$, formula (4.36) becomes

$$e^{-at}\|v(t)\| \le \overline{M}\sqrt{\eta} + NP \int_0^t e^{a(\sigma-1)s} e^{-a\sigma s} \|v(s)\|^{\sigma} ds.$$

Note that by (4.38), using [99, Lemma 2.1], one can get

$$e^{-at}\|v(t)\| \le \frac{1}{\left[(\overline{M}\sqrt{\eta})^{1-\sigma} - (\sigma-1)NP \int_0^t e^{a(\sigma-1)s}ds \right]^{\frac{1}{\sigma-1}}}$$

$$= \frac{1}{\left[(\overline{M}\sqrt{\eta})^{1-\sigma} - NP \frac{e^{a(\sigma-1)t}-1}{a} \right]^{\frac{1}{\sigma-1}}}.$$

Thus,

$$\|v(t)\| \le \frac{e^{at}}{\left[(\overline{M}\sqrt{\eta})^{1-\sigma} - NP \frac{e^{a(\sigma-1)t}-1}{a} \right]^{\frac{1}{\sigma-1}}}.$$

Following (4.39), we have $\|v(t)\|^2 < \delta, \ \forall \, t \in \Omega$. The proof is finished. $\qquad\square$

Theorem 4.12. *The solution of (4.23) with (I_1) is finite time stable with respect to $\{0, \Omega, \eta, \delta, \tau\}$ provided that F satisfies (H_8) with $m = 1$ and the following inequalities hold:*

$$\left[\overline{M} e^{at} + NL \frac{e^{at} - 1}{a} \right] \sqrt{\eta} < \sqrt{\delta}, \ \forall \, t \in [0, \tau], \tag{4.40}$$

and

$$\left[\overline{M} + NL \frac{1 - e^{-a\tau}}{a} \right] e^{(NLe^{-a\tau} + a)t - NLe^{-a\tau}\tau} \sqrt{\eta} < \sqrt{\delta}, \ \forall \, t \in (\tau, T]. \tag{4.41}$$

Proof. Note that F satisfies (H_8) with $m = 1$, then (4.36) can be translated into

$$e^{-at} \|v(t)\| \leq \overline{M} \sqrt{\eta} + NLe^{-a\tau} \int_0^t e^{-a(s-\tau)} \|v(s - \tau)\| ds.$$

Note that $e^{-at} \|v(t)\| = e^{-at} \|\psi(t)\|, \ t \in [-\tau, 0]$.

By using [96, Lemma 2.8]:

(i) For $0 \leq t \leq \tau$,

$$e^{-at} \|v(t)\| \leq \overline{M} \sqrt{\eta} + NL \int_0^t e^{-as} \|\psi(s - \tau)\| ds$$

$$< \left[\overline{M} - NL \frac{e^{-at} - 1}{a} \right] \sqrt{\eta}.$$

Thus,

$$\|v(t)\| < \left[\overline{M} e^{at} + NL \frac{e^{at} - 1}{a} \right] \sqrt{\eta}.$$

By (4.40), we have $\|v(t)\|^2 < \delta, \ \forall \, t \in \Omega$.

(ii) For $\tau < t \leq T$,

$$e^{-at} \|v(t)\| \leq \left[\overline{M} \sqrt{\eta} + NL \int_0^\tau e^{-as} \|\psi(s - \tau)\| ds \right] e^{NLe^{-a\tau}(t-\tau)}$$

$$< \left[\overline{M} - NL \frac{e^{-a\tau} - 1}{a} \right] e^{NLe^{-a\tau}(t-\tau)} \sqrt{\eta}.$$

Thus,

$$\|v(t)\| < \left[\overline{M} + NL \frac{1 - e^{-a\tau}}{a} \right] e^{(NLe^{-a\tau} + a)t - NLe^{-a\tau}\tau} \sqrt{\eta}.$$

By (4.41), we have $\|v(t)\|^2 < \delta, \ \forall \, t \in \Omega$. The proof is finished. $\square$

Theorem 4.13. *The solution of (4.23) with (I_1) is finite time stable with respect to $\{0, \Omega, \eta, \delta, \tau\}$ if F satisfies (H_8), $0 < m < 1$, and the following inequalities hold:*

$$\left[\overline{M}\sqrt{\eta} + NL\frac{1 - e^{-a\tau}}{a}(\sqrt{\eta})^m \right]e^{at} < \sqrt{\delta}, \ \forall\, t \in [0, \tau], \tag{4.42}$$

and

$$\left[(\overline{M}\sqrt{\eta})^{1-m} + NLe^{-a\tau}\frac{e^{-a(m-1)\tau} - e^{a(m-1)(t-\tau)}}{a} \right]^{\frac{1}{1-m}} e^{at}$$
$$< \sqrt{\delta}, \ \forall\, t \in (\tau, T]. \tag{4.43}$$

Proof. Note that with (H_8) with $0 < m < 1$, (4.36) becomes

$$e^{-at}\|v(t)\| \leq \overline{M}\sqrt{\eta} + NLe^{-a\tau}\int_0^t e^{a(m-1)(s-\tau)}e^{-am(s-\tau)}\|v(s-\tau)\|^m ds.$$

Note that $e^{-at}\|v(t)\| = e^{-at}\|\psi(t)\|$, $t \in [-\tau, 0]$.

By using [98, Lemma 3]:

(i) For $0 \leq t \leq \tau$,

$$e^{-at}\|v(t)\| \leq \overline{M}\sqrt{\eta} + NL\int_0^\tau e^{-as}\|\psi(s-\tau)\|^m ds$$
$$< \overline{M}\sqrt{\eta} + NL\frac{1 - e^{-a\tau}}{a}(\sqrt{\eta})^m,$$

that is,

$$\|v(t)\| < \left[\overline{M}\sqrt{\eta} + NL\frac{1 - e^{-a\tau}}{a}(\sqrt{\eta})^m \right]e^{at}.$$

By (4.42), we have $\|v(t)\|^2 < \delta, \forall\, t \in \Omega$.

(ii) For $\tau < t \leq T$,

$$e^{-at}\|v(t)\| \leq \left[(\overline{M}\sqrt{\eta})^{1-m} + (1-m)NLe^{-a\tau}\int_0^t e^{a(m-1)(s-\tau)}ds \right]^{\frac{1}{1-m}}$$
$$= \left[(\overline{M}\sqrt{\eta})^{1-m} + NLe^{-a\tau}\frac{e^{-a(m-1)\tau} - e^{a(m-1)(t-\tau)}}{a} \right]^{\frac{1}{1-m}},$$

that is,

$$\|v(t)\| < \left[(\overline{M}\sqrt{\eta})^{1-m} + NLe^{-a\tau}\frac{e^{-a(m-1)\tau} - e^{a(m-1)(t-\tau)}}{a} \right]^{\frac{1}{1-m}} e^{at}.$$

By (4.43), we obtain $\|v(t)\|^2 < \delta, \forall\, t \in \Omega$. The proof is completed. $\qquad\square$

Theorem 4.14. *If $F(t, v(t), v(t - \tau)) = f(t, v(t))$, $\|f(t, v(t))\| \leq \varphi(\|v(t)\|)$, the solution of (4.23) with (I_1) is finite time stable with respect to $\{0, \Omega, \eta, \delta, \tau\}$ provided that*

$$e^{at}\Psi^{-1}\left(\Psi(\overline{M}\sqrt{\eta}) + \int_0^t N\omega(s)ds\right) < \sqrt{\delta}, \qquad (4.44)$$

where $\omega(t)\varphi(e^{-at}\|v(t)\|) = e^{-at}\varphi(\|v(t)\|)$ and $\Psi(u) = \int_{\mu_0}^{\mu}\frac{ds}{\varphi(s)}$; $\Psi^{-1}(\cdot)$ means the inverse function of $\Psi(\cdot)$.

Proof. Note that $\|f(t, v(t))\| \leq \varphi(\|v(t)\|)$, (4.36) becomes

$$e^{-at}\|v(t)\| \leq \overline{M}\sqrt{\eta} + N\int_0^t e^{-as}\varphi(\|v(s)\|)ds$$

$$= \overline{M}\sqrt{\eta} + N\int_0^t \omega(s)\varphi(e^{-as}\|v(s)\|)ds.$$

By Lemma 4.3,

$$e^{-at}\|v(t)\| \leq \Psi^{-1}\left(\Psi(\overline{M}\sqrt{\eta}) + \int_0^t N\omega(s)ds\right),$$

which implies

$$\|v(t)\| \leq e^{at}\Psi^{-1}\left(\Psi(\overline{M}\sqrt{\eta}) + \int_0^t N\omega(s)ds\right).$$

By (4.44), we get $\|v(t)\|^2 < \delta, \forall\, t \in \Omega$. $\qquad\qquad\square$

4.2.2.2 Impulsive condition (I_2) holds

Theorem 4.15. *The solution of (4.22) with (I_2) is finite time stable with respect to $\{0, \Omega, \eta, \delta, \tau\}$ provided that*

$$e^{at}(Ne^{a\tau} + Q)\left(\sqrt{\eta} + \int_{-\tau}^0 \|\psi'(s) - A\psi(s)\|ds\right) < \sqrt{\delta}, \;\forall\, t \in \Omega, \quad (4.45)$$

where $Q = \sum_{j=1}^r\left(\prod_{k=j+1}^r(1 + \|C_k\|e^{-\|\check{B}\|\tau})\right)e^{-\|\check{B}\|\tau}e^{-at_j}\|d_j\|.$

Proof. Taking the norm for (4.27) via Lemma 4.6, we obtain

$$\|v(t)\|$$

$$\leq \|H(t, -\tau)\|\|\psi(-\tau)\| + \int_{-\tau}^0 \|H(t, s)\|\|\psi'(s) - A\psi(s)\|ds$$

$$\leq \left\{\left(\prod_{-\tau < t_j \leq t}(1 + \|C_j\|e^{-\|\check{B}\|\tau})\right)e^{-\|\check{B}\|\tau}e^{a(t+\tau)}\right.$$

$$+ \sum_{-\tau < t_j \le t} \left(\prod_{t_j < t_k \le t} (1 + \|C_k\| e^{-\|\check{B}\|\tau}) \right) e^{-\|\check{B}\|\tau} e^{a(t-t_j)} \|d_j\| \right\} \|\psi(-\tau)\|$$

$$+ \int_{-\tau}^{0} \left\{ \left(\prod_{s < t_j \le t} (1 + \|C_j\| e^{-\|\check{B}\|\tau}) \right) e^{-\|\check{B}\|\tau} e^{a(t-s)} \right.$$

$$+ \sum_{s < t_j \le t} \left(\prod_{t_j < t_k \le t} (1 + \|C_k\| e^{-\|\check{B}\|\tau}) \right) e^{-\|\check{B}\|\tau} e^{a(t-t_j)} \|d_j\| \right\} \|\psi'(s) - A\psi(s)\| ds$$

$$\le N e^{a(t+\tau)} \left(\sqrt{\eta} + \int_{-\tau}^{0} \|\psi'(s) - A\psi(s)\| ds \right)$$

$$+ \sum_{j=1}^{r} \left(\prod_{k=j+1}^{r} (1 + \|C_k\| e^{-\|\check{B}\|\tau}) \right) e^{-\|\check{B}\|\tau} e^{a(t-t_j)} \|d_j\|$$

$$\times \left(\sqrt{\eta} + \int_{-\tau}^{0} \|\psi'(s) - A\psi(s)\| ds \right)$$

$$= e^{at} (N e^{a\tau} + Q) \left(\sqrt{\eta} + \int_{-\tau}^{0} \|\psi'(s) - A\psi(s)\| ds \right). \tag{4.46}$$

Linking (4.45) and (4.46), we obtain $\|v(t)\|^2 < \delta, \ \forall \, t \in \Omega$. $\qquad \square$

Remark 4.2. Denoting $d_i = 0$, $i = 1, 2, \cdots$, Theorem 4.15 reduces to Theorem 4.7.

Theorem 4.16. *The solution of (4.22) with (I_2) is finite time stable with respect to $\{0, \Omega, \eta, \delta, \tau\}$ provided that*

$$e^{bt} (N_1 e^{b\tau} + Q_1) \left(\sqrt{\eta} + \int_{-\tau}^{0} \|\psi'(s) - A\psi(s)\| ds \right) < \sqrt{\delta}, \ \forall \, t \in \Omega, \quad (4.47)$$

where $b = \alpha(A) + \alpha(\check{B})$, $N_1 = \left(\prod_{j=1}^{r} (1 + \|C_j\| e^{-\alpha(\check{B})\tau}) \right) e^{-\alpha(\check{B})\tau}$, and

$$Q_1 = \sum_{j=1}^{r} \left(\prod_{k=j+1}^{r} (1 + \|C_k\| e^{-\alpha(\check{B})\tau}) \right) e^{-\alpha(\check{B})\tau} e^{-bt_j} \|d_j\|.$$

Proof. Taking the norm for (4.27) via Lemma 4.6, we obtain

$$\|v(t)\| \le \|H(t, -\tau)\| \|\psi(-\tau)\| + \int_{-\tau}^{0} \|H(t, s)\| \|\psi'(s) - A\psi(s)\| ds$$

$$\le N_1 e^{b(t+\tau)} \left(\sqrt{\eta} + \int_{-\tau}^{0} \|\psi'(s) - A\psi(s)\| ds \right)$$

$$+ \sum_{j=1}^{r} \left(\prod_{k=j+1}^{r} (1 + \|C_k\| e^{-\alpha(\check{B})\tau}) \right) e^{-\alpha(\check{B})\tau} e^{b(t-t_j)} \|d_j\|$$

$$\times \left(\sqrt{\eta} + \int_{-\tau}^{0} \|\psi'(s) - A\psi(s)\| ds \right)$$

$$= e^{bt}(N_1 e^{b\tau} + Q_1)\left(\sqrt{\eta} + \int_{-\tau}^{0} \|\psi'(s) - A\psi(s)\| ds \right). \tag{4.48}$$

Linking (4.47) and (4.48), we obtain $\|v(t)\|^2 < \delta, \ \forall\, t \in \Omega.$ □

Theorem 4.17. *The solution of (4.23) with (I_2) is finite time stable with respect to $\{0, \Omega, \eta, \delta, \tau\}$ if F satisfies (H_7), $\sigma = 1$, and*

$$(Ne^{a\tau} + Q)\left(\sqrt{\eta} + \int_{-\tau}^{0} \|\psi'(s) - A\psi(s)\| ds \right) e^{(PN+a)t + \frac{PQ(e^{at}-1)}{a}}$$

$$< \sqrt{\delta}, \ \forall\, t \in \Omega, \tag{4.49}$$

where Q is defined in Theorem 4.15.

Proof. By Lemma 4.8, any solution of (4.23) with (I_2) has the form

$$v(t) = H(t, -\tau)\psi(-\tau) + \int_{-\tau}^{0} H(t, s)[\psi'(s) - A\psi(s)]ds$$

$$+ \sum_{\gamma=0}^{i-1} \int_{t_\gamma}^{t_{\gamma+1}} H(t, s)F(s, v(s), v(s - \tau))ds$$

$$+ \int_{t_i}^{t} H(t, s)F(s, v(s), v(s - \tau))ds. \tag{4.50}$$

Combining Lemma 4.6 with (4.50), we have

$$\|v(t)\| \le \|H(t, -\tau)\| \|\psi(-\tau)\| + \int_{-\tau}^{0} \|H(t, s)\| \|\psi'(s) - A\psi(s)\| ds$$

$$+ \int_{0}^{t} \|H(t, s)\| \|F(s, v(s), v(s - \tau))\| ds$$

$$\le e^{at}(Ne^{a\tau} + Q)\left(\sqrt{\eta} + \int_{-\tau}^{0} \|\psi'(s) - A\psi(s)\| ds \right)$$

$$+ \int_{0}^{t} \left[\left(\prod_{s < t_j \le t} (1 + \|C_j\| e^{-\|\check{B}\|\tau}) \right) e^{-\|\check{B}\|\tau} e^{a(t-s)} + \sum_{s < t_j \le t} \left(\prod_{t_j < t_k \le t} \right. \right.$$

$$\left. \left. (1 + \|C_k\| e^{-\|\check{B}\|\tau}) \right) e^{-\|\check{B}\|\tau} e^{a(t-t_j)} \|d_j\| \right] \|F(s, v(s), v(s - \tau))\| ds$$

$$\le e^{at}(Ne^{a\tau} + Q)\left(\sqrt{\eta} + \int_{-\tau}^{0} \|\psi'(s) - A\psi(s)\| ds \right)$$

$$+ \int_{0}^{t} e^{at}(Ne^{-as} + Q)\|F(s, v(s), v(s - \tau))\| ds. \tag{4.51}$$

Because F satisfies $F(t, v(t), v(t - \tau)) = f(t, v(t))$, $\|f(t, v(t))\| \leq P\|v(t)\|$, (4.51) yields

$$e^{-at}\|v(t)\| \leq (Ne^{a\tau} + Q)\left(\sqrt{\eta} + \int_{-\tau}^{0} \|\psi'(s) - A\psi(s)\|ds\right) + P\int_{0}^{t} (N + Qe^{as})e^{-as}\|v(s)\|ds.$$

By using the well-known classical Gronwall inequality, we have

$$e^{-at}\|v(t)\| \leq (Ne^{a\tau} + Q)\left(\sqrt{\eta} + \int_{-\tau}^{0} \|\psi'(s) - A\psi(s)\|ds\right)e^{P\int_{0}^{t}(N+Qe^{as})ds}$$

$$= (Ne^{a\tau} + Q)\left(\sqrt{\eta} + \int_{-\tau}^{0} \|\psi'(s) - A\psi(s)\|ds\right)e^{PNt+\frac{PQ(e^{at}-1)}{a}},$$

which implies that $\|v(t)\|^2 < \delta$ via (4.49), $t \in \Omega$. The proof is finished. $\quad\square$

Theorem 4.18. *If F satisfies (H_7) and $0 < \sigma < 1$, then the solution of (4.23) with (I_2) is finite time stable with respect to $\{0, \Omega, \eta, \delta, \tau\}$ provided that*

$$e^{at}\left\{\left[(Ne^{a\tau} + Q)\left(\sqrt{\eta} + \int_{-\tau}^{0} \|\psi'(s) - A\psi(s)\|ds\right)\right]^{1-\sigma}\right.$$
$$\left. - PN\frac{e^{a(\sigma-1)t} - 1}{a} + \frac{1-\sigma}{\sigma}PQ\frac{e^{a\sigma t} - 1}{a}\right\}^{\frac{1}{1-\sigma}} < \sqrt{\delta}, \ \forall\, t \in \Omega. \quad (4.52)$$

Proof. For $0 < \sigma < 1$, formula (4.51) becomes

$$e^{-at}\|v(t)\| \leq (Ne^{a\tau} + Q)\left(\sqrt{\eta} + \int_{-\tau}^{0} \|\psi'(s) - A\psi(s)\|ds\right)$$
$$+ P\int_{0}^{t} (Ne^{a(\sigma-1)s} + Qe^{a\sigma s})e^{-a\sigma s}\|v(s)\|^{\sigma}ds.$$

By applying Lemma [98, Lemma 2], one can get

$$e^{-at}\|v(t)\| \leq \left\{\left[(Ne^{a\tau} + Q)\left(\sqrt{\eta} + \int_{-\tau}^{0} \|\psi'(s) - A\psi(s)\|ds\right)\right]^{1-\sigma}\right.$$
$$\left. + (1-\sigma)P\int_{0}^{t} (Ne^{a(\sigma-1)s} + Qe^{a\sigma s})ds\right\}^{\frac{1}{1-\sigma}}$$
$$= \left\{\left[(Ne^{a\tau} + Q)\left(\sqrt{\eta} + \int_{-\tau}^{0} \|\psi'(s) - A\psi(s)\|ds\right)\right]^{1-\sigma}\right.$$
$$\left. - PN\frac{e^{a(\sigma-1)t} - 1}{a} + \frac{1-\sigma}{\sigma}PQ\frac{e^{a\sigma t} - 1}{a}\right\}^{\frac{1}{1-\sigma}}.$$

By (4.52), we have $\|v(t)\|^2 < \delta$, $\forall\, t \in \Omega$. The proof is completed. $\square$

Theorem 4.19. *If F satisfies (H_8) and $m = 1$, the solution of (4.23) with (I_2) is finite time stable with respect to $\{0, \Omega, \eta, \delta, \tau\}$ provided that*

$$
e^{at}\left[(Ne^{a\tau} + Q)\left(\sqrt{\eta} + \int_{-\tau}^{0} \|\psi'(s) - A\psi(s)\|ds\right)\right.
$$
$$
\left. + (N\frac{1 - e^{-at}}{a} + Qt)L\sqrt{\eta}\right] < \sqrt{\delta},\ \forall\, t \in [0, \tau],\qquad (4.53)
$$

and

$$
\left[(Ne^{a\tau} + Q)\left(\sqrt{\eta} + \int_{-\tau}^{0} \|\psi'(s) - A\psi(s)\|ds\right)\right.
$$
$$
\left. + (N\frac{1 - e^{-a\tau}}{a} + Qt)L\sqrt{\eta}\right]e^{(LNe^{-a\tau}+a)t+LQ\frac{e^{a(t-\tau)}-1}{a}}
$$
$$
< \sqrt{\delta},\ \forall\, t \in (\tau, T].\qquad (4.54)
$$

Proof. Note that for $m = 1$, (4.51) becomes

$$
e^{-at}\|v(t)\| \le (Ne^{a\tau} + Q)\left(\sqrt{\eta} + \int_{-\tau}^{0} \|\psi'(s) - A\psi(s)\|ds\right)
$$
$$
+ L\int_{0}^{t}(Ne^{-a\tau} + Qe^{a(s-\tau)})e^{-a(s-\tau)}\|v(s - \tau)\|ds,
$$

and note that $e^{-at}\|v(t)\| = e^{-at}\|\psi(t)\|$, $t \in [-\tau, 0]$.

By using [96, Lemma 2.8]:

(i) For $0 \le t \le \tau$,

$$
e^{-at}\|v(t)\| \le (Ne^{a\tau} + Q)\left(\sqrt{\eta} + \int_{-\tau}^{0} \|\psi'(s) - A\psi(s)\|ds\right)
$$
$$
+ L\int_{0}^{t}(Ne^{-as} + Q)\|\psi(s - \tau)\|ds
$$
$$
= (Ne^{a\tau} + Q)\left(\sqrt{\eta} + \int_{-\tau}^{0} \|\psi'(s) - A\psi(s)\|ds\right)
$$
$$
+ (N\frac{1 - e^{-at}}{a} + Qt)L\sqrt{\eta},
$$

that is,

$$
\|v(t)\| < e^{at}\left[(Ne^{a\tau} + Q)\left(\sqrt{\eta} + \int_{-\tau}^{0} \|\psi'(s) - A\psi(s)\|ds\right)\right.
$$
$$
\left. + (N\frac{1 - e^{-at}}{a} + Qt)L\sqrt{\eta}\right].
$$

By (4.53), one gets $\|v(t)\|^2 < \delta$, $\forall\, t \in \Omega$.

(ii) For $\tau < t \leq T$,

$$
\begin{aligned}
e^{-at}\|v(t)\| &\leq \left[(Ne^{a\tau} + Q)\left(\sqrt{\eta} + \int_{-\tau}^{0} \|\psi'(s) - A\psi(s)\|ds\right)\right.\\
&\quad \left. + (N\frac{1 - e^{-a\tau}}{a} + Q\tau)L\sqrt{\eta}\right]e^{L\int_{\tau}^{t}(Ne^{-a\tau} + Qe^{a(s-\tau)})ds}\\
&= \left[(Ne^{a\tau} + Q)\left(\sqrt{\eta} + \int_{-\tau}^{0} \|\psi'(s) - A\psi(s)\|ds\right)\right.\\
&\quad \left. + (N\frac{1 - e^{-a\tau}}{a} + Q\tau)L\sqrt{\eta}\right]e^{LNe^{-a\tau}t + LQ\frac{e^{a(t-\tau)} - 1}{a}},
\end{aligned}
$$

that is,

$$
\begin{aligned}
\|v(t)\| &< \left[(Ne^{a\tau} + Q)\left(\sqrt{\eta} + \int_{-\tau}^{0} \|\psi'(s) - A\psi(s)\|ds\right)\right.\\
&\quad \left. + (N\frac{1 - e^{-a\tau}}{a} + Q\tau)L\sqrt{\eta}\right]e^{(LNe^{-a\tau} + a)t + LQ\frac{e^{a(t-\tau)} - 1}{a}}.
\end{aligned}
$$

By (4.54), we have $\|v(t)\|^2 < \delta$, $\forall\, t \in \Omega$. The proof is finished. $\qquad\square$

Theorem 4.20. *The solution of (4.23) with (I_2) is finite time stable with respect to $\{0, \Omega, \eta, \delta, \tau\}$ if F satisfies (H_8), $0 < m < 1$, and the following inequalities hold:*

$$
\begin{aligned}
e^{at}&\left[(Ne^{a\tau} + Q)\left(\sqrt{\eta} + \int_{-\tau}^{0} \|\psi'(s) - A\psi(s)\|ds\right)\right.\\
&\quad \left. + (N\frac{1 - e^{-a\tau}}{a} + Q\tau)L(\sqrt{\eta})^m\right] < \sqrt{\delta}, \quad \forall\, t \in [0, \tau],
\end{aligned}
\tag{4.55}
$$

and

$$
\begin{aligned}
e^{at}&\left\{\left[(Ne^{a\tau} + Q)\left(\sqrt{\eta} + \int_{-\tau}^{0} \|\psi'(s) - A\psi(s)\|ds\right)\right]^{1-m}\right.\\
&\quad \left. - LNe^{-a\tau}\frac{e^{a(m-1)(t-\tau)} - e^{-a(m-1)\tau}}{a} + \frac{1 - m}{m}LQ\frac{e^{am(t-\tau)} - e^{-am\tau}}{a}\right\}^{\frac{1}{1-m}}\\
&< \sqrt{\delta}, \quad \forall\, t \in (\tau, T].
\end{aligned}
\tag{4.56}
$$

Proof. For $0 < m < 1$, formula (4.51) becomes

$$
\begin{aligned}
e^{-at}&\|v(t)\|\\
&\leq (Ne^{a\tau} + Q)\left(\sqrt{\eta} + \int_{-\tau}^{0} \|\psi'(s) - A\psi(s)\|ds\right)
\end{aligned}
$$

$$+ L \int_0^t (N e^{-a\tau} e^{a(m-1)(s-\tau)} + Q e^{am(s-\tau)}) e^{-am(s-\tau)} \|v(s-\tau)\|^m ds.$$

Note that $e^{-at} \|v(t)\| = e^{-at} \|\psi(t)\|$, $t \in [-\tau, 0]$.

By using [98, Lemma 3]:

(i) For $0 \le t \le \tau$,

$$e^{-at} \|v(t)\| \le (N e^{a\tau} + Q)\left(\sqrt{\eta} + \int_{-\tau}^0 \|\psi'(s) - A\psi(s)\| ds \right)$$

$$+ L \int_0^t (N e^{-as} + Q) \|\psi(s-\tau)\|^m ds$$

$$= (N e^{a\tau} + Q)\left(\sqrt{\eta} + \int_{-\tau}^0 \|\psi'(s) - A\psi(s)\| ds \right)$$

$$+ (N \frac{1 - e^{-a\tau}}{a} + Qt) L (\sqrt{\eta})^m.$$

Thus,

$$\|v(t)\| < e^{at}\left[(N e^{a\tau} + Q)\left(\sqrt{\eta} + \int_{-\tau}^0 \|\psi'(s) - A\psi(s)\| ds \right) \right.$$

$$\left. + (N \frac{1 - e^{-a\tau}}{a} + Q\tau) L (\sqrt{\eta})^m \right].$$

Note that due to (4.55), we have $\|v(t)\|^2 < \delta$, $\forall\, t \in \Omega$.

(ii) For $\tau < t \le T$,

$$e^{-at} \|v(t)\|$$

$$\le \left\{ \left[(N e^{a\tau} + Q)\left(\sqrt{\eta} + \int_{-\tau}^0 \|\psi'(s) - A\psi(s)\| ds \right) \right]^{1-m} \right.$$

$$\left. + (1-m) L \int_0^t (N e^{-a\tau} e^{a(m-1)(s-\tau)} + Q e^{am(s-\tau)}) ds \right\}^{\frac{1}{1-m}}$$

$$= \left\{ \left[(N e^{a\tau} + Q)\left(\sqrt{\eta} + \int_{-\tau}^0 \|\psi'(s) - A\psi(s)\| ds \right) \right]^{1-m} \right.$$

$$\left. - L N e^{-a\tau} \frac{e^{a(m-1)(t-\tau)} - e^{-a(m-1)\tau}}{a} + \frac{1-m}{m} L Q \frac{e^{am(t-\tau)} - e^{-am\tau}}{a} \right\}^{\frac{1}{1-m}}.$$

Thus,

$$\|v(t)\| < e^{at}\left\{ \left[(N e^{a\tau} + Q)\left(\sqrt{\eta} + \int_{-\tau}^0 \|\psi'(s) - A\psi(s)\| ds \right) \right]^{1-m} \right.$$

$$- LNe^{-a\tau} \frac{e^{a(m-1)(t-\tau)} - e^{-a(m-1)\tau}}{a}$$

$$+ \frac{1-m}{m} LQ \frac{e^{am(t-\tau)} - e^{-am\tau}}{a} \Big\}^{\frac{1}{1-m}},$$

which implies that $\|v(t)\|^2 < \delta$ by (4.56), $t \in \Omega$. The proof is completed. $\quad\square$

4.2.2.3 Numerical examples and discussion

In this section, we give numerical examples to demonstrate the validity of our theoretical results. We also provide some discussion based on the results in the literature and present advantages of finding the explicit representation of solutions of impulsive delay systems.

Example 4.3. Consider the following linear delay system with impulses:

$$\begin{cases} v'(t) = Av(t) + Bv(t - 0.4), \ t \in \Omega = [0, 3.5], \ t \notin \mathscr{T}, \\ v(t) = (0.3, 0.5)^{\top}, \ -0.4 \leq t \leq 0, \\ \Delta v(t_i) = C_i v(t_i), \ t_i \in \mathscr{T}, \end{cases}$$

where we set $T = 3.5$, $\tau = 0.4$, $\mathscr{T} = \{1, 2, 3, \cdots\}$, and

$$A = \begin{pmatrix} 0.3 & 0 \\ 0 & 0.4 \end{pmatrix}, \ B = \begin{pmatrix} 0.6 & 0 \\ 0 & 0.4 \end{pmatrix}, \ C_i = \begin{pmatrix} 0.1 & 0 \\ 0 & -0.1 \end{pmatrix},$$

$$i = 1, 2, 3, \cdots. \tag{4.57}$$

Obviously, $\Omega \cap \mathscr{T} = \{1, 2, 3\}$ and $AB = BA$, $AC_i = C_i A$, $BC_i = C_i B$, for $i = 1, 2, 3$, $r = 3$, $\|\psi(t)\|_C^2 = 0.25 < \eta = 0.3$.

By calculation,

$$\check{B} = \begin{pmatrix} 0.5321 & 0 \\ 0 & 0.3409 \end{pmatrix}, \quad a = 0.9321, \quad N = 1.0206, \quad \overline{M} = 1.2251,$$

$$\int_{-\tau}^{0} \|\psi'(s) - A\psi(s)\| ds = 0.08.$$

By calculation, we find that the requirements of δ in the above theorems are different to ensure that the system is stable on $[0, 3.5]$. In fact, Theorem 4.7 and Theorem 4.16 ($d_i = 0$) require $\sqrt{\delta} > 70.11$, and Theorem 4.8 requires $\sqrt{\delta} > 55.41$. We also note that [100, Theorem 3.5] ($\alpha = 1$) requires $\sqrt{\delta} > 1652.52$ and [100, Theorem 3.6] ($\alpha = 1$) requires $\sqrt{\delta} > 635.05$.

Example 4.4. Consider the following impulsive delay system with a nonlinearity independent on delay:

$$\begin{cases} v'(t) = Av(t) + Bv(t - 0.4) + f(t, v(t)), \ t \in [0, 3.5], \ t \notin \mathscr{T}, \\ v(t) = (0.3, 0.5)^\top, \ -0.4 \le t \le 0, \\ \Delta v(t_i) = C_i v(t_i), \ t_i \in \mathscr{T}, \end{cases}$$

where we set $T = 3.5$, $\tau = 0.4$, $r = 3$, A, B, C_i, $i = 1, 2, 3$, are given in (4.57), and

$$f(t, v(t)) = \begin{pmatrix} 0.2v_1(t)\sin t \\ -0.1v_2(t)\sin t \end{pmatrix}.$$

We can select $P = 0.2$ such that $\| f(t, v(t)) \| \le P \| v(t) \|$.

Letting $\Omega = [0, 3.5]$, by computer calculation, Theorem 4.9 requires $\sqrt{\delta} > 113.19$, Theorem 4.10 requires $\sqrt{\delta} > 69.61$, Theorem 4.11 requires $\sqrt{\delta} > 210.44$, Theorem 4.17 ($d_i = 0$) requires $\sqrt{\delta} > 143.23$, and Theorem 4.18 ($d_i = 0$) requires $\sqrt{\delta} > 85.99$.

Example 4.5. Consider the following impulsive delay system with a nonlinearity dependent on delay:

$$\begin{cases} v'(t) = Av(t) + Bv(t - 0.4) + g(t, v(t - 0.4)), \ t \in [0, 3.5], \ t \notin \mathscr{T}, \\ v(t) = (0.3, 0.5)^\top, \ -0.4 \le t \le 0, \\ \Delta v(t_i) = C_i v(t_i) + d_i, \ t_i \in \mathscr{T}, \end{cases}$$

where we set $T = 3.5$, $\tau = 0.4$, $r = 3$, A, B, C_i are given in (4.57), and

$$d_i = \begin{pmatrix} 0.4 \\ 0.8 \end{pmatrix}, \ i = 1, 2, 3, \quad g(t, v(t - 0.4)) = \begin{pmatrix} 0.2v_1(t - 0.4)^{1/2} \\ 0.5v_2(t - 0.4)^{1/2} \end{pmatrix}.$$

We can select $L = 0.5$ such that $\| g(t, v(t - 0.4)) \| \le L \| v(t - 0.4) \|^{1/2}$.

Letting $\Omega = [0, 3.5]$, by computer calculation, Theorem 4.12 requires $\sqrt{\delta} > 187.63$, Theorem 4.13 requires $\sqrt{\delta} > 86.69$, Theorem 4.19($d_i = 0$) requires $\sqrt{\delta} > 266.27$, and Theorem 4.20($d_i = 0$) requires $\sqrt{\delta} > 104.87$.

4.2.2.4 Conclusion

This section deals with delay differential systems with impulsive conditions and linear parts defined by permutable matrices. The solutions are expressed with the aid of a special function called the impulsive delayed matrix exponential. The sufficient conditions of finite time stability are derived by applying the properties of the impulsive delayed matrix exponential and Gronwall's integral inequalities. The results in this part are motivated from [101].

4.3 Controllability

4.3.1 Controllability of impulsive delay differential systems

Consider the relative controllability of the following impulsive semilinear delay differential systems:

$$\begin{cases} v'(t) = Av(t) + Bv(t - \tau) + f(t, v(t)) + Du(t), \ t \in \Omega, \ t \notin \mathscr{T}, \ \tau > 0, \\ \Delta v(t_i) = C_i v(t_i), \ t_i \in \mathscr{T} = \{t_1, t_2, \cdots, t_k\}, \\ v(t) = \psi(t), \ -\tau \leq t \leq 0, \end{cases}$$

$$(4.58)$$

where $A, B, C_i, D \in \mathbb{R}^{n \times n}$, $AB = BA$, $AC_i = C_i A$, $BC_i = C_i B$ for each $i \in \{1, 2, \cdots, k\}$, $v \in C^1([-\tau, 0], \mathbb{R}^n) \cup \left[PC(\Omega, \mathbb{R}^n) \cap (\cup_{i=0}^k C^1(t_i, t_{i+1})) \right]$ (here $t_0 = 0$ and $t_{k+1} = T$), $\psi \in C_\tau^1$, $f \in C(\Omega \times \mathbb{R}^n, \mathbb{R}^n)$, $\Omega := [0, T]$, $T > 0$, $0 < t_1 < t_2 < \cdots < t_k < T$, and the control function $u(\cdot)$ takes values from $L^2(\Omega, \mathbb{R}^n)$.

Definition 4.4. (see [38, Definition 4]) System (4.58) is called relatively controllable if for an arbitrary initial vector function $\psi \in C^1([-\tau, 0], \mathbb{R}^n)$, the final state of the vector $v_1 \in \mathbb{R}^n$ at time T, there exists a control $u \in L^2(\Omega, \mathbb{R}^n)$ such that system (4.58) has a solution $v \in C^1([-\tau, 0], \mathbb{R}^n) \cup \left[PC(\Omega, \mathbb{R}^n) \cap (\cup_{i=0}^k C^1(t_i, t_{i+1})) \right]$ (here $t_0 = 0$ and $t_{k+1} = T$) that satisfies $v(T) = v_1$.

For any solution $v \in C^1([-\tau, 0], \mathbb{R}^n) \cup \left[PC(\Omega, \mathbb{R}^n) \cap (\cup_{i=0}^k C^1(t_i, t_{i+1})) \right]$ (here $t_0 = 0$ and $t_{k+1} = T$) of (4.58), from Corollary 4.1, we obtain

$$v(t) = Y(t, -\tau)\psi(-\tau) + \int_{-\tau}^{0} Y(t, s)[\psi'(s) - A\psi(s)]ds$$

$$+ \int_0^t Y(t, s)[f(s, v(s)) + Du(s)]ds. \tag{4.59}$$

4.3.1.1 Relative controllability of linear systems

In this section, we investigate the relative controllability of the linear impulsive delay controlled system

$$\begin{cases} v'(t) = Av(t) + Bv(t - \tau) + Du(t), \ t \in \Omega, \ t \notin \mathscr{T}, \ \tau > 0, \\ \Delta v(t_i) = C_i v(t_i), \ t_i \in \mathscr{T}, \\ v(t) = \psi(t), \ -\tau \leq t \leq 0, \end{cases} \tag{4.60}$$

using the impulsive delay Cauchy matrix introduced. Next we construct a suitable control function for (4.58), which means that we give a condition necessary and sufficient for $u \in L^2(\Omega, \mathbb{R}^n)$ to lead the solution of (4.58) to v_1 at time T.

We apply Krasnoselskii's fixed point theorem (see [82]) to show that (4.58) is also relatively controllable under suitable conditions.

The impulsive delay Grammian matrix, an extension of the classical Grammian matrix for linear differential systems, is as follows:

$$W_\tau[0, T] = \int_0^T Y(T, s) D D^\top Y^\top(T, s) ds. \tag{4.61}$$

Theorem 4.21. *System (4.60) is relatively controllable if and only if $W_\tau[0, T]$ is nonsingular.*

Proof. First we establish sufficiency. Since $W_\tau[0, T]$ is nonsingular, its inverse $W_\tau[0, T]$ is well defined. For any final state $v_1 \in \mathbb{R}^n$ one can select a control function as follows:

$$u(t) = D^\top V^\top(T, t) W_\tau^{-1}[0, T] \tilde{\eta},$$

where

$$\tilde{\eta} = v_1 - Y(T, -\tau) \psi(-\tau) - \int_{-\tau}^0 Y(T, s)[\psi'(s) - A\psi(s)] ds.$$

Then

$$\begin{aligned}
v(T) &= Y(T, -\tau) \psi(-\tau) + \int_{-\tau}^0 Y(T, s)[\psi'(s) - A\psi(s)] ds \\
&\quad + \int_0^T Y(T, s) D u(s) ds \\
&= Y(T, -\tau) \psi(-\tau) + \int_{-\tau}^0 Y(T, s)[\psi'(s) - A\psi(s)] ds \\
&\quad + \int_0^T Y(T, s) D D^\top Y^\top(T, s) W_\tau^{-1}[0, T] \tilde{\eta} ds \\
&= v_1.
\end{aligned}$$

We argue by contradiction to prove our necessity result. Assume $W_\tau[0, T]$ is singular, i.e., there exists at least one nonzero state $\tilde{v} \in \mathbb{R}^n$ such that

$$\tilde{v}^\top W_\tau[0, T] \tilde{v} = 0.$$

One obtains

$$\begin{aligned}
0 = \tilde{v}^\top W_\tau[0, T] \tilde{v} &= \int_0^T \tilde{v}^\top Y(T, s) D D^\top Y^\top(T, s) \tilde{v} ds \\
&= \int_0^T \|\tilde{v}^\top Y(T, s) D\|^2 ds,
\end{aligned}$$

which implies

$$\tilde{v}^\top Y(T,s)D = \mathbf{0}^\top, \quad \forall s \in \Omega.$$

Since system (4.60) is relatively controllable, according to Definition 4.4, there exists a control $u_1(t)$ that drives the initial state to zero at T, i.e.,

$$v(T) = Y(T,-\tau)\psi(-\tau) + \int_{-\tau}^{0} Y(T,s)[\psi'(s) - A\psi(s)]ds$$
$$+ \int_{0}^{T} Y(T,s)Du_1(s)ds = \mathbf{0}. \tag{4.62}$$

Similarly, there also exists a control $u_2(t)$ that drives the initial state to the (nonzero) state $\tilde{v}$ at T, i.e.,

$$v(T) = Y(T,-\tau)\psi(-\tau) + \int_{-\tau}^{0} Y(T,s)[\psi'(s) - A\psi(s)]ds$$
$$+ \int_{0}^{T} Y(T,s)Du_2(s)ds$$
$$= \tilde{v}. \tag{4.63}$$

Then from (4.62) and (4.63), we have

$$\tilde{v} = \int_{0}^{T} Y(T,s)D[u_2(s) - u_1(s)]ds. \tag{4.64}$$

Multiplying both sides of (4.64) by $\tilde{v}^\top$, we obtain

$$\tilde{v}^\top \tilde{v} = \int_{0}^{T} \tilde{v}^\top Y(T,s)D[u_2(s) - u_1(s)]ds = 0.$$

Thus $\tilde{v} = \mathbf{0}$, which conflicts with $\tilde{v}$ being nonzero. Thus, the impulsive delay Grammian matrix $W_\tau[0,T]$ is nonsingular. The proof is finished. $\qquad\square$

4.3.1.2 Relative controllability of semilinear systems

We assume the following:

(H_W): The operator $W : L^2(\Omega, \mathbb{R}^n) \to \mathbb{R}^n$ defined by

$$Wu = \int_{0}^{T} Y(T,s)Du(s)ds$$

has an inverse operator W^{-1} which takes values in $L^2(\Omega, \mathbb{R}^n)/ker\,W$. Then we set

$$M = \|W^{-1}\|_{L_b(\mathbb{R}^n, L^2(\Omega, \mathbb{R}^n)/ker\,W)}.$$

From [87, Remark 3.3], we know

$$M = \sqrt{\|W_\tau^{-1}[0,T]\|}. \tag{4.65}$$

(H_F): The function $f : \Omega \times \mathbb{R}^n \to \mathbb{R}^n$ is continuous and there exist a constant $q > 1$ and $L_f(\cdot) \in L^q(\Omega, \mathbb{R}^+)$ such that

$$\|f(t, v) - f(t, \mu)\| \le L_f(t)\|v - \mu\|, \ v, \mu \in \mathbb{R}^n.$$

Theorem 4.22. *Suppose that (H_W) and (H_F) are satisfied. Then system (4.58) is relatively controllable provided that*

$$c\left[1 + \frac{d\|D\|M}{a}(e^{aT} - 1)\right] < 1, \tag{4.66}$$

where $d = \left(\prod_{j=1}^{k}(1 + \|C_j\|e^{-\|\check{B}\|\tau})\right)e^{-\|\check{B}\|\tau}$ *and*

$$c = \left(\prod_{j=1}^{k}(1 + \|C_j\|e^{-\|\check{B}\|\tau})\right)e^{-\|\check{B}\|\tau}[\frac{1}{ap}(e^{apT} - 1)]^{\frac{1}{p}}\|L_f\|_{L^q(\Omega,\mathbb{R}^+)},$$

$$\frac{1}{p} + \frac{1}{q} = 1, \ p, q > 1.$$

Proof. Let $v_1 \in \mathbb{R}^n$ be the final state. Using hypothesis (H_W) for arbitrary $v(\cdot) \in PC(\Omega, \mathbb{R}^n)$, we define the control function $u_v(t)$ by

$$u_v(t) = W^{-1}\left(v_1 - Y(T, -\tau)\psi(-\tau) - \int_{-\tau}^{0} Y(T, s)[\psi'(s) - A\psi(s)]ds\right.$$
$$\left. - \int_{0}^{T} Y(T, s)f(s, v(s))ds\right)(t), \ t \in \Omega. \tag{4.67}$$

We show that, using this control, the operator $\mathcal{F} : PC(\Omega, \mathbb{R}^n) \to PC(\Omega, \mathbb{R}^n)$ defined by

$$(\mathcal{F}v)(t) = Y(t, -\tau)\psi(-\tau) + \int_{-\tau}^{0} Y(t, s)[\psi'(s) - A\psi(s)]ds$$
$$+ \int_{0}^{t} Y(t, s)f(s, v(s))ds + \int_{0}^{t} Y(t, s)Du_v(s)ds$$

has a fixed point v, which is a mild solution of (4.58).

We check that $(\mathcal{F}v)(T) = v_1$, which means that u_v steers system (4.58) from $(\mathcal{F}v)(0)$ to v_1 in finite time T. This implies system (4.58) is relatively controllable on Ω.

For each positive number r, let $\mathcal{B}_r = \{v \in PC(\Omega, \mathbb{R}^n) : \|v\|_{PC} \leq r\}$ (a bounded, closed, and convex set of $PC(\Omega, \mathbb{R}^n)$). Set $\overline{N} = \sup_{t \in \Omega} \|f(t, 0)\|$.

We divide the proof into several steps.

Step 1. We claim that there exists a positive number r such that $\mathcal{F}(\mathcal{B}_r) \subseteq \mathcal{B}_r$. From (H_F) and Hölder's inequality, we obtain

$$\int_0^t e^{a(t-s)} L_f(s)ds \leq \left(\int_0^t e^{ap(t-s)}ds \right)^{\frac{1}{p}} \left(\int_0^t L_f^q(s)ds \right)^{\frac{1}{q}}$$

$$\leq \left[\frac{1}{ap}(e^{apt} - 1) \right]^{\frac{1}{p}} \|L_f\|_{L^q(\Omega, \mathbb{R}^+)}$$

and

$$\int_0^t e^{a(t-s)} \|f(s, 0)\|ds \leq \overline{N} \int_0^t e^{a(t-s)}ds = \frac{\overline{N}}{a}(e^{at} - 1).$$

From (4.67), using (H_W) and (H_F), we have

$$\|u_v(t)\| \leq \|W^{-1}\|_{L_b(\mathbb{R}^n, L^2(\Omega, \mathbb{R}^n)/\ker W)} \bigg(\|v_1\| + \|Y(T, -\tau)\|\|\psi(-\tau)\|$$

$$+ \int_{-\tau}^0 \|Y(T, s)\|\|\psi'(s) - A\psi(s)\|ds + \int_0^T \|Y(T, s)\|\|f(s, v(s))\|ds \bigg)$$

$$\leq M \bigg(\|v_1\| + \bigg(\prod_{j=1}^k (1 + \|C_j\|e^{-\|\check{B}\|\tau}) \bigg) e^{-\|\check{B}\|\tau} e^{a(T+\tau)} \|\psi(-\tau)\|$$

$$+ \bigg(\prod_{j=1}^k (1 + \|C_j\|e^{-\|\check{B}\|\tau}) \bigg) e^{-\|\check{B}\|\tau} \int_{-\tau}^0 e^{a(T-s)} \|\psi'(s) - A\psi(s)\|ds$$

$$+ \bigg(\prod_{j=1}^k (1 + \|C_j\|e^{-\|\check{B}\|\tau}) \bigg) e^{-\|\check{B}\|\tau}$$

$$\times \int_0^T e^{a(T-s)} (\|f(s, v(s)) - f(s, 0)\| + \|f(s, 0)\|)ds \bigg)$$

$$\leq M \bigg(\|v_1\| + \bigg(\prod_{j=1}^k (1 + \|C_j\|e^{-\|\check{B}\|\tau}) \bigg) e^{-\|\check{B}\|\tau} e^{a(T+\tau)} \|\psi(-\tau)\|$$

$$+ \bigg(\prod_{j=1}^k (1 + \|C_j\|e^{-\|\check{B}\|\tau}) \bigg) e^{-\|\check{B}\|\tau} \int_{-\tau}^0 e^{a(T-s)} \|\psi'(s) - A\psi(s)\|ds$$

$$+ \bigg(\prod_{j=1}^k (1 + \|C_j\|e^{-\|\check{B}\|\tau}) \bigg) e^{-\|\check{B}\|\tau}$$

$$\times \int_0^T e^{a(T-s)}(L_f(s)\|v(s)\| + \|f(s,0)\|)ds\Bigg)$$

$$\leq M\Bigg(\|v_1\| + \Bigg(\prod_{j=1}^k (1 + \|C_j\|e^{-\|\check{B}\|\tau})\Bigg)e^{-\|\check{B}\|\tau}e^{a(T+\tau)}\|\psi(-\tau)\|$$

$$+ \Bigg(\prod_{j=1}^k (1 + \|C_j\|e^{-\|\check{B}\|\tau})\Bigg)e^{-\|\check{B}\|\tau}\int_{-\tau}^0 e^{a(T-s)}\|\psi'(s) - A\psi(s)\|ds$$

$$+ \Bigg(\prod_{j=1}^k (1 + \|C_j\|e^{-\|\check{B}\|\tau})\Bigg)e^{-\|\check{B}\|\tau}[\frac{1}{ap}(e^{apT}-1)]^{\frac{1}{p}}$$

$$\times \|L_f\|_{L^q(\Omega,\mathbb{R}^+)}\|v\|_{PC}$$

$$+ \Bigg(\prod_{j=1}^k (1 + \|C_j\|e^{-\|\check{B}\|\tau})\Bigg)e^{-\|\check{B}\|\tau}\frac{\overline{N}}{a}(e^{aT}-1)\Bigg)$$

$$\leq M\|v_1\| + Mb + Mc\|v\|_{PC},$$

where

$$\mathfrak{b} = \Bigg(\prod_{j=1}^k (1 + \|C_j\|e^{-\|\check{B}\|\tau})\Bigg)e^{-\|\check{B}\|\tau}e^{a(T+\tau)}\|\psi(-\tau)\|$$

$$+ \Bigg(\prod_{j=1}^k (1 + \|C_j\|e^{-\|\check{B}\|\tau})\Bigg)e^{-\|\check{B}\|\tau}\int_{-\tau}^0 e^{a(T-s)}\|\psi'(s) - A\psi(s)\|ds$$

$$+ \Bigg(\prod_{j=1}^k (1 + \|C_j\|e^{-\|\check{B}\|\tau})\Bigg)e^{-\|\check{B}\|\tau}\frac{\overline{N}}{a}(e^{aT}-1).$$

From (H_W) and (H_F) we have

$$\|(\mathcal{F}v)(t)\|$$

$$\leq \|Y(t,-\tau)\|\|\psi(-\tau)\| + \int_{-\tau}^0 \|Y(t,s)\|\|\psi'(s) - A\psi(s)\|ds$$

$$+ \int_0^t \|Y(t,s)\|\|f(s,v(s))\|ds + \int_0^t \|Y(t,s)\|\|D\|\|u_v(s)\|ds$$

$$\leq \Bigg(\prod_{j=1}^k (1 + \|C_j\|e^{-\|\check{B}\|\tau})\Bigg)e^{-\|\check{B}\|\tau}e^{a(t+\tau)}\|\psi(-\tau)\|$$

$$+ \int_{-\tau}^0 \Bigg(\prod_{s<t_j\leq t}(1 + \|C_j\|e^{-\|\check{B}\|\tau})\Bigg)e^{-\|\check{B}\|\tau}e^{a(t-s)}\|\psi'(s) - A\psi(s)\|ds$$

$$+ \int_0^t \left(\prod_{s < t_j \le t} (1 + \|C_j\| e^{-\|\check{B}\|\tau}) \right) e^{-\|\check{B}\|\tau}$$

$$\times\, e^{a(t-s)} (L_f(s)\|v(s)\| + \|f(s,0)\|)\,ds$$

$$+ \int_0^t \left(\prod_{s < t_j \le t} (1 + \|C_j\| e^{-\|\check{B}\|\tau}) \right)$$

$$\times\, e^{-\|\check{B}\|\tau} e^{a(t-s)} \|D\|[M\|v_1\| + Mb + Mc\|v\|_{PC}]\,ds$$

$$\le \mathfrak{b} + c\|v\|_{PC} + \left(\prod_{j=1}^{k} (1 + \|C_j\| e^{-\|\check{B}\|\tau}) \right) e^{-\|\check{B}\|\tau} \|D\|$$

$$\times\, [M\|v_1\| + Mb + Mc\|v\|_{PC}] \int_0^t e^{a(t-s)}\,ds$$

$$\le \mathfrak{b}\left[1 + \left(\prod_{j=1}^{k} (1 + \|C_j\| e^{-\|\check{B}\|\tau}) \right) e^{-\|\check{B}\|\tau} \frac{\|D\|M}{a}(e^{at} - 1) \right]$$

$$+ \left(\prod_{j=1}^{k} (1 + \|C_j\| e^{-\|\check{B}\|\tau}) \right) e^{-\|\check{B}\|\tau} \frac{\|D\|M}{a}(e^{at} - 1)\|v_1\|$$

$$+ c\left[1 + \left(\prod_{j=1}^{k} (1 + \|C_j\| e^{-\|\check{B}\|\tau}) \right) e^{-\|\check{B}\|\tau} \frac{\|D\|M}{a}(e^{at} - 1) \right]\|v\|_{PC}$$

$$\le \mathfrak{b}\left[1 + \frac{d\|D\|M}{a}(e^{aT} - 1) \right] + \frac{d\|D\|M}{a}(e^{aT} - 1)\|v_1\|$$

$$+ c\left[1 + \frac{d\|D\|M}{a}(e^{aT} - 1) \right] r = r,$$

where

$$r = \frac{\mathfrak{b}\left[1 + \frac{d\|D\|M}{a}(e^{aT} - 1) \right] + \frac{d\|D\|M}{a}(e^{aT} - 1)\|v_1\|}{1 - c\left[1 + \frac{d\|D\|M}{a}(e^{aT} - 1) \right]}.$$

Hence, we obtain $\mathcal{F}(\mathcal{B}_r) \subseteq \mathcal{B}_r$ for such an r.

Now, we define operators $\mathcal{F}_1$ and $\mathcal{F}_2$ on $\mathcal{B}_r$ as

$$(\mathcal{F}_1 v)(t) = Y(t, -\tau)\psi(-\tau) + \int_{-\tau}^{0} Y(t, s)[\psi'(s) - A\psi(s)]\,ds$$

$$+ \int_0^t Y(t, s) D u_v(s)\,ds$$

and

$$(\mathcal{F}_2 v)(t) = \int_0^t Y(t,s) f(s, v(s)) ds,$$

for $t \in \Omega$, respectively.

Step 2. We claim that $\mathcal{F}_1$ is a contraction mapping.

Let $v, \gamma \in \mathcal{B}_r$. From (H_W) and (H_F), for each $t \in \Omega$, we have

$$\|u_v(t) - u_\gamma(t)\|$$

$$\leq M \int_0^T \|Y(T,s)\| \|f(s, v(s)) - f(s, \gamma(s))\| ds$$

$$\leq M \int_0^T \left(\prod_{s < t_j \leq T} (1 + \|C_j\| e^{-\|\check{B}\|\tau}) \right) e^{-\|\check{B}\|\tau} e^{a(T-s)} L_f(s)(\|v(s) - \gamma(s)\|) ds$$

$$\leq M \left(\prod_{j=1}^k (1 + \|C_j\| e^{-\|\check{B}\|\tau}) \right) e^{-\|\check{B}\|\tau} \left[\frac{1}{ap}(e^{apT} - 1) \right]^{\frac{1}{p}}$$

$$\times \|L_f\|_{L^q(\Omega, \mathbb{R}^+)} \|v - \gamma\|_{PC}$$

$$\leq Mc \|v - \gamma\|_{PC}.$$

Thus,

$$\|(\mathcal{F}_1 v)(t) - (\mathcal{F}_1 \gamma)(t)\|$$

$$\leq \int_0^t \|Y(t,s)\| \|D\| \|u_v(s) - u_\gamma(s)\| ds$$

$$\leq \int_0^t \left(\prod_{s < t_j \leq t} (1 + \|C_j\| e^{-\|\check{B}\|\tau}) \right) e^{-\|\check{B}\|\tau} e^{a(t-s)} ds \|D\| Mc \|v - \gamma\|_{PC}$$

$$\leq \frac{d\|D\|Mc}{a}(e^{aT} - 1)\|v - \gamma\|_{PC},$$

so we obtain

$$\|\mathcal{F}_1 v - \mathcal{F}_1 \gamma\|_{PC} \leq \overline{P} \|v - \gamma\|_{PC},$$

where $\overline{P} = \frac{d\|D\|Mc}{a}(e^{aT} - 1)$. From (4.66), we have $\overline{P} < 1$, so $\mathcal{F}_1$ is a contraction.

Step 3. We claim that $\mathcal{F}_2 : \mathcal{B}_r \to PC(\Omega, \mathbb{R}^n)$ is a compact and continuous operator.

Let $v_n \in \mathcal{B}_r$ with $v_n \to v$ in $\mathcal{B}_r$. Using (H_F), we have $f(s, v_n(s)) \to f(s, v(s))$ in $PC(\Omega, \mathbb{R}^n)$ and thus using the Lebesgue dominated convergence theorem we have

$$\|(\mathcal{F}_2 v_n)(t) - (\mathcal{F}_2 v)(t)\|$$

$$\leq \int_0^t \|Y(t,s)\| \|f(s,v_n(s)) - f(s,v(s))\| ds$$

$$\leq \left(\prod_{j=1}^k (1 + \|C_j\| e^{-\|\check{B}\|\tau}) \right) e^{-\|\check{B}\|\tau} \int_0^t e^{a(t-s)} \|f(s,v_n(s)) - f(s,v(s))\| ds$$

$$\to 0 \quad \text{as} \quad n \to 0,$$

which implies that $\mathcal{F}_2$ is continuous on $\mathcal{B}_r$.

To check the compactness of $\mathcal{F}_2 : \mathcal{B}_r \to PC(\Omega, \mathbb{R}^n)$, we prove that $\mathcal{F}_2(\mathcal{B}_r)$ is equicontinuous and uniformly bounded. In fact, for any $v \in \mathcal{B}_r$, $t_z < t \leq t + h \leq t_{z+1}$, $z = 0, 1, \cdots, k$,

$$(\mathcal{F}_2 v)(t+h) - (\mathcal{F}_2 v)(t)$$

$$= \int_0^{t+h} Y(t+h,s) f(s,v(s)) ds - \int_0^t Y(t,s) f(s,v(s)) ds$$

$$= \int_t^{t+h} Y(t+h,s) f(s,v(s)) ds$$

$$+ \int_0^t (e^{A(t+h-s)} - e^{A(t-s)}) X(t,s+\tau) f(s,v(s)) ds$$

$$+ \int_0^t e^{A(t+h-s)} (X(t+h,s+\tau) - X(t,s+\tau)) f(s,v(s)) ds.$$

Let

$$S_1 = \int_t^{t+h} Y(t+h,s) f(s,v(s)) ds,$$

$$S_2 = \int_0^t e^{A(t+h-s)} (X(t+h,s+\tau) - X(t,s+\tau)) f(s,v(s)) ds,$$

$$S_3 = \int_0^t (e^{A(t+h-s)} - e^{A(t-s)}) X(t,s+\tau) f(s,v(s)) ds$$

$$= \int_0^t (e^{Ah} - E) Y(t,s) f(s,v(s)) ds.$$

From above, we see that

$$\|(\mathcal{F}_2 v)(t+h) - (\mathcal{F}_2 v)(t)\| \leq \|S_1\| + \|S_2\| + \|S_3\|.$$

Now, we only need to check $\|S_i\| \to 0$ as $h \to 0$, $i = 1, 2, 3$. Clearly,

$$\|S_1\|$$

$$\leq \int_t^{t+h} \|Y(t+h,s)\| \|f(s,v(s))\| ds$$

$$\leq \int_t^{t+h} \left(\prod_{s < t_j \leq t+h} (1 + \|C_j\| e^{-\|\check{B}\|\tau}) \right) e^{-\|\check{B}\|\tau} e^{a(t+h-s)} (L_f(s)\|v(s)\|$$

$$+ \|f(s,0)\|) ds$$

$$\leq e^{-\|\check{B}\|\tau} [\frac{1}{ap}(e^{aph} - 1)]^{\frac{1}{p}} \|L_f\|_{L^q(\Omega,\mathbb{R}^+)} \|v\|_{PC} + e^{-\|\check{B}\|\tau} \frac{\overline{N}}{a}(e^{ah} - 1)$$

$$\to 0 \quad \text{as} \quad h \to 0,$$

$$\|S_2\|$$

$$\leq \int_0^t \|e^{A(t+h-s)}\| \|X(t+h, s+\tau) - X(t, s+\tau)\| \|f(s, v(s))\| ds$$

$$\leq e^{\|A\|T} \int_0^t \left\| e_\tau^{\check{B}(t+h-\tau-s)} - e_\tau^{\check{B}(t-\tau-s)} + \sum_{s < t_j \leq t+h} C_j e_\tau^{\check{B}(t+h-\tau-t_j)} X(t_j, s+\tau) \right.$$

$$\left. - \sum_{s < t_j \leq t} C_j e_\tau^{\check{B}(t-\tau-t_j)} X(t_j, s+\tau) \right\| (L_f(s)\|v(s)\| + \|f(s,0)\|) ds$$

$$\leq e^{\|A\|T} \int_0^t \left\| e_\tau^{\check{B}(t+h-\tau-s)} - e_\tau^{\check{B}(t-\tau-s)} \right.$$

$$\left. + \sum_{s < t_j \leq t} C_j (e_\tau^{\check{B}(t+h-\tau-t_j)} - e_\tau^{\check{B}(t-\tau-t_j)}) X(t_j, s+\tau) \right\| (L_f(s)\|v(s)\|$$

$$+ \|f(s,0)\|) ds$$

$$\leq e^{\|A\|T} \int_0^t \|e_\tau^{\check{B}(t+h-\tau-s)} - e_\tau^{\check{B}(t-\tau-s)}\| (L_f(s)\|v(s)\| + \|f(s,0)\|) ds$$

$$+ \sum_{j=1}^k e^{\|A\|T} \int_0^t \|C_j(e_\tau^{\check{B}(t+h-\tau-t_j)} - e_\tau^{\check{B}(t-\tau-t_j)}) X(t_j, s+\tau)\| L_f(s)\|v(s)\| ds$$

$$+ \sum_{j=1}^k e^{\|A\|T} \int_0^t \|C_j(e_\tau^{\check{B}(t+h-\tau-t_j)} - e_\tau^{\check{B}(t-\tau-t_j)}) X(t_j, s+\tau)\| \|f(s,0)\| ds$$

$$\leq e^{\|A\|T} \|L_f\|_{L^q(\Omega,\mathbb{R}^+)} r \left(\int_0^t \|e_\tau^{\check{B}(t+h-\tau-s)} - e_\tau^{\check{B}(t-\tau-s)}\|^p ds \right)^{\frac{1}{p}}$$

$$+ e^{\|A\|T} \overline{N} \int_0^t \|e_\tau^{\check{B}(t+h-\tau-s)} - e_\tau^{\check{B}(t-\tau-s)}\| ds$$

$$+ \sum_{j=1}^k e^{\|A\|T} \|L_f\|_{L^q(\Omega,\mathbb{R}^+)} r$$

$$\times \left(\int_0^t \|C_j(e_\tau^{\check{B}(t+h-\tau-t_j)} - e_\tau^{\check{B}(t-\tau-t_j)}) X(t_j, s+\tau)\|^p ds \right)^{\frac{1}{p}}$$

$$+ \sum_{j=1}^{k} e^{\|A\|T} \overline{N} \int_0^t \|C_j (e_\tau^{\check{B}(t+h-\tau-t_j)} - e_\tau^{\check{B}(t-\tau-t_j)}) X(t_j, s + \tau)\| ds$$

$$\to 0 \quad \text{as} \quad h \to 0,$$

and

$$\|S_3\|$$

$$\leq \int_0^t \|e^{Ah} - I\| \|Y(t,s)\| \|f(s, v(s))\| ds$$

$$\leq \|e^{Ah} - I\|$$

$$\times \int_0^t \left(\prod_{s < t_j \leq t} (1 + \|C_j\| e^{-\|\check{B}\|\tau}) \right) e^{-\|\check{B}\|\tau} e^{a(t-s)} (L_f(s) \|v(s)\| + \|f(s,0)\|) ds$$

$$\leq \|e^{Ah} - I\| d \left(\left[\frac{1}{ap} (e^{apt} - 1) \right]^{\frac{1}{p}} \|L_f\|_{L^q(\Omega, \mathbb{R}^+)} r + \frac{\overline{N}}{a} (e^{at} - 1) \right)$$

$$\to 0 \quad \text{as} \quad h \to 0.$$

As a result, we immediately obtain

$$\|(\mathcal{F}_2 v)(t + h) - (\mathcal{F}_2 v)(t)\| \to 0 \quad \text{as} \quad h \to 0,$$

for all $v \in \mathcal{B}_r$. Therefore, $\mathcal{F}_2(\mathcal{B}_r)$ is equicontinuous in $PC(\Omega, \mathbb{R}^n)$.

Next, repeating the above computations, we have

$$\|(\mathcal{F}_2 v)(t)\|$$

$$\leq \int_0^t \|Y(t,s)\| \|f(s, v(s))\| ds$$

$$\leq \int_0^t \left(\prod_{s < t_j \leq t} (1 + \|C_j\| e^{-\|\check{B}\|\tau}) \right) e^{-\|\check{B}\|\tau} e^{a(t-s)} (L_f(s) \|v(s)\| + \|f(s,0)\|) ds$$

$$\leq d \left[\frac{1}{ap} (e^{apT} - 1) \right]^{\frac{1}{p}} \|L_f\|_{L^q(\Omega, \mathbb{R}^+)} r + \frac{d\overline{N}}{a} (e^{aT} - 1).$$

Hence $\mathcal{F}_2(\mathcal{B}_r)$ is uniformly bounded. From the PC-type Arzela–Ascoli theorem (see [102, Theorem 2.1]), $\mathcal{F}_2(\mathcal{B}_r)$ is relatively compact in $PC(\Omega, \mathbb{R}^n)$. Thus, $\mathcal{F}_2 : \mathcal{B}_r \to PC(\Omega, \mathbb{R}^n)$ is a compact and continuous operator.

Hence, using Krasnoselskii's fixed point theorem (see [82]), $\mathcal{F}$ has a fixed point v on $\mathcal{B}_r$. Obviously, v is a solution of system (4.58) satisfying $v(T) = v_1$. The boundary condition $v(t) = \psi(t)$, $-\tau \leq t \leq 0$, holds from (4.59). The proof is complete. $\qquad \square$

Remark 4.3. Now we make a comparison with [87]. Let $C_i = \Theta$, $i = 1, 2, \cdots, k$. Then system (4.58) reduces to [87, (3)]. For relative controllability results for the linear system, one has $Y(\cdot, \cdot) = e^{A(\cdot - \cdot)} e^{\check{B}(\cdot - \tau - \cdot)}_\tau$ and the Grammian matrix (4.61) which is the same as in [87, (10)]. For the nonlinear system, set $C_i = \Theta$, $a = \overline{N}$, $c = e^{-\|\check{B}\|\tau} M_2$, $d = e^{-\|\check{B}\|\tau}$, and then (4.66) becomes

$$e^{-\|\check{B}\|\tau} M_2 \left[1 + \frac{e^{-\|\check{B}\|\tau} \|D\| M}{\overline{N}} (e^{\overline{N}T} - 1) \right] < 1. \tag{4.68}$$

From (4.68) with $M_2 \left[1 + \frac{M}{N} (e^{\overline{N}t_1} - 1) \|C\| \right] < 1$ ([87, (23)]), one sees that the factor (4.68) improves [87, (23)] since $e^{-\|\check{B}\|\tau}$ is in (4.68).

4.3.1.3 Numerical examples and discussion

Example 4.6. Set $T = 1$. Consider the following semilinear impulsive delay differential controlled system:

$$\begin{cases} v'(t) = Av(t) + Bv(t - 0.5) + f(t, v(t)) + Du(t), \ t \in \Omega := [0, 1], \ t \notin \mathscr{T}, \\ \Delta v(t_i) = C_i v(t_i), \ t_i \in \mathscr{T} := \{0.5\}, \\ \psi(t) = (0.2, 0.3)^\top, \ -0.5 \le t \le 0, \end{cases} \tag{4.69}$$

where we set $k = 1$, $\tau = 0.5$, $\mathscr{T} = \{0.5\}$, and

$$A = \begin{pmatrix} 0.2 & 0 \\ 0 & 0.2 \end{pmatrix}, \ B = \begin{pmatrix} 0.2 & 0.1 \\ 0 & 0.3 \end{pmatrix}, \ D = \begin{pmatrix} 1 & 0 \\ 0 & 1 \end{pmatrix},$$

$$C_i = \begin{pmatrix} 0.3 & 0.1 \\ 0 & 0.4 \end{pmatrix}, \ i = 1, \ f(t, v(t)) = \begin{pmatrix} 0.2t v_1(t) \\ 0.3t v_2(t) \end{pmatrix}.$$

Clearly $\|A\| = 0.2$, $\|\check{B}\| = 0.2715$, $\|D\| = 1$, and $\|C_i\| = 0.4$ $(i = 1)$. Also,

$$AB = \begin{pmatrix} 0.04 & 0.02 \\ 0 & 0.06 \end{pmatrix} = BA, \ AC_i = \begin{pmatrix} 0.06 & 0.02 \\ 0 & 0.08 \end{pmatrix} = C_i A,$$

$$BC_i = \begin{pmatrix} 0.06 & 0.06 \\ 0 & 0.12 \end{pmatrix} = C_i B, \ i = 1,$$

$$\check{B} = e^{-A\tau} B = \begin{pmatrix} 1.8221 & 0.3644 \\ 0 & 1.4577 \end{pmatrix}.$$

Clearly $a = \|A\| + \|\check{B}\| = 0.4715$.

Now we use (4.65) to obtain M. For this purpose, we need to obtain $W_\tau[0, T]$ and then $W_\tau[0, T]^{-1}$. The delay Grammian matrix has the following explicit form:

$$W_\tau[0, T] = \int_0^T Y(T, s) D D^\top Y^\top(T, s) ds$$
$$= \int_0^1 e^{A(1-s)} X(1, s+0.5) X^\top(1, s+0.5) e^{A^\top(1-s)} ds$$
$$= W_1 + W_2,$$

where

$$W_1 =$$
$$\int_0^{0.5} e^{A(1-s)} \left[I + \check{B}(0.5 - s) + C_1 \right] \times \left[I + \check{B}^\top(0.5 - s) + C_1^\top \right] e^{A^\top(1-s)} ds$$

and

$$W_2 = \int_{0.5}^1 e^{A(1-s)} I^2 e^{A^\top(1-s)} ds.$$

Note

$$W_1 = \begin{pmatrix} 1.2371 & 0.1230 \\ 0.1230 & 1.4622 \end{pmatrix}, \quad W_2 = \begin{pmatrix} 0.5535 & 0 \\ 0 & 0.5535 \end{pmatrix}.$$

Therefore, we obtain

$$W_\tau[0, 1] = \begin{pmatrix} 1.7906 & 0.1230 \\ 0.1230 & 2.0157 \end{pmatrix}, \quad W_\tau^{-1}[0, 1] = \begin{pmatrix} 0.5608 & -0.0342 \\ -0.0342 & 0.4982 \end{pmatrix},$$

and

$$M = \sqrt{\|W_\tau^{-1}[0, 1]\|} = \sqrt{0.5266} = 0.7257.$$

For any $v, \mu \in \mathbb{R}^n$,

$$\|f(t, v) - f(t, \mu)\| = \max\{0.2t\|v_1 - \mu_1\|, 0.3t\|v_2 - \mu_2\|\}$$
$$\leq 0.3t \max\{\|v_1 - \mu_1\|, \|v_2 - \mu_2\|\} = 0.3t\|v - \mu\|.$$

Now we set $L_f(t) = 0.3t$, $L_f \in L^q(\Omega, \mathbb{R}^+)$ with $p = q = 2$. Note $\|L_f\|_{L^2(\Omega, \mathbb{R}^+)}$ $= (\int_0^1 (0.3s)^2 ds)^{\frac{1}{2}} = 0.1732,$

$$d = \left(\prod_{j=1}^k (1 + \|C_j\| e^{-\|\check{B}\|\tau}) \right) e^{-\|\check{B}\|\tau} = (1 + \|C_1\| e^{-0.5\|\check{B}\|})^2 e^{-0.5\|\check{B}\|} = 1.3681,$$

$$c = \left(\prod_{j=1}^{k} (1 + \|C_j\| e^{-\|\check{B}\|\tau}) \right) e^{-\|\check{B}\|\tau} [\frac{1}{ap}(e^{apT} - 1)]^{\frac{1}{p}} \|L_f\|_{L^q(\Omega,\mathbb{R}^+)}$$

$$= d[\frac{1}{ap}(e^{apT} - 1)]^{\frac{1}{p}} \|L_f\|_{L^q(\Omega,\mathbb{R}^+)} = 0.3055,$$

and

$$c[1 + \frac{d\|D\|M}{a}(e^{aT} - 1)] = 0.3055[1 + \frac{1.3681 \times 0.7257}{0.4715}(e^{0.4715} - 1)]$$
$$= 0.6930 < 1.$$

Thus all the conditions of Theorem 4.22 are satisfied, so (4.69) is relatively controllable on [0, 1].

Example 4.7. In Example 4.6 we keep the parameters unchanged except f and let

$$f(t, v(t)) = \begin{pmatrix} 0.3t^2 v_1(t) \sin t \\ 0.4t v_2(t) \cos 2t \end{pmatrix}. \tag{4.70}$$

For any $t \in \Omega$ and $v, \mu \in \mathbb{R}^n$,

$$\|f(t, v) - f(t, \mu)\| = \max\{0.3t^2 \|(v_1 - \mu_1) \sin t\|, 0.4t \|(v_2 - \mu_2) \cos 2t\|\}$$
$$\leq 0.4t \max\{\|v_1 - \mu_1\|, \|v_2 - \mu_2\|\}$$
$$= 0.4t \|v - \mu\|.$$

Set $L_f(t) = 0.4t$, $L_f \in L^q(\Omega, \mathbb{R}^+)$ with $p = q = 2$. Note

$$\|L_f\|_{L^2(\Omega,\mathbb{R}^+)} = \left(\int_0^1 (0.4s)^2 ds \right)^{\frac{1}{2}} = 0.2309,$$

$$c = d\left[\frac{1}{ap}(e^{apT} - 1) \right]^{\frac{1}{p}} \|L_f\|_{L^q(\Omega,\mathbb{R}^+)} = 0.4073,$$

and

$$c\left[1 + \frac{d\|D\|M}{a}(e^{aT} - 1) \right] = 0.4073\left[1 + \frac{1.3681 \times 0.7257}{0.4715}(e^{0.4715} - 1) \right]$$
$$= 0.9239 < 1.$$

From Theorem 4.22, we know (4.69) with (4.70) is relatively controllable on [0, 1].

Example 4.8. Consider the relative controllability of system (4.69) (with $f(t, v(t)) \equiv 0$) on Ω, where A, B, C_i, D are defined in Example 4.6.

According to Theorem 4.21, we know that system (4.69) is relatively controllable when $f(t, v(t)) = \mathbf{0}$. Further, one can get

$$\tilde{\eta} = v_1 - Y(1, -0.5)\psi(-0.5) - \int_{-0.5}^{0} Y(1, s)[\psi'(s) - A\psi(s)]ds$$

$$= v_1 - \begin{pmatrix} -0.0444 \\ -0.0630 \end{pmatrix}.$$

By using the form of the control $u(t)$, we arrive at

$$u(t) = D^\top V^\top(T, t)W_\tau^{-1}[0, T]\tilde{\eta} = X^\top(1, t + 0.5)e^{A^\top(1-t)}W_\tau^{-1}[0, 1]\tilde{\eta}$$

$$= \begin{cases} \left[E + \check{B}^\top(0.5 - t) + C_1^\top\right]e^{A^\top(1-t)}W_\tau^{-1}[0, 1]\tilde{\eta}, & 0 \leq t < 0.5, \\ Ie^{A^\top(1-t)}W_\tau^{-1}[0, 1]\tilde{\eta}, & 0.5 \leq t < 1, \end{cases}$$

where $\check{B}$ and $W_\tau^{-1}[0, 1]$ are given in Example 4.6.

4.3.1.4 Conclusion

The main contribution of this paper is to introduce a relative controllability method for impulsive delay controlled systems with linear parts defined by permutable matrices. We use the impulsive delayed Cauchy matrix to construct the impulsive delay Grammian matrix. We give a sufficient and necessary condition to examine when a linear delay controlled system is relatively controllable. Also, we apply a fixed point method to establish a relative controllability result for impulsive semilinear delay controlled systems. The results in this part are motivated from [103].

Chapter 5

Fractional delay systems

5.1 Finite time stability and controllability for Caputo type

5.1.1 Finite time stability for Caputo type

In this part, we study finite time stability of the linear homogeneous fractional delay differential equations

$$
\begin{cases}
({}^{C}D_{0+}^{\alpha} y)(x) = By(x - \tau), \ y(x) \in \mathbb{R}^{n}, \ x \in J := [0, T], \ \tau > 0, \\
y(x) = \varphi(x), \ -\tau \leq x \leq 0,
\end{cases} \tag{5.1}
$$

and the linear nonhomogeneous fractional delay differential equations

$$
\begin{cases}
({}^{C}D_{0+}^{\alpha} y)(x) = By(x - \tau) + f(x), \ x \in J, \ \tau > 0, \ f \in C(J, \mathbb{R}^{n}), \\
y(x) = \varphi(x), \ -\tau \leq x \leq 0,
\end{cases} \tag{5.2}
$$

where ${}^{C}D_{0+}^{\alpha} y(\cdot)$ is the Caputo derivative of order $\alpha \in (0, 1)$ and of lower limit zero, $B \in \mathbb{R}^{n \times n}$ denotes a constant matrix, $T = k^{*}\tau$ for a fixed $k^{*} \in \Lambda := \{1, 2, \cdots\}$, τ is a fixed delay time, and $\varphi(\cdot)$ is an arbitrary continuously differentiable vector function, i.e., $\varphi \in C_{\tau}^{1} := C^{1}([-\tau, 0], \mathbb{R}^{n})$.

We denote $\|y\| = \sum_{i=1}^{n} |y_{i}|$ and $\|A\| = \max\limits_{1 \leq j \leq n} \sum_{i=1}^{n} |a_{ij}|$, which are the Euclidean vector norm and matrix norm, respectively; y_{i} and a_{ij} are the elements of the vector y and the matrix A, respectively. We denote by $C(J, \mathbb{R}^{n})$ the Banach space of a vector-valued continuous function from $J \to \mathbb{R}^{n}$ endowed with the norm $\|x\| = \max\limits_{t \in J} \|x(t)\|$ for a norm $\|\cdot\|$ on $\mathbb{R}^{n}$. We introduce a space $C^{1}(J, \mathbb{R}^{n}) = \{x \in C(J, \mathbb{R}^{n}) : x' \in C(J, \mathbb{R}^{n})\}$. In addition, we note $\|\varphi\| = \max\limits_{x \in [-\tau, 0]} \|\varphi(x)\|$.

Definition 5.1. (see [104]) The Caputo derivative of order $0 < \alpha < 1$ for a function $y : [0, \infty) \to \mathbb{R}$ can be written as $({}^{C}D_{0+}^{\alpha} y)(x) = \frac{1}{\Gamma(1-\alpha)} \int_{0}^{x} \frac{y'(t)}{(x-t)^{\alpha}} dt$, $x > 0$.

Definition 5.2. (see [104]) The fractional integral of order $\alpha \in (0, 1)$ with the lower limit zero for a function $y : [0, \infty) \to \mathbb{R}$ is defined as $\mathbb{I}^{\alpha} y(x) = \frac{1}{\Gamma(\alpha)} \int_{0}^{x} \frac{y(t)}{(x-t)^{1-\alpha}} dt$, $x > 0$, provided the right side is pointwise defined on $[0, \infty)$, where $\Gamma(\cdot)$ is the Gamma function.

Stability and Controls Analysis for Delay Systems. https://doi.org/10.1016/B978-0-32-399792-8.00011-6

Copyright © 2023 Elsevier Inc. All rights reserved.

Definition 5.3. (see [104]) The Riemann–Liouville derivative of order $\alpha \in (0, 1)$ for a function $y : [0, \infty) \to \mathbb{R}$ can be written as $(^{RL}D_{0+}^{\alpha} y)(x) = \frac{1}{\Gamma(1-\alpha)} \frac{d}{dx} \int_0^x \frac{y(t)}{(x-t)^{\alpha}} dt, x > 0$.

Definition 5.4. (see [104]) The Caputo derivative of order $\alpha \in (0, 1)$ for a function $y : [0, \infty) \to \mathbb{R}$ can be written as $(^{C}D_{0+}^{\alpha} y)(x) = (^{RL}D_{0+}^{\alpha} y)(x) - \frac{y(0)}{\Gamma(1-\alpha)} x^{-\alpha}, x > 0$.

Definition 5.5. (see [38, Definition 4]) Mittag-Leffler functions $E_{\alpha}(z)$ and $E_{\alpha,\alpha}(z)$ are defined by

$$E_{\alpha}(x) = \sum_{k=0}^{\infty} \frac{x^k}{\Gamma(k\alpha + 1)}, \quad E_{\alpha,\beta}(x) = \sum_{k=0}^{\infty} \frac{x^k}{\Gamma(k\alpha + \beta)}, \quad x \in \mathbb{R},$$

and α, β are positive real numbers.

Next, we introduce the concept of delayed Mittag-Leffler-type matrix function, which is an analogy of the delayed exponential matrix e_{τ}^{Bt} in [1].

Definition 5.6. The delayed Mittag-Leffler-type matrix function $\mathbb{E}_{\tau}^{Bx^{\alpha}} : \mathbb{R} \to \mathbb{R}^{n \times n}$ is defined by

$$\mathbb{E}_{\tau}^{Bx^{\alpha}} = \begin{cases} \Theta, & -\infty < x < -\tau, \\ I, & -\tau \leq x \leq 0, \\ I + B \dfrac{x^{\alpha}}{\Gamma(\alpha + 1)} + B^2 \dfrac{(x - \tau)^{2\alpha}}{\Gamma(2\alpha + 1)} + \cdots + B^k \dfrac{(x - (k-1)\tau)^{k\alpha}}{\Gamma(k\alpha + 1)}, \\ \qquad\qquad (k - 1)\tau \leq x \leq k\tau, \ k \in \Lambda. \end{cases}$$

$$(5.3)$$

Definition 5.7. The delayed two-parameter Mittag-Leffler-type matrix $\mathbb{E}_{\tau,\alpha}^{B,\alpha} : \mathbb{R} \to \mathbb{R}^{n \times n}$ is defined by

$$\mathbb{E}_{\tau,\alpha}^{Bx^{\alpha}} = \begin{cases} \Theta, & -\infty < x \leq -\tau, \\ I \dfrac{(\tau + x)^{\alpha - 1}}{\Gamma(\alpha)}, & -\tau < x \leq 0, \\ I \dfrac{(\tau + x)^{\alpha - 1}}{\Gamma(\alpha)} + B \dfrac{x^{2\alpha - 1}}{\Gamma(\alpha + \alpha)} + \cdots + B^k \dfrac{(x - (k-1)\tau)^{(k+1)\alpha - 1}}{\Gamma(k\alpha + \alpha)}, \\ \qquad\qquad (k - 1)\tau < x \leq k\tau, \ k \in \Lambda. \end{cases}$$

$$(5.4)$$

Lemma 5.1. *Letting* $(k - 1)\tau \leq x \leq k\tau$ *and* $k \in \Lambda$, *we have*

$$\int_{(k-1)\tau}^{x} (x - t)^{-\alpha} (t - (k-1)\tau)^{k\alpha - 1} dt = (x - (k-1)\tau)^{(k-1)\alpha} \mathbb{B}[1 - \alpha, k\alpha],$$

where $\mathbb{B}[\xi, \eta] = \int_0^1 s^{\xi-1}(1-s)^{\eta-1}ds$ *is a Beta function.*

Proof. Integration by parts gives

$$\int_{(k-1)\tau}^x (x-t)^{-\alpha}(t-(k-1)\tau)^{k\alpha-1}dt$$

$$= \int_0^{x-(k-1)\tau}(x-(k-1)\tau)^{-\alpha}(1-\frac{z}{x-(k-1)\tau})^{-\alpha}z^{k\alpha-1}dz$$

$$= (x-(k-1)\tau)^{(k-1)\alpha}\mathbb{B}[1-\alpha, k\alpha].$$

The proof is completed. $\square$

Lemma 5.2. *Letting* $(k-1)\tau < x \le k\tau$, $0 \le s \le t$, *and letting* $k \in \Lambda$ *be a fixed number, we have*

$$\int_{(k-1)\tau+s}^x (x-t)^{-\alpha}(t-(k-1)\tau-s)^{k\alpha-1}dt$$

$$= (x-(k-1)\tau-s)^{(k-1)\alpha}\mathbb{B}[1-\alpha, k\alpha].$$

Proof. Integration by parts gives

$$\int_{(k-1)\tau+s}^x (x-t)^{-\alpha}(t-(k-1)\tau-s)^{k\alpha-1}dt$$

$$= \int_0^{x-(k-1)\tau-s}(x-(k-1)\tau-s-z)^{-\alpha}z^{k\alpha-1}dz$$

(where $t-(k-1)\tau-s=z$)

$$= \int_0^{x-(k-1)\tau-s}(x-(k-1)\tau-s)^{-\alpha}\left(1-\frac{z}{x-(k-1)\tau-s}\right)^{-\alpha}z^{k\alpha-1}dz$$

$$= \int_0^1 (x-(k-1)\tau-s)^{(k-1)\alpha}(1-y)^{-\alpha}y^{k\alpha-1}dy$$

(where $y(x-(k-1)\tau-s)=z$)

$$= (x-(k-1)\tau-s)^{(k-1)\alpha}\mathbb{B}[1-\alpha, k\alpha].$$

The proof is completed. $\square$

Lemma 5.3. *Letting* $(k-1)\tau < x \le k\tau$, $0 \le s \le t$, *and letting* $k \in \Lambda$ *be a fixed number, we have*

$$\int_s^x (x-t)^{-\alpha}\mathbb{E}_{\tau,\alpha}^{B(t-\tau-s)^\alpha}dt$$

$$= \int_s^x (x-t)^{-\alpha}I\frac{(t-s)^{\alpha-1}}{\Gamma(\alpha)}dt + \int_{\tau+s}^x (x-t)^{-\alpha}B\frac{(t-\tau-s)^{2\alpha-1}}{\Gamma(2\alpha)}dt$$

$$+ \cdots + \int_{(k-1)\tau+s}^{x} (x-t)^{-\alpha} B^{k-1} \frac{(t-(k-1)\tau-s)^{k\alpha-1}}{\Gamma(k\alpha)} dt. \qquad (5.5)$$

Proof. For arbitrary $x \in ((k-1)\tau, k\tau]$, $0 \le s \le t$, $k \in \Lambda$ is a fixed number. We adopt mathematical induction to prove our result.

(i) For $k=1$, $0 < x \le \tau$, using $\mathbb{E}_{\tau,\alpha}^{B \cdot \alpha}$ defined in (5.4), we have

$$\int_{s}^{x} (x-t)^{-\alpha} \mathbb{E}_{\tau,\alpha}^{B(t-\tau-s)^{\alpha}} dt = \int_{s}^{x} (x-t)^{-\alpha} I \frac{(t-s)^{\alpha-1}}{\Gamma(\alpha)} dt.$$

(ii) For $k=2$, $\tau < x \le 2\tau$, using $\mathbb{E}_{\tau,\alpha}^{B \cdot \alpha}$ defined in (5.4) again, we have

$$\int_{s}^{x} (x-t)^{-\alpha} \mathbb{E}_{\tau,\alpha}^{B(t-\tau-s)^{\alpha}} dt$$

$$= \int_{s}^{\tau+s} (x-t)^{-\alpha} \mathbb{E}_{\tau,\alpha}^{B(t-\tau-s)^{\alpha}} dt + \int_{\tau+s}^{x} (x-t)^{-\alpha} \mathbb{E}_{\tau,\alpha}^{B(t-\tau-s)^{\alpha}} dt$$

$$= \int_{s}^{\tau+s} (x-t)^{-\alpha} I \frac{(t-s)^{\alpha-1}}{\Gamma(\alpha)} dt$$

$$\quad + \int_{\tau+s}^{x} (x-t)^{-\alpha} \left(I \frac{(t-s)^{\alpha-1}}{\Gamma(\alpha)} + B \frac{(t-\tau-s)^{2\alpha-1}}{\Gamma(2\alpha)} \right) dt$$

$$= \int_{s}^{x} (x-t)^{-\alpha} I \frac{(t-s)^{\alpha-1}}{\Gamma(\alpha)} dt + \int_{\tau+s}^{x} (x-t)^{-\alpha} B \frac{(t-\tau-s)^{2\alpha-1}}{\Gamma(2\alpha)} dt.$$

(iii) Supposing $k=N$, $(N-1)\tau < x \le N\tau$, and $N \in \Lambda$, the following relation holds:

$$\int_{s}^{x} (x-t)^{-\alpha} \mathbb{E}_{\tau,\alpha}^{B(t-\tau-s)^{\alpha}} dt$$

$$= \int_{s}^{x} (x-t)^{-\alpha} I \frac{(t-s)^{\alpha-1}}{\Gamma(\alpha)} dt + \int_{\tau+s}^{x} (x-t)^{-\alpha} B \frac{(t-\tau-s)^{2\alpha-1}}{\Gamma(2\alpha)} dt$$

$$\quad + \cdots + \int_{(N-1)\tau+s}^{x} (x-t)^{-\alpha} B^{N-1} \frac{(t-(N-1)\tau-s)^{N\alpha-1}}{\Gamma(N\alpha)} dt.$$

Next, for $k=N+1$, $N\tau < x \le (N+1)\tau$, we have

$$\int_{s}^{x} (x-t)^{-\alpha} \mathbb{E}_{\tau,\alpha}^{B(t-\tau-s)^{\alpha}} dt$$

$$= \int_{s}^{\tau+s} (x-t)^{-\alpha} I \frac{(t-s)^{\alpha-1}}{\Gamma(\alpha)} dt$$

$$\quad + \int_{\tau+s}^{2\tau+s} (x-t)^{-\alpha} \left(I \frac{(t-s)^{\alpha-1}}{\Gamma(\alpha)} + B \frac{(t-\tau-s)^{2\alpha-1}}{\Gamma(2\alpha)} \right) dt$$

$$+ \cdots + \int_{N\tau+s}^{x} (x-t)^{-\alpha} \left(I \frac{(t-s)^{\alpha-1}}{\Gamma(\alpha)} + \cdots + B^N \frac{(t-N\tau-s)^{(N+1)\alpha-1}}{\Gamma((N+1)\alpha)} \right) dt$$

$$= \int_{s}^{x} (x-t)^{-\alpha} I \frac{(t-s)^{\alpha-1}}{\Gamma(\alpha)} dt + \int_{\tau+s}^{x} (x-t)^{-\alpha} B \frac{(t-\tau-s)^{2\alpha-1}}{\Gamma(2\alpha)} dt$$

$$+ \cdots + \int_{N\tau+s}^{x} (x-t)^{-\alpha} B^N \frac{(t-N\tau-s)^{(N+1)\alpha-1}}{\Gamma((N+1)\alpha)} dt,$$

which implies that (5.5) holds for any $(k-1)\tau < x \le k\tau$ and $k \in \Lambda$ is a fixed number. The proof is completed. $\qquad\square$

Lemma 5.4. *For any $x \in [(k-1)\tau, k\tau]$ and $k \in \Lambda$, we have*

$$\left\| \mathbb{E}_{\tau}^{Bx^{\alpha}} \right\| \le E_{\alpha}(\|B\|x^{\alpha}).$$

Proof. By formula (5.3), one has

$$\|\mathbb{E}_{\tau}^{Bx^{\alpha}}\| \le 1 + \|B\| \frac{x^{\alpha}}{\Gamma(\alpha+1)} + \|B\|^2 \frac{x^{2\alpha}}{\Gamma(2\alpha+1)} + \cdots + \|B\|^k \frac{x^{k\alpha}}{\Gamma(k\alpha+1)}$$

$$\le \sum_{k=0}^{\infty} \frac{(\|B\|x^{\alpha})^k}{\Gamma(k\alpha+1)} = E_{\alpha}(\|B\|x^{\alpha}).$$

The proof is completed. $\qquad\square$

Lemma 5.5. *For any $x \in ((k-1)\tau, k\tau]$, $k \in \Lambda$ being a fixed number, and $0 \le t < x$, we have*
   (i) For $0 \le t < x - (k-1)\tau$ and $k \in \Lambda$, we have

$$\left\| \mathbb{E}_{\tau,\alpha}^{B(x-\tau-t)^{\alpha}} \right\| \le \sum_{i=1}^{k} \|B\|^{i-1} \frac{(x-(i-1)\tau-t)^{i\alpha-1}}{\Gamma((i-1)\alpha+\alpha)}.$$

   (ii) For $x - (j-1)\tau \le t < x - (j-2)\tau$ and $j = 2, 3, \cdots k$, we have

$$\left\| \mathbb{E}_{\tau,\alpha}^{B(x-\tau-t)^{\alpha}} \right\| \le \sum_{i=2}^{j} \|B\|^{i-2} \frac{(x-(i-2)\tau-t)^{(i-1)\alpha-1}}{\Gamma((i-2)\alpha+\alpha)}.$$

Proof. By formula (5.4), one has
   (i) For $0 \le t < x - (k-1)\tau$ and $k \in \Lambda$, we have

$$\left\| \mathbb{E}_{\tau,\alpha}^{B(x-\tau-t)^{\alpha}} \right\|$$

$$\le \frac{(x-t)^{\alpha-1}}{\Gamma(\alpha)} + \|B\| \frac{(x-\tau-t)^{2\alpha-1}}{\Gamma(\alpha+\alpha)} + \cdots + \|B\|^{k-1} \frac{(x-(k-1)\tau-t)^{k\alpha-1}}{\Gamma((k-1)\alpha+\alpha)}$$

$$\leq \frac{(x-t)^{\alpha-1}}{\Gamma(\alpha)} + \sum_{i=2}^{k} \|B\|^{i-1} \frac{(x-(i-1)\tau-t)^{i\alpha-1}}{\Gamma((i-1)\alpha+\alpha)}$$

$$\leq \sum_{i=1}^{k} \|B\|^{i-1} \frac{(x-(i-1)\tau-t)^{i\alpha-1}}{\Gamma((i-1)\alpha+\alpha)}.$$

(ii) For $x-(j-1)\tau \leq t < x-(j-2)\tau$ and $j = 2, 3, \cdots, k$, we have

$$\left\| \mathbb{E}_{\tau,\alpha}^{B(x-\tau-t)^{\alpha}} \right\|$$

$$\leq \frac{(x-t)^{\alpha-1}}{\Gamma(\alpha)} + \|B\| \frac{(x-\tau-t)^{2\alpha-1}}{\Gamma(\alpha+\alpha)} + \cdots$$

$$+ \|B\|^{j-2} \frac{(x-(j-2)\tau-t)^{(j-1)\alpha-1}}{\Gamma((j-2)\alpha+\alpha)}$$

$$\leq \frac{(x-t)^{\alpha-1}}{\Gamma(\alpha)} + \sum_{i=3}^{j} \|B\|^{i-2} \frac{(x-(i-2)\tau-t)^{(i-1)\alpha-1}}{\Gamma((i-2)\alpha+\alpha)}$$

$$\leq \sum_{i=2}^{j} \|B\|^{i-2} \frac{(x-(i-2)\tau-t)^{(i-1)\alpha-1}}{\Gamma((i-2)\alpha+\alpha)}.$$

The proof is completed. $\qquad\qquad\qquad\qquad\qquad\qquad\qquad\qquad\qquad\quad\square$

Lemma 5.6. *For any $x \in ((k-1)\tau, k\tau]$ and $k \in \Lambda$ being a fixed number, we have*

$$\left\| \int_{-\tau}^{0} \mathbb{E}_{\tau}^{B(x-\tau-t)^{\alpha}} \varphi'(t)dt \right\| \leq E_{\alpha}(\|B\|x^{\alpha}) \int_{-\tau}^{0} \|\varphi'(t)\|dt.$$

Proof. According to the definition of $\mathbb{E}_{\tau}^{B,\alpha}$, we obtain

$$\left\| \int_{-\tau}^{0} \mathbb{E}_{\tau}^{B(x-\tau-t)^{\alpha}} \varphi'(t)dt \right\|$$

$$= \left\| \int_{-\tau}^{x-k\tau} \left(I + B\frac{(x-\tau-t)^{\alpha}}{\Gamma(\alpha+1)} + \cdots + B^{k}\frac{(x-k\tau-t)^{k\alpha}}{\Gamma(k\alpha+1)} \right) \varphi'(t)dt \right.$$

$$+ \int_{x-k\tau}^{0} \left(I + B\frac{(x-\tau-t)^{\alpha}}{\Gamma(\alpha+1)} + \cdots + B^{k-1}\frac{(x-(k-1)\tau-t)^{(k-1)\alpha}}{\Gamma((k-1)\alpha+1)} \right) \varphi'(t)dt \right\|$$

$$\leq \left(\int_{-\tau}^{0} \left(1 + \|B\|\frac{x^{\alpha}}{\Gamma(\alpha+1)} + \cdots + \|B\|^{k-1}\frac{x^{(k-1)\alpha}}{\Gamma((k-1)\alpha+1)} \right) \right.$$

$$+ \int_{-\tau}^{x-k\tau} \|B\|^{k}\frac{x^{k\alpha}}{\Gamma(k\alpha+1)} \right) \|\varphi'(t)\|dt$$

$$
\leq \left(I + \frac{\|B\| x^{\alpha}}{\Gamma(\alpha+1)} + \cdots + \frac{\|B\|^{k-1} x^{(k-1)\alpha}}{\Gamma((k-1)\alpha+1)} \right) \int_{-\tau}^{0} \|\varphi'(t)\| dt
$$

$$
+ \frac{\|B\|^{k} x^{k\alpha}}{\Gamma(k\alpha+1)} \int_{-\tau}^{x-k\tau} \|\varphi'(t)\| dt
$$

$$
\leq \left(I + \frac{\|B\| x^{\alpha}}{\Gamma(\alpha+1)} + \cdots + \frac{\|B\|^{k-1} x^{(k-1)\alpha}}{\Gamma((k-1)\alpha+1)} + \frac{\|B\|^{k} x^{k\alpha}}{\Gamma(k\alpha+1)} \right) \int_{-\tau}^{0} \|\varphi'(t)\| dt
$$

$$
\leq \sum_{k=0}^{\infty} \frac{(\|B\| x^{\alpha})^{k}}{\Gamma(k\alpha+1)} \int_{-\tau}^{0} \|\varphi'(t)\| dt
$$

$$
\leq E_{\alpha}(\|B\| x^{\alpha}) \int_{-\tau}^{0} \|\varphi'(t)\| dt.
$$

The proof is completed. $\qquad\square$

Lemma 5.7. *For any $x \in ((k-1)\tau, k\tau]$, $k \in \Lambda$, $j = 2, 3, \cdots, k$, and $\alpha \geq \frac{1}{2}$, we have*

$$
\int_{0}^{x-(k-1)\tau} \left\| \mathbb{E}_{\tau,\alpha}^{B(x-\tau-t)^{\alpha}} \right\| \|\phi(t)\| dt + \int_{x-(k-1)\tau}^{x} \left\| \mathbb{E}_{\tau,\alpha}^{B(x-\tau-t)^{\alpha}} \right\| \|\phi(t)\| dt
$$

$$
\leq E_{\alpha,\alpha}(\|B\| x^{\alpha}) \int_{0}^{x} (x-t)^{\alpha-1} \|\phi(t)\| dt, \quad \phi \in C(J, \mathbb{R}^{n}).
$$

Proof. According to the definition of $\mathbb{E}_{\tau,\alpha}^{B\cdot\alpha}$, we obtain

$$
\int_{0}^{x-(k-1)\tau} \left\| \mathbb{E}_{\tau,\alpha}^{B(x-\tau-t)^{\alpha}} \right\| \|\phi(t)\| dt + \int_{x-(k-1)\tau}^{x} \left\| \mathbb{E}_{\tau,\alpha}^{B(x-\tau-t)^{\alpha}} \right\| \|\phi(t)\| dt
$$

$$
= \int_{0}^{x-(k-1)\tau} \left(\sum_{i=1}^{k} \|B\|^{i-1} \frac{(x-(i-1)\tau-t)^{i\alpha-1}}{\Gamma((i-1)\alpha+\alpha)} \right) \|\phi(t)\| dt
$$

$$
+ \sum_{j=2}^{k} \int_{x-(j-1)\tau}^{x-(j-2)\tau} \left(\sum_{i=2}^{j} \|B\|^{i-2} \frac{(x-(i-2)\tau-t)^{(i-1)\alpha-1}}{\Gamma((i-2)\alpha+\alpha)} \right) \|\phi(t)\| dt
$$

$$
\leq \int_{0}^{x-(k-1)\tau} \left(\sum_{i=1}^{k} \|B\|^{i-1} \frac{(x-t)^{i\alpha-1}}{\Gamma((i-1)\alpha+\alpha)} \right) \|\phi(t)\| dt
$$

$$
+ \int_{x-(k-1)\tau}^{x} \left(\sum_{i=2}^{k} \|B\|^{i-2} \frac{(x-t)^{(i-1)\alpha-1}}{\Gamma((i-2)\alpha+\alpha)} \right) \|\phi(t)\| dt
$$

$$
\leq \left(\int_{0}^{x} \sum_{i=2}^{k} \|B\|^{i-2} \frac{(x-t)^{(i-1)\alpha-1}}{\Gamma((i-2)\alpha+\alpha)} \right.
$$

$$+ \int_0^{x-(k-1)\tau} \|B\|^{k-1} \frac{(x-t)^{k\alpha-1}}{\Gamma((k-1)\alpha+\alpha)} \Bigg) \|\phi(t)\| dt$$

$$\leq \int_0^x (x-t)^{\alpha-1} E_{\alpha,\alpha}(\|B\|(x-t)^\alpha)\|\phi(t)\| dt$$

$$\leq E_{\alpha,\alpha}(\|B\|x^\alpha) \int_0^x (x-t)^{\alpha-1}\|\phi(t)\| dt,$$

where we use the monotonic property of $f(\cdot) = \cdot^{k\alpha-1}$ for $\alpha \geq \frac{1}{2}$, which implies that $(x-(k-1)\tau-t)^{k\alpha-1} \leq (x-t)^{k\alpha-1}$, $x \in ((k-1)\tau, k\tau]$. The proof is completed. $\qquad\square$

5.1.1.1 Representation of solutions for linear systems

In this section, we seek explicit formula of solutions to fractional delay system by adopting the classical ideas to find solutions of linear fractional ODEs.

Theorem 5.1. *For the delayed Mittag-Leffler-type matrix $\mathbb{E}_\tau^{Bx^\alpha} : \mathbb{R} \to \mathbb{R}^{n\times n}$, one has*

$$({}^C D_{0+}^\alpha \mathbb{E}_\tau^{Bt^\alpha})(x) = B\mathbb{E}_\tau^{B(x-\tau)^\alpha}, \tag{5.6}$$

i.e., $\mathbb{E}_\tau^{Bx^\alpha}$ is a solution of $({}^C D_{0+}^\alpha y)(x) = By(x-\tau)$, which satisfies initial conditions $\mathbb{E}_\tau^{Bx^\alpha} = I$, $-\tau \leq x \leq 0$.

Proof. For arbitrary $x \in (-\infty, -\tau]$, $\mathbb{E}_\tau^{Bx^\alpha} = \mathbb{E}_\tau^{B(x-\tau)^\alpha} = \Theta$. Obviously, (5.6) holds. For arbitrary $x \in [-\tau, 0]$, $\mathbb{E}_\tau^{Bx^\alpha} = I$ and $\mathbb{E}_\tau^{B(x-\tau)^\alpha} = \Theta$. Note that ${}^C D_{0+}^\alpha I = \Theta = B\Theta$. Thus, (5.6) holds.

For arbitrary $x \in [(k-1)\tau, k\tau]$, $k \in \Lambda$, we adopt mathematical induction to prove our result.

(*i*) For $k = 1$, $0 \leq x \leq \tau$, we have

$$y(x) = \mathbb{E}_\tau^{Bx^\alpha} = I + \frac{Bx^\alpha}{\Gamma(\alpha+1)}, \quad y'(x) = \frac{\alpha Bx^{\alpha-1}}{\Gamma(\alpha+1)}. \tag{5.7}$$

Next, having the Caputo fractional differentiated expression of $\mathbb{E}_\tau^{B\cdot\alpha}$ via (5.7) and Lemma 5.1, we obtain

$$({}^C D_{0+}^\alpha \mathbb{E}_\tau^{Bt^\alpha})(x) = \frac{\alpha B}{\Gamma(\alpha+1)\Gamma(1-\alpha)} \int_0^x (x-t)^{-\alpha} t^{\alpha-1} dt$$

$$= \frac{\alpha B\Gamma(1-\alpha)\Gamma(\alpha)}{\Gamma(\alpha+1)\Gamma(1-\alpha)} = B. \tag{5.8}$$

(*ii*) For $k = 2$, $\tau \leq x \leq 2\tau$, we have

$$y(x) = \mathbb{E}_\tau^{Bx^\alpha} = I + \frac{Bx^\alpha}{\Gamma(\alpha+1)} + \frac{B^2(x-\tau)^{2\alpha}}{\Gamma(2\alpha+1)},$$

$$y'(x) = \frac{\alpha B x^{\alpha-1}}{\Gamma(\alpha+1)} + \frac{2\alpha B^2 (x-\tau)^{2\alpha-1}}{\Gamma(2\alpha+1)}. \tag{5.9}$$

Next, having the Caputo fractional differentiated expression of $\mathbb{E}_\tau^{B,\alpha}$ via (5.8), (5.9), and Lemma 5.1, we obtain

$$\begin{aligned}
(^C D_{0^+}^\alpha \mathbb{E}_\tau^{Bt^\alpha})(x) &= B + \frac{2\alpha B^2}{\Gamma(1-\alpha)\Gamma(2\alpha+1)} \int_\tau^x (x-t)^{-\alpha}(t-\tau)^{2\alpha-1} dt \\
&= B + \frac{2\alpha B^2 (x-\tau)^\alpha}{\Gamma(1-\alpha)} \frac{\Gamma(1-\alpha)\Gamma(2\alpha)}{\Gamma(\alpha+1)\Gamma(2\alpha+1)} \\
&= B + \frac{B^2 (x-\tau)^\alpha}{\Gamma(\alpha+1)}.
\end{aligned}$$

(iii) Supposing $k = n$, $(n-1)\tau \leq x \leq n\tau$, and $n \in \Lambda$, the following relation holds:

$$\begin{aligned}
(^C D_{0^+}^\alpha \mathbb{E}_\tau^{Bt^\alpha})(x) = B &+ \frac{B^2(x-\tau)^\alpha}{\Gamma(\alpha+1)} + \frac{B^3(x-2\tau)^{2\alpha}}{\Gamma(2\alpha+1)} + \cdots \\
&+ \frac{B^n(x-(n-1)\tau)^{(n-1)\alpha}}{\Gamma((n-1)\alpha+1)}.
\end{aligned}$$

Next, for $k = n+1$, $n\tau \leq x \leq (n+1)\tau$, by elementary computation, one obtains

$$\begin{aligned}
y'(x) = \frac{\alpha B x^{\alpha-1}}{\Gamma(\alpha+1)} &+ \frac{2\alpha B^2 (x-\tau)^{2\alpha-1}}{\Gamma(2\alpha+1)} + \cdots \\
&+ \frac{(n+1)\alpha B^{n+1}(x-n\tau)^{(n+1)\alpha-1}}{\Gamma((n+1)\alpha+1)}.
\end{aligned} \tag{5.10}$$

Having the Caputo fractional differentiated expression of $\mathbb{E}_\tau^{B,\alpha}$ via (5.10) and Lemma 5.1, we obtain

$$\begin{aligned}
(^C D_{0^+}^\alpha & \mathbb{E}_\tau^{Bt^\alpha})(x) \\
&= \frac{\alpha B}{\Gamma(1-\alpha)} \int_0^x (x-t)^{-\alpha} \frac{t^{\alpha-1}}{\Gamma(\alpha+1)} dt \\
&\quad + \frac{2\alpha B^2}{\Gamma(1-\alpha)} \int_\tau^x (x-t)^{-\alpha} \frac{(t-\tau)^{2\alpha-1}}{\Gamma(2\alpha+1)} dt \\
&\quad + \cdots + \frac{(n+1)\alpha B^{n+1}}{\Gamma(1-\alpha)} \int_{n\tau}^x (x-t)^{-\alpha} \frac{(t-n\tau)^{(n+1)\alpha-1}}{\Gamma((n+1)\alpha+1)} dt \\
&= B + \frac{B^2(x-\tau)^\alpha}{\Gamma(\alpha+1)} + \frac{B^3(x-2\tau)^{2\alpha}}{\Gamma(2\alpha+1)} + \cdots + \frac{B^{n+1}(x-n\tau)^{n\alpha}}{\Gamma(n\alpha+1)},
\end{aligned}$$

which implies that (5.6) holds for any $(k-1)\tau \leq x \leq k\tau$ and $k \in \Lambda$. The proof is completed. $\square$

Theorem 5.2. *A solution $y \in C([-\tau, T], \mathbb{R}^n)$ of (5.1) can be expressed by the following formula:*

$$y(x) = \mathbb{E}_\tau^{Bx^\alpha} \varphi(-\tau) + \int_{-\tau}^{0} \mathbb{E}_\tau^{B(x-\tau-s)^\alpha} \varphi'(s)ds.$$

Proof. Let matrix $Y_0(x) = \mathbb{E}_\tau^{Bx^\alpha}$ satisfy Theorem 5.1. We should search any solution of (5.1) satisfying initial conditions $y(x) = \varphi(x)$, $-\tau \le x \le 0$, in the form

$$y(x) = Y_0(x)c + \int_{-\tau}^{0} Y_0(x - \tau - s)z(s)ds. \tag{5.11}$$

In this formula, c is a vector of unknown constant and $z(\cdot)$ is an unknown continuously differentiable vector function. The matrix $Y_0(x)$ is a solution of (5.1). Then, for arbitrary c and $z(\cdot)$, formula (5.11) is a solution of (5.1). Therefore, we choose c and $z(\cdot)$ satisfying initial conditions, $y(x) = \varphi(x)$, $-\tau \le x \le 0$.

Let us assume $x = -\tau$. It follows from (5.3) that $Y_0(-\tau) = I$, $Y_0(-2\tau - s) = \Theta$, $-\tau < s \le 0$, and $Y_0(-2\tau - s) = I$, $s = -\tau$. Thus, $y(-\tau) = \varphi(-\tau) = c$, and formula (5.11) takes the form

$$y(x) = \mathbb{E}_\tau^{Bx^\alpha} \varphi(-\tau) + \int_{-\tau}^{0} \mathbb{E}_\tau^{B(x-\tau-s)^\alpha} z(s)ds.$$

Since $-\tau \le x \le 0$, one should divide the interval into two subintervals:

(i) For the interval $-\tau \le s \le x$, so $-\tau \le x - \tau - s \le x$, the delayed Mittag-Leffler-type matrix is equal to $\mathbb{E}_\tau^{B(x-\tau-s)^\alpha} = I$, $-\tau \le s \le x$.

(ii) For the interval $x \le s \le 0$, so $x - \tau \le x - \tau - s \le -\tau$, the delayed Mittag-Leffler-type matrix is equal to

$$\mathbb{E}_\tau^{B(x-\tau-s)^\alpha} = \begin{cases} \Theta, & x < s \le 0, \\ I, & s = x. \end{cases}$$

Thus on the interval $-\tau \le x \le 0$, we have

$$\varphi(x) = \varphi(-\tau) + \int_{-\tau}^{x} z(s)ds. \tag{5.12}$$

Having differentiated on (5.12), we obtain $z(x) = \varphi'(x)$. The proof is completed. $\square$

Next, we seek the explicit formula of solutions to linear nonhomogeneous fractional delay differential equations by adopting the classical ideas to find solutions of linear fractional ODEs. To achieve our aim, we need to find a special solution to (5.2). The next theorem will show how to find this special solution.

Theorem 5.3. *A solution $\overline{y} \in C([-\tau, T], \mathbb{R}^n)$ of (5.2) satisfying initial conditions $y(x) = 0$, $x \in [-\tau, 0]$ has the form*

$$\overline{y}(x) = \int_0^x \mathbb{E}_{\tau,\alpha}^{B(x-\tau-t)^\alpha} f(t)dt, \quad x > 0.$$

Proof. By using the method of variation of constants, any solution of nonhomogeneous system $\overline{y}(x)$ should satisfy the form

$$\overline{y}(x) = \int_0^x \mathbb{E}_{\tau,\alpha}^{B(x-\tau-t)^\alpha} c(t)dt, \tag{5.13}$$

where $c(t)$, $0 \le t \le x$, is an unknown vector function and $\overline{y}(0) = 0$.

Having Caputo fractional differentiation on both sides of (5.13), we obtain the following cases:

(i) For $0 < x \le \tau$, according to (5.2), we have

$$({}^C D_{0+}^\alpha \overline{y})(x) = B\overline{y}(x - \tau) + f(x) = B \int_0^{x-\tau} \mathbb{E}_{\tau,\alpha}^{B(x-2\tau-t)^\alpha} c(t)dt + f(x)$$

$$= f(x),$$

where we note that $\mathbb{E}_{\tau,\alpha}^{B(x-2\tau-\cdot)^\alpha} = \Theta$ due to (5.4).

According to Definition 5.4 and Lemma 5.2, we have

$$({}^C D_{0+}^\alpha \overline{y})(x) = ({}^{RL} D_{0+}^\alpha \overline{y})(x)$$

$$= \frac{1}{\Gamma(1-\alpha)} \frac{d}{dx} \int_0^x (x-t)^{-\alpha} \left(\int_0^t \mathbb{E}_{\tau,\alpha}^{B(t-\tau-s)^\alpha} c(s)ds \right) dt$$

$$= \frac{1}{\Gamma(1-\alpha)} \frac{d}{dx} \int_0^x c(s) \int_s^x (x-t)^{-\alpha} \mathbb{E}_{\tau,\alpha}^{B(t-\tau-s)^\alpha} dsdt$$

$$= \frac{1}{\Gamma(1-\alpha)} \frac{d}{dx} \int_0^x c(s) \left(\int_s^x (x-t)^{-\alpha} I \frac{(t-s)^{\alpha-1}}{\Gamma(\alpha)} dt \right) ds$$

$$= \frac{1}{\Gamma(1-\alpha)} \frac{d}{dx} \int_0^x \frac{\mathbb{B}[1-\alpha,\alpha]}{\Gamma(\alpha)} c(s)ds$$

$$= c(x).$$

Hence, we obtain $c(x) = f(x)$.

(ii) For $k\tau < x \le (k+1)\tau$ and $k \in \Lambda$, according to (5.2), we have

$$({}^C D_{0+}^\alpha \overline{y})(x) = B\overline{y}(x - \tau) + f(x)$$

$$= B \int_0^{x-\tau} \mathbb{E}_{\tau,\alpha}^{B(x-2\tau-t)^\alpha} c(t)dt + f(x)$$

$$= B \left(\int_0^{x-\tau} \frac{(x-\tau-t)^{\alpha-1}}{\Gamma(\alpha)} c(t)dt \right.$$

$$+ \int_0^{x-2\tau} B \frac{(x - 2\tau - t)^{2\alpha-1}}{\Gamma(2\alpha)} c(t)dt$$

$$+ \cdots + \int_0^{x-k\tau} B^{k-1} \frac{(x - k\tau - t)^{k\alpha-1}}{\Gamma(k\alpha)} c(t)dt \Bigg) + f(x).$$

According to Definition 5.4, we have

$$({}^C D_{0+}^{\alpha} \overline{y})(x) = ({}^{RL} D_{0+}^{\alpha} \overline{y})(x)$$

$$= \frac{1}{\Gamma(1-\alpha)} \frac{d}{dx} \int_0^x (x-t)^{-\alpha} \left(\int_0^t \mathbb{E}_{\tau,\alpha}^{B(t-\tau-s)^{\alpha}} c(s)ds \right) dt$$

$$= \frac{1}{\Gamma(1-\alpha)} \frac{d}{dx} \int_0^x \int_0^t (x-t)^{-\alpha} \mathbb{E}_{\tau,\alpha}^{B(t-\tau-s)^{\alpha}} c(s)ds\,dt$$

$$= \frac{1}{\Gamma(1-\alpha)} \frac{d}{dx} \int_0^x c(s) \left(\int_s^x (x-t)^{-\alpha} \mathbb{E}_{\tau,\alpha}^{B(t-\tau-s)^{\alpha}} dt \right) ds.$$

According to Lemmas 5.2 and 5.3, we obtain

$$({}^C D_{0+}^{\alpha} \overline{y})(x) = ({}^{RL} D_{0+}^{\alpha} \overline{y})(x)$$

$$= \frac{1}{\Gamma(1-\alpha)} \frac{d}{dx} \int_0^x c(s) \left(\int_s^x (x-t)^{-\alpha} \mathbb{E}_{\tau,\alpha}^{B(t-\tau-s)^{\alpha}} dt \right) ds$$

$$= \frac{d}{dx} \int_0^x c(s)ds$$

$$+ \frac{1}{\Gamma(1-\alpha)} \frac{d}{dx} \int_0^{x-\tau} c(s) \left(\int_{\tau+s}^x (x-t)^{-\alpha} B \frac{(t - \tau - s)^{2\alpha-1}}{\Gamma(2\alpha)} dt \right) ds + \cdots$$

$$+ \frac{1}{\Gamma(1-\alpha)} \frac{d}{dx} \int_0^{x-k\tau} c(s) \left(\int_{k\tau+s}^x (x-t)^{-\alpha} B^k \frac{(t - k\tau - s)^{(k+1)\alpha-1}}{\Gamma((k+1)\alpha)} dt \right) ds$$

$$= c(x) + B \int_0^{x-\tau} \frac{(x - \tau - s)^{\alpha-1}}{\Gamma(\alpha)} c(s)ds + B^2 \int_0^{x-2\tau} \frac{(x - 2\tau - s)^{2\alpha-1}}{\Gamma(2\alpha)} c(s)ds$$

$$+ \cdots + B^k \int_0^{x-k\tau} \frac{(x - k\tau - s)^{k\alpha-1}}{\Gamma(k\alpha)} c(s)ds$$

$$= B \int_0^{x-\tau} \mathbb{E}_{\tau,\alpha}^{B(x-2\tau-t)^{\alpha}} c(t)dt + f(x).$$

Hence, we obtain $c(x) = f(x)$. The proof is completed. $\qquad\square$

It is obvious that any solution y of (5.2) can be written as a sum $y(x) = y_0(x) + \overline{y}(x)$, where $y_0(x)$ is a solution of (5.1) satisfying condition $y(x) = \varphi(x)$, $-\tau \leq x \leq 0$, and $\overline{y}(x)$ is a solution of (5.2) satisfying $y(0) = 0$. The following theorem presents the construction of formulas of solutions to (5.2). The proof is straightforward, so we omit it here.

Theorem 5.4. *A solution* $y \in C([-\tau, T], \mathbb{R}^n)$ *of* (5.2) *has the form*

$$y(x) = \mathbb{E}_\tau^{Bx^\alpha} \varphi(-\tau) + \int_{-\tau}^0 \mathbb{E}_\tau^{B(x-\tau-t)^\alpha} \varphi'(t)dt + \int_0^x \mathbb{E}_{\tau,\alpha}^{B(x-\tau-t)^\alpha} f(t)dt.$$

5.1.1.2 Existence of solutions for nonlinear system

In this section, we study the existence of solutions of the following nonlinear fractional delay differential equations:

$$\begin{cases} (^C D_{0^+}^\alpha y)(x) = By(x - \tau) + f(x, y(x)), \ x \in J, \ \tau > 0, \ f \in C(J \times \mathbb{R}^n, \mathbb{R}^n), \\ y(x) = \varphi(x), \ -\tau \le x \le 0. \end{cases}$$

$$(5.14)$$

Definition 5.8. A function $y \in C([-\tau, T], \mathbb{R}^n)$ is called the solution of (5.14) if it satisfies the following integral equation:

$$y(x) = \mathbb{E}_\tau^{Bx^\alpha} \varphi(-\tau) + \int_{-\tau}^0 \mathbb{E}_\tau^{B(x-\tau-t)^\alpha} \varphi'(t)dt + \int_0^x \mathbb{E}_{\tau,\alpha}^{B(x-\tau-t)^\alpha} f(t, y(t))dt.$$

$$(5.15)$$

We introduce the following assumptions:

$[H_1]$ $f \in C(J, \mathbb{R}^n)$ and there exists a $\widetilde{K} > 0$ such that $\| f(x, y) - f(x, z)\| \le \widetilde{K} \|y - z\|$ for any $x \in J$ and all $y, z \in \mathbb{R}^n$.

$[H_2]$ $x = \widetilde{K} \left(\sum_{i=1}^k \frac{\|B\|^{i-1}}{\Gamma(i\alpha+1)} (T - (i - 1)\tau)^{i\alpha} \right) < 1$, where $k \in \Lambda$ is a fixed number.

Denote $M := \int_{-\tau}^0 \|\varphi'(t)\|dt$. Now we are ready to present the following existence and uniqueness result.

Theorem 5.5. *Assume that* $[H_1]$ *and* $[H_2]$ *hold. Then* (5.14) *has a unique solution* $y \in C([-\tau, T], \mathbb{R}^n)$.

Proof. Define an operator $\Re : C([-\tau, T], \mathbb{R}^n) \to C([-\tau, T], \mathbb{R}^n)$ by

$$(\Re y)(x) = \mathbb{E}_\tau^{Bx^\alpha} \varphi(-\tau) + \int_{-\tau}^0 \mathbb{E}_\tau^{B(x-\tau-t)^\alpha} \varphi'(t)dt$$

$$+ \int_0^x \mathbb{E}_{\tau,\alpha}^{B(x-\tau-t)^\alpha} f(t, y(t))dt. \tag{5.16}$$

It is obvious that $\Re$ is well defined due to $[H_1]$. Next, we check that $\Re$ is a contraction mapping.

For any $y, z \in C([-\tau, T], \mathbb{R}^n)$, by Lemma 5.5, we obtain

$$\|(\Re y)(x) - (\Re z)(x)\| \le \widetilde{K} \left(\sum_{i=1}^k \frac{\|B\|^{i-1}}{\Gamma(i\alpha + 1)} (x - (i - 1)\tau)^{i\alpha} \right) \|y - z\|,$$

which implies that

$$\|\Re y - \Re z\| \le x\|y - z\|.$$

By $[H_2]$, one can apply the contraction mapping principle to complete the proof.
$\square$

Next, we give another existence and uniqueness result based on the below locally Lipschitz condition.

$[H_1']$ $f \in C([0, \infty), \mathbb{R}^n)$. For every $r > 0$ and $x_1 > 0$ there exists a constant $K := K(x_1, r)$ such that $\|f(x, y) - f(x, z)\| \le K\|y - z\|$ for all $x \in [0, x_1]$ and $y, z \in B_r = \{y \in \mathbb{R}^n : \|y\| \le r\}$.

Theorem 5.6. *Assume that $[H_1']$ holds. Then* (5.14) *has a unique solution* $y \in C([-\tau, x_m), \mathbb{R}^n)$, *where* $x_m \le \infty$. *Further, if* $x_m < \infty$, *then* $\lim_{x \to x_m} \|y(x)\| = \infty$.

Proof. Consider $\Re : C([-\tau, x_m), \mathbb{R}^n) \to C([-\tau, x_m), \mathbb{R}^n)$ defined in (5.16) again.

Step 1. A prior estimate.

For arbitrary $x \in [-\tau, 0]$, we have $\|y(x)\| = \|\varphi(x)\|$. Choose a subinterval $I_1 = [0, x_1]$. Let $\mu(x) = \sum_{i=1}^{k} \frac{\|B\|^{i-1}}{\Gamma(i\alpha+1)}(x - (i-1)\tau)^{i\alpha}$, where k comes from set $\max\{k \in \Lambda : k\tau \le x_1\}$, $r_1 = (\|\varphi\| + M)E_\alpha(\|B\|x_1^\alpha) + 1$, and $N_1 = \sup_{x \in I_1} \|f(x, 0)\|$. For any $y \in C([0, x_1], \mathbb{R}^n)$ with $\{\|y(x)\| : x \in [0, x_1]\} \le r_1$, we have

$$\|(\Re y)(x)\|$$

$$\le \|\mathbb{E}_\tau^{Bx^\alpha}\|\|\varphi(-\tau)\| + \left\|\int_{-\tau}^{0} \mathbb{E}_\tau^{B(x-\tau-t)^\alpha}\varphi'(t)dt\right\|$$

$$+ \int_0^x \left\|\mathbb{E}_{\tau,\alpha}^{B(x-\tau-t)^\alpha}\right\|\|f(t, y(t))\|dt$$

$$\le (\|\varphi(-\tau)\| + M)E_\alpha(\|B\|x^\alpha) + \int_0^x \left\|\mathbb{E}_{\tau,\alpha}^{B(x-\tau-t)^\alpha}\right\|\|f(t, y(t)) - f(t, 0)\|dt$$

$$+ \int_0^x \left\|\mathbb{E}_{\tau,\alpha}^{B(x-\tau-t)^\alpha}\right\|\|f(t, 0)\|dt$$

$$\le (\|\varphi(-\tau)\| + M)E_\alpha(\|B\|x_1^\alpha) + K\|y\|\int_0^x \left\|\mathbb{E}_{\tau,\alpha}^{B(x-\tau-t)^\alpha}\right\|dt$$

$$+ N_1\int_0^x \left\|\mathbb{E}_{\tau,\alpha}^{B(x-\tau-t)^\alpha}\right\|dt$$

$$\le (\|\varphi\| + M)E_\alpha(\|B\|x_1^\alpha) + (Kr_1 + N_1)\mu(x). \tag{5.17}$$

Choose $l_1 = \min\{x_1, x_1^*\}$, where x_1^* comes from set $\left\{x_1^* : \mu(x_1^*) = \frac{1}{Kr_1 + N_1}\right\}$. It follows from (5.17) that $\|(\Re y)(x)\| \le r_1$, for all $x \in [0, l_1]$.

Step 2. Local existence and uniqueness of solution.

For $x \in [0, l_1]$ and $y, z \in B_r$, use Lemma 5.5 to derive

$$\|(\Re y)(x) - (\Re z)(x)\| \le \int_0^x \|\mathbb{E}_{\tau,\alpha}^{B(x-\tau-t)^\alpha}\| \|f(x, y(t)) - f(x, z(t))\| dt$$

$$\le K\left(\int_0^x \|\mathbb{E}_{\tau,\alpha}^{B(x-\tau-t)^\alpha}\| dt\right) \|y - z\|$$

$$\le K\left(\sum_{i=1}^{k} \frac{\|B\|^{i-1}}{\Gamma(i\alpha+1)}(l_1 - (i-1)\tau)^{i\alpha}\right) \|y - z\|,$$

where k comes from set $\max\{k \in \Lambda : k\tau \le l_1\}$.

Choosing l_1 satisfying $\sum_{i=1}^{k} \frac{\|B\|^{i-1}}{\Gamma(i\alpha+1)}(l_1 - (i-1)\tau)^{i\alpha} < \frac{1}{2K}$, we can obtain

$$\|\Re y - \Re z\| \le \frac{1}{2}\|y - z\|.$$

This implies that $\Re$ has a fixed point on $[0, l_1]$, which reduces to a solution.

Step 3. Extension of solution.

Next, we extend the solution for $x \ge l_1$ by solving the fixed point problem, $z = \Re z$, where $\Re$ is given by

$$(\Re z)(x) = \mathbb{E}_\tau^{Bx^\alpha} \varphi(-\tau) + \int_{-\tau}^0 \mathbb{E}_\tau^{B(x-\tau-t)^\alpha} \varphi'(t)dt$$

$$+ \int_0^x \mathbb{E}_{\tau,\alpha}^{B(x-\tau-t)^\alpha} f(t, z(t))dt, \, x \in [0, l_2].$$

Set $z(x) = y(x)$ for $x \in [0, l_2]$, where l_2 is defined as follows. Take $x_2 > l_1$ and define $I_2 = [0, x_2]$, $\mu(x) = \sum_{i=1}^{k} \frac{\|B\|^{i-1}}{\Gamma(i\alpha+1)}(x - (i-1)\tau)^{i\alpha}$, where k comes from set $\max\{k \in \Lambda : k\tau \le x_2\}$, $r_2 = (\|\varphi\| + M)E_\alpha(\|B\|x_2^\alpha) + 1$, and $N_2 = \sup_{x \in I_2} \|f(x, 0)\|$.

Choose $l_2 = l_1 + \min\{x_2 - l_1, x_2^*\}$, where x_2^* comes from set $\left\{x_2^* : \mu(x_2^*) = \frac{1}{Kr_2 + N_2}\right\}$. For $x \in [l_1, l_2]$, we obtain $\|(\Re y)(x)\| \le r_2$. Repeating this procedure, we get the maximal interval of existence of solution. Thus (5.14) has a unique solution $y \in C([-\tau, x_m), \mathbb{R}^n)$.

Now we verify that $\lim_{x \to x_m} \|y(x_m)\| = \infty$ for x_m is finite. If not, then there exists a sequence $\{l_n\}$ converging to x_m and a finite positive number r such that $\|y(l_n)\| \le r$ for all n. Taking n sufficiently large, so that l_n is infinitesimally close to x_m, one can use the previous arguments to extend the solution beyond x_m, which is a contradiction. The proof is completed. $\qquad\square$

5.1.1.3 Finite time stability results for Caputo type

In this section, we show finite time stability results by using delayed Mittag-Leffler-type matrices.

Definition 5.9. (see [24]) Eq. (5.1) is called finite time stable with respect to $\{0, J, \tau, \delta, \kappa\}$ if and only if $\|\varphi\| < \delta$ implies $\|y(x)\| < \kappa$, $\forall\, x \in J$, where $\varphi(x)$, $-\tau \le x \le 0$, is the initial time of observation, δ, κ are real positive numbers, and $\delta < \kappa$.

Theorem 5.7. *Suppose $M = \int_{-\tau}^{0} \|\varphi'(s)\| ds < \infty$. If*

$$E_\alpha(\|B\|x^\kappa) < \frac{\kappa}{\delta + M}, \quad \forall\, x \in J, \tag{5.18}$$

then (5.1) is finite time stable with respect to $\{0, J, \tau, \delta, \kappa\}$.

Proof. According to Theorem 5.2, the solution of (5.1) can be expressed in the following form:

$$y(x) = \mathbb{E}_\tau^{Bx^\alpha}\varphi(-\tau) + \int_{-\tau}^{0} \mathbb{E}_\tau^{B(x-\tau-s)^\alpha}\varphi'(s)ds. \tag{5.19}$$

By Lemma 5.4 and the properties of norm $\|\cdot\|$, via (5.18) we have

$$\|y(x)\| \le \|\varphi(-\tau)\|\|\mathbb{E}_\tau^{Bx^\alpha}\| + \int_{-\tau}^{0} \|\varphi'(s)\|\|\mathbb{E}_\tau^{B(x-\tau-s)^\alpha}\| ds$$

$$\le \delta E_\alpha(\|B\|x^\alpha) + E_\alpha(\|B\|x^\alpha)\int_{-\tau}^{0} \|\varphi'(s)\| ds$$

$$\le E_\alpha(\|B\|x^\alpha)(\delta + M) < \kappa, \quad \forall\, x \in J,$$

where we use $\|\mathbb{E}_\tau^{B(x-\tau-s)^\alpha}\| \le \|\mathbb{E}_\tau^{Bx^\alpha}\|$. The proof is completed. $\qquad\square$

Theorem 5.8. *Suppose k, α are two constants satisfying $\alpha \le \frac{1}{k}$, $k \in \Lambda$. If*

$$E_\alpha(\|B\|x^\alpha) + E_\alpha(\|B\|\tau^\alpha) < \frac{\kappa}{\delta}, \quad \forall\, x \in J, \tag{5.20}$$

then (5.1) is finite time stable with respect to $\{0, J, \tau, \delta, \kappa\}$.

Proof. By integration by parts, one can change expression (5.19) to the form

$$y(x) = \mathbb{E}_\tau^{B(x-\tau)^\alpha}\varphi(0) + \int_{-\tau}^{0}\sum_{i=1}^{k} \frac{i\alpha B^i(x-i\tau-s)^{i\alpha-1}}{\Gamma(i\alpha+1)}\varphi(s)ds,$$

where $\mathbb{E}_\tau^{B(x-\tau-s)^\alpha} = \sum_{i=0}^{k} B^i \frac{(x-i\tau-s)^{i\alpha}}{\Gamma(i\alpha+1)}$ and
$\frac{d(\mathbb{E}_\tau^{B(x-\tau-s)^\alpha})}{ds} = -\sum_{i=1}^{k} \frac{i\alpha B^i (x-i\tau-s)^{i\alpha-1}}{\Gamma(i\alpha+1)}$.

By Lemma 5.4 and the properties of norm $\|\cdot\|$, via (5.20) we have

$$\|y(x)\| \leq \|\varphi(0)\| \|\mathbb{E}_\tau^{Bx^\alpha}\| + \sum_{i=1}^{k} \left\| \int_{-\tau}^{0} \frac{i\alpha B^i (x-i\tau-s)^{i\alpha-1}}{\Gamma(i\alpha+1)} ds \right\| \|\varphi\|$$

$$\leq \|\varphi(0)\| \|\mathbb{E}_\tau^{Bx^\alpha}\|$$

$$+ \sum_{i=1}^{k} \left\| \frac{B^i}{\Gamma(i\alpha+1)} ((x-(i-1)\tau)^{i\alpha} - (x-i\tau)^{i\alpha}) \right\| \|\varphi\|$$

$$\leq \|\varphi(0)\| \|\mathbb{E}_\tau^{Bx^\alpha}\| + \|\varphi\| \sum_{i=1}^{k} \left\| \frac{B^i}{\Gamma(i\alpha+1)} \tau^{i\alpha} \right\|$$

$$\leq \|\varphi(0)\| \|\mathbb{E}_\tau^{Bx^\alpha}\| + \|\varphi\| \sum_{i=1}^{\infty} \frac{\|B\|^i}{\Gamma(i\alpha+1)} \tau^{i\alpha}$$

$$\leq \delta E_\alpha(\|B\| x^\alpha) + \delta E_\alpha(\|B\| \tau^\alpha) < \kappa, \quad \forall\, x \in J,$$

where we use inequality $a^\theta - b^\theta \leq (a-b)^\theta, a > b > 0, 0 < \theta \leq 1$. The proof is completed. $\qquad\square$

Next, we continue to study the finite time stability of (5.14). We give the basic definition of finite time stable as follows.

Definition 5.10. (see [24]) Let y be a solution of (5.14). We say (5.14) is finite time stable with respect to $\{0, J, \tau, \delta, \eta\}$ If and only if $\|\varphi\| < \delta$ implies $\|y(x)\| < \eta, \forall\, x \in J$, where $\varphi(x), \ -\tau \leq x \leq 0$ is the initial time of observation, δ, η are real positive numbers, and $\delta < \eta$.

We impose the following assumptions:

[H_3] There exists an $\omega(\cdot) \in C(J, \mathbb{R}^+)$ such that $\|f(x,y)\| \leq \omega(x)$, for $x \in J$ and $y \in \mathbb{R}^n$.

[H_4] There exists a $\psi(\cdot) \in L^q(J, \mathbb{R}^+), \frac{1}{q} = 1 - \frac{1}{p}, p > 1$, such that

$\|f(x,y)\| \leq \psi(x)$ for $x \in J, y \in \mathbb{R}^n$ and $Q(x) := \left(\int_0^x \psi(t)^q dt\right)^{\frac{1}{q}} < \infty$.

[H_5] There exists a $L > 0$ such that $\|f(x,y)\| \leq L\|y\|, x \in J$, and $y \in \mathbb{R}^n$.

Theorem 5.9. *Assume that* [H_1]–[H_3] *hold. For a fixed* $k \in \Lambda$, *if*

$$(\delta + M) E_\alpha(\|B\| x^\alpha) + \|\omega\| \left(\sum_{i=1}^{k} \frac{\|B\|^{i-1}}{\Gamma(i\alpha+1)} (x-(i-1)\tau)^{i\alpha} \right)$$

$$< \eta, \quad \forall\, x \in J, \tag{5.21}$$

then (5.14) *is finite time stable with respect to* $\{0, J, \tau, \delta, \eta\}$.

Proof. By $[H_1]$, $[H_2]$, and Theorem 5.5, we know (5.14) has a unique solution $y \in C([-\tau, T], \mathbb{R}^n)$. By Lemmas 5.4, 5.5, and 5.6 and the properties of norm $\|\cdot\|$, via (5.15) we have

$$\|y(x)\|$$

$$\leq \|E_\tau^{Bx^\alpha} \varphi(-\tau)\| + \left\| \int_{-\tau}^0 \mathbb{E}_\tau^{B(x-\tau-t)^\alpha} \varphi'(t)dt \right\| + \int_0^x \left\| \mathbb{E}_{\tau,\alpha}^{B(x-\tau-t)^\alpha} \right\| \omega(t)dt$$

$$\leq (\|\varphi(-\tau)\| + M) E_\alpha(\|B\|x^\alpha)$$

$$+ \|\omega\| \int_0^{x-(k-1)\tau} \sum_{i=1}^k \|B\|^{i-1} \frac{(x-(i-1)\tau - t)^{i\alpha-1}}{\Gamma((i-1)\alpha + \alpha)} dt$$

$$+ \|\omega\| \sum_{j=2}^k \int_{x-(j-1)\tau}^{x-(j-2)\tau} \left(\sum_{i=2}^j \|B\|^{i-2} \frac{(x-(i-2)\tau - t)^{(i-1)\alpha-1}}{\Gamma((i-2)\alpha + \alpha)} \right) dt$$

$$\leq (\delta + M) E_\alpha(\|B\|x^\alpha)$$

$$+ \|\omega\| \sum_{i=1}^k \frac{\|B\|^{i-1}}{\Gamma(i\alpha + 1)} ((x-(i-1)\tau)^{i\alpha} - ((k-i)\tau)^{i\alpha})$$

$$+ \|\omega\| \sum_{j=2}^k$$

$$\times \left(\sum_{i=2}^j \frac{\|B\|^{i-2}}{\Gamma((i-1)\alpha + 1)} \left[((j-i+1)\tau)^{\alpha(i-1)} - ((j-i)\tau)^{\alpha(i-1)} \right] \right)$$

$$\leq (\delta + M) E_\alpha(\|B\|x^\alpha) + \|\omega\| \left(\sum_{i=1}^k \frac{\|B\|^{i-1}}{\Gamma(i\alpha + 1)} (x-(i-1)\tau)^{i\alpha} \right) < \eta.$$

Due to (5.21), the proof is completed. $\qquad\square$

Theorem 5.10. *Assume that $[H_1]$, $[H_2]$, $[H_4]$, and $\alpha > 1 - \frac{1}{p}$ ($p > 1$) hold. For a fixed $k \in \Lambda$, if*

$$(\delta + M) E_\alpha(\|B\|x^\alpha) + Q(x) \sum_{i=1}^k \left(\frac{\|B\|^{i-1}}{\Gamma(i\alpha)} \cdot \frac{(x-(i-1)\tau)^{i\alpha-1+\frac{1}{p}}}{(pi\alpha - p + 1)^{\frac{1}{p}}} \right)$$

$$< \eta, \ \forall \, x \in J, \tag{5.22}$$

then (5.14) is finite time stable with respect to $\{0, J, \tau, \delta, \eta\}$.

Proof. By Lemmas 5.4, 5.5, and 5.6 and the properties of norm $\| \cdot \|$, via (5.15) and (5.22) we have

$$\|y(x)\|$$

$$\leq \|\varphi(-\tau)\| \|\mathbb{E}_\tau^{Bx^\alpha}\| + E_\alpha(\|B\|x^\alpha) \int_{-\tau}^0 \|\varphi'(t)\| dt + \int_0^x \frac{(x-t)^{\alpha-1}}{\Gamma(\alpha)} \psi(t) dt$$

$$+ \cdots + \int_0^{x-(k-1)\tau} \|B\|^{k-1} \frac{(x-(k-1)\tau-t)^{k\alpha-1}}{\Gamma(k\alpha)} \psi(t) dt$$

$$\leq (\|\varphi(-\tau)\| + M) E_\alpha(\|B\|x^\alpha)$$

$$+ \sum_{i=1}^k \frac{\|B\|^{i-1}}{\Gamma(i\alpha)} \int_0^{x-(i-1)\tau} (x-(i-1)\tau-t)^{i\alpha-1} \psi(t) dt$$

$$\leq (\|\varphi(-\tau)\| + M) E_\alpha(\|B\|x^\alpha)$$

$$+ \sum_{i=1}^k \frac{\|B\|^{i-1}}{\Gamma(i\alpha)} \left(\int_0^{x-(i-1)\tau} (x-(i-1)\tau-t)^{p(i\alpha-1)} dt \right)^{\frac{1}{p}}$$

$$\times \left(\int_0^{x-(i-1)\tau} \psi(t)^q dt \right)^{\frac{1}{q}}$$

$$\leq (\|\varphi(-\tau)\| + M) E_\alpha(\|B\|x^\alpha)$$

$$+ \sum_{i=1}^k \frac{\|B\|^{i-1}}{\Gamma(i\alpha)} \left(\int_0^{x-(i-1)\tau} (x-(i-1)\tau-t)^{p(i\alpha-1)} dt \right)^{\frac{1}{p}} \left(\int_0^x \psi(t)^q dt \right)^{\frac{1}{q}}$$

$$\leq (\delta + M) E_\alpha(\|B\|x^\alpha) + Q(x) \sum_{i=1}^k \left(\frac{\|B\|^{i-1}}{\Gamma(i\alpha)} \cdot \frac{(x-(i-1)\tau)^{i\alpha-1+\frac{1}{p}}}{(pi\alpha - p + 1)^{\frac{1}{p}}} \right) < \eta.$$

The proof is completed. $\qquad\square$

Theorem 5.11. *Assume that $[H_1]$–$[H_3]$ and $\alpha \geq \frac{1}{2}$ hold. For a fixed $k \in \Lambda$, if*

$$(\delta + M) E_\alpha(\|B\|x^\alpha) + \frac{\|\omega\|}{\alpha} x^\alpha E_{\alpha,\alpha}(\|B\|x^\alpha) < \eta, \ \forall \ x \in J, \qquad (5.23)$$

then (5.14) is finite time stable with respect to $\{0, J, \tau, \delta, \eta\}$.

Proof. By Lemmas 5.4, 5.5, 5.6, and 5.7 and the properties of norm $\| \cdot \|$, via (5.15) and (5.23) we have

$$\|y(x)\|$$

$$\leq (\|\varphi(-\tau)\| + M) E_\alpha(\|B\|x^\alpha) + E_{\alpha,\alpha}(\|B\|x^\alpha) \int_0^x (x-t)^{\alpha-1} \omega(t) dt$$

$$\leq (\|\varphi(-\tau)\| + M) E_\alpha(\|B\|x^\alpha) + \|\omega\| E_{\alpha,\alpha}(\|B\|x^\alpha) \int_0^x (x-t)^{\alpha-1} dt$$

$$\leq (\delta + M) E_\alpha(\|B\|x^\alpha) + \frac{\|\omega\|}{\alpha} x^\alpha E_{\alpha,\alpha}(\|B\|x^\alpha) < \eta.$$

The proof is completed. $\qquad\qquad\square$

Theorem 5.12. *Assume that* $[H_1]$, $[H_2]$, $[H_5]$, *and* $\alpha \geq \frac{1}{2}$ *hold. For a fixed* $k \in \Lambda$, *if*

$$E_\alpha(\|B\|x^\alpha) E_\alpha(L\Gamma(\alpha) E_{\alpha,\alpha}(\|B\|x^\alpha)x^\alpha) < \frac{\eta}{\delta + M}, \ \forall \, x \in J, \qquad (5.24)$$

then (5.14) *is finite time stable with respect to* $\{0, J, \tau, \delta, \eta\}$.

Proof. For $x \in J$, by Lemmas 5.4, 5.5, 5.6, and 5.7 and the properties of norm $\|\cdot\|$, we have

$$\|y(x)\|$$

$$\leq \|\varphi(-\tau)\| \|\mathbb{E}_\tau^{Bx^\alpha}\| + E_\alpha(\|B\|x^\alpha) \int_{-\tau}^0 \|\varphi'(t)\| dt$$

$$+ \int_0^x \left\| \mathbb{E}_{\tau,\alpha}^{B(x-\tau-t)^\alpha} \right\| \|f(t, y(t))\| dt$$

$$\leq (\|\varphi(-\tau)\| + M) E_\alpha(\|B\|x^\alpha) + L E_{\alpha,\alpha}(\|B\|x^\alpha) \int_0^x (x-t)^{\alpha-1} \|y(t)\| dt.$$

According to (5.24), we have

$$\|y(x)\|$$

$$\leq (\|\varphi(-\tau)\| + M) E_\alpha(\|B\|x^\alpha) + L E_{\alpha,\alpha}(\|B\|x^\alpha) \int_0^x (x-t)^{\alpha-1} \|y(x)\| dt$$

$$\leq (\delta + M) E_\alpha(\|B\|x^\alpha) E_\alpha(L\Gamma(\alpha) E_{\alpha,\alpha}(\|B\|x^\alpha)x^\alpha) < \eta.$$

The proof is completed. $\qquad\qquad\square$

5.1.1.4 Numerical examples and discussion

Example 5.1. Let $\alpha = 0.2$, $\tau = 0.2$, $T = 0.8$, $k = 4$, $B = \begin{pmatrix} 0.2 & 0 \\ 0 & 0.8 \end{pmatrix}$. Consider

$$\begin{cases} {}^c D_{0+}^{0.2} y(x) = B y(x - 0.2), & y(x) = (y_1(x), y_2(x))^\top, \ x \in [0, 0.8], \\ \varphi(x) = (0.1, 0.2)^\top, & -0.2 \leq x \leq 0. \end{cases}$$

$$(5.25)$$

By Theorem 5.2, the solution of (5.25) can be expressed by the following explicit form:

$$y(x) = \mathbb{E}_{0.2}^{Bx^{0.2}} \varphi(-0.2)$$

$$= \begin{cases} \left(I + B\frac{x^{0.2}}{\Gamma(1.2)}\right)\begin{pmatrix} 0.1 \\ 0.2 \end{pmatrix}, & 0 \le x \le 0.2, \\[2ex] \left(I + B\frac{x^{0.2}}{\Gamma(1.2)} + B^2\frac{(x-0.2)^{0.4}}{\Gamma(1.4)}\right)\begin{pmatrix} 0.1 \\ 0.2 \end{pmatrix}, & 0.2 < x \le 0.4, \\[2ex] \left(I + B\frac{x^{0.2}}{\Gamma(1.2)} + B^2\frac{(x-0.2)^{0.4}}{\Gamma(1.4)} + B^3\frac{(x-0.4)^{0.6}}{\Gamma(1.6)}\right)\begin{pmatrix} 0.1 \\ 0.2 \end{pmatrix}, & 0.4 < x \le 0.6, \\[2ex] \left(I + B\frac{x^{0.2}}{\Gamma(1.2)} + B^2\frac{(x-0.2)^{0.4}}{\Gamma(1.4)} + B^3\frac{(x-0.4)^{0.6}}{\Gamma(1.6)} \right. \\[1ex] \left. \quad + B^4\frac{(x-0.6)^{0.8}}{\Gamma(1.8)}\right)\begin{pmatrix} 0.1 \\ 0.2 \end{pmatrix}, & 0.6 < x \le 0.8. \end{cases}$$

By calculation, $\|\varphi\| = 0.3$, $M = \int_{-0.2}^{0} \|\Theta\| ds = 0$, $E_{0.2}(0.8^{1.2}) = 4.155$, $E_{0.2}(\|B\|0.2^{0.2}) = E_{0.2}(0.8 \times 0.2^{0.2}) = 2.4851$.

Concerning the definition of finite time stability (see Definition 5.9), we need to find a certain threshold κ which makes the state $\|y(x)\|$ of system (5.1) not exceed κ during a fixed finite time interval J. On the one hand, we can use an explicit formula of solution to (5.25) via numerical simulation to find a corresponding $\kappa = 1.1307$ for a fixed $T = 0.8$. On the other hand, by checking the conditions in Theorems 5.7 and 5.8 and Theorem 4.2 in [24] for the finite time interval $[0, 0.8]$, we get the relative optimal threshold $\kappa = 1.29$.

Example 5.2. Let $\alpha = 0.6$, $\tau = 0.2$, $T = 0.6$, and $k = 3$. Consider

$$\begin{cases} {}^{c}D_{0^+}^{0.6} y(x) = By(x - 0.2) + f(x, y(x)), & x \in [0, 0.6], \\ \varphi(x) = (x, 2x)^{\top}, & -0.2 \le x \le 0, \end{cases} \tag{5.26}$$

where

$$B = \begin{pmatrix} 0.3 & 0 \\ 0 & 0.5 \end{pmatrix}, \qquad f(x, y(x)) = \begin{pmatrix} x^2\frac{|y_1(x)|}{1+|y_1(x)|} \\ x^2\frac{|y_2(x)|}{1+|y_2(x)|} \end{pmatrix}.$$

A solution of (5.26) can be expressed in the following form:

$$y(x) = \begin{pmatrix} y_1(x) \\ y_2(x) \end{pmatrix} = \mathbb{E}_{0.2}^{Bx^{0.6}} \varphi(-0.2) + \int_{-0.2}^{0} \mathbb{E}_{0.2}^{B(x-0.2-s)^{0.6}} \begin{pmatrix} 1 \\ 2 \end{pmatrix} ds$$

$$+ \int_{0}^{x} \mathbb{E}_{0.2,0.6}^{B(x-0.2-t)^{0.6}} \begin{pmatrix} x^2 \frac{|y_1(x)|}{1+|y_1(x)|} \\ x^2 \frac{|y_2(x)|}{1+|y_2(x)|} \end{pmatrix} dt.$$

Obviously, $\|f(x,y) - f(x,z)\| \le 2x^2\|y-z\|$ for $y, z \in \mathbb{R}^n$ and $\|f(x,y)\| < 2x^2$, $\forall\, x \in J$. Now put $p = 2$, $q = 2$, $K = L = 0.72$, $x = 0.7078$, and $\omega(x) = \psi(x) = 2x^2$.

By calculation, one has $\|\varphi\|_C = 0.6$, $M = \int_{-0.2}^{0} \|\varphi'(t)\| dt = 0.6$, $E_{0.6}(\|B\|T^{0.6}) = 1.572$, $E_{0.6,0.6}(\|B\|T^{0.6}) = 1.2692$, $E_\alpha(\|B\|T^\alpha)E_\alpha(L\Gamma(\alpha)E_{\alpha,\alpha}(\|B\|T^\alpha)T^\alpha) = 4.2827$, $\|\omega\| = 0.72$, $Q(T) = 0.2494$. Next, $\sum\limits_{i=1}^{3} \frac{\|B\|^{i-1}}{\Gamma(i\alpha+1)}(T-(i-1)\tau)^{i\alpha} = 0.9831$ and $\sum\limits_{i=1}^{3} \frac{\|B\|^{i-1}(T-(i-1)\tau)^{i\alpha-1+\frac{1}{p}}}{\Gamma(i\alpha)(pi\alpha-p+1)^{\frac{1}{p}}} = 1.6896$.

Concerning the definition of finite time stability (see Definition 5.10), we need to find a certain threshold η which makes the state $\|y(x)\|$ of (5.14) not exceed η on J. By checking the required conditions in Theorems 5.9, 5.10, 5.11, and 5.12 on $[0, 0.6]$, we get a relatively optimal threshold $\eta = 2.33$.

5.1.1.5 Conclusion

We give sufficient conditions for the finite time stability results derived based on the delayed matrix exponential approach. We demonstrate the validity of the designed method and provide some discussion by using a numerical example. The results in this part are motivated from [14,105].

5.1.2 Relative controllability results for Caputo type

Mahmudov [106] introduced the concept of delay perturbation of matrix function $X_{\tau,\alpha,\alpha}^{A,B}(t)$, $\alpha, \alpha > 0$, of two-parameter Mittag-Leffler-type generated by A and B as follows:

$$X_{\tau,\alpha,\beta}^{A,B}(t) = \begin{cases} \Theta, & -\tau \le t < 0, \\ I, & t = 0, \\ \sum\limits_{i=0}^{\infty}\sum\limits_{j=0}^{p-1} Q_{i+1}(j\tau)\dfrac{(t-j\tau)^{i\alpha+\beta-1}}{\Gamma(i\alpha+\beta)}, & (p-1)\tau < t \le p\tau, \end{cases}$$

$$(5.27)$$

where the brief form of $Q_{i+1}(j\tau)$ is

j	$Q_1(j\tau)$	$Q_2(j\tau)$	$Q_3(j\tau)$	$Q_4(j\tau)$	$\cdots$	$Q_{p+1}(j\tau)$
0	I	A	A^2	A^3	$\cdots$	A^p
1	Θ	B	$AB+BA$	$A(AB+BA)+BA^2$	$\cdots$	$\cdots$
2	Θ	Θ	B^2	$AB^2+B(AB+BA)$	$\cdots$	$\cdots$
3	Θ	Θ	Θ	B^3	$\cdots$	$\cdots$
$\cdots$	$\cdots$	$\cdots$	$\cdots$	$\cdots$	$\cdots$	$\cdots$
p	Θ	Θ	Θ	Θ	$\cdots$	B^p

and $X^{A,B}_{\tau,\alpha,\beta}(t)$ can also be written in (5.28) if A and B are permutable matrices, i.e., if $AB = BA$,

$$
X^{A,B}_{\tau,\alpha,\beta}(t) = \begin{cases}
\Theta, & -\tau \leqslant t < 0, \\
I, & t = 0, \\
\displaystyle\sum_{i=0}^{\infty} A^i \frac{t^{i\alpha+\beta-1}}{\Gamma(i\alpha+\beta)} + \sum_{i=1}^{\infty} \binom{i}{1} A^{i-1} B \frac{(t-\tau)^{i\alpha+\beta-1}}{\Gamma(i\alpha+\beta)} + \cdots \\
\quad + \displaystyle\sum_{i=p-1}^{\infty} \binom{i}{p-1} A^{i-p+1} B^{p-1} \frac{(t-(p-1)\tau)^{i\alpha+\beta-1}}{\Gamma(i\alpha+\beta)}, \\
(p-1)\tau < t \leqslant p\tau.
\end{cases}
$$

$$(5.28)$$

In general, the condition of $AB = BA$ is strong and special for the controlled system since we always avoid putting such restriction to consider controllability problems. The flexible coefficient of matrix B will be used to adjust the control function and deal with more difficulty from the main part A. For example, if the linear system generalized by A is not stable, then we can use a suitable B to make that the perturbation system is stable. In such case, it is impossible to require that $AB = BA$.

Next, $X^{A,B}_{\tau,\alpha,\beta}$ in (5.27) does generalize (5.28) in [1] and gives an explicit way to compute the delay perturbation of matrix functions in the case of nonpermutable matrices. Based on (5.27), Mahmudov [106] studied the nonhomogeneous fractional order delay system

$$
\begin{cases}
(^C D^\alpha_{-\tau^+} v)(t) = Av(t) + Bv(t-\tau) + g(t), \ t \in J, \ \tau > 0, \\
v(t) = \psi(t), \ -\tau \leqslant t \leqslant 0,
\end{cases}
\tag{5.29}
$$

and obtained the expression of the solution of (5.29) by using $X^{A,B}_{\tau,\alpha,\beta}(t)$. We have

$$
v(t) = X^{A,B}_{\tau,\alpha,1}(t+\tau)\psi(-\tau) + \int_{-\tau}^{0} X^{A,B}_{\tau,\alpha,\alpha}(t-s)[(^C D^\alpha_{-\tau^+}\psi)(s) - A\psi(s)]ds
$$

$$+ \int_0^t X_{\tau,\alpha,\alpha}^{A,B}(t-s)g(s)ds, \tag{5.30}$$

where $(^C D_{-\tau^+}^\alpha v)(\cdot)$ is the Caputo fractional derivative, $\alpha \in (0, 1)$, A, $B \in \mathbb{R}^{n \times n}$, $g \in C([-\tau, \tau_1], \mathbb{R}^n)$, $J := [0, \tau_1]$, $\tau_1 > 0$.

Now, we investigate the relative controllability of the following semilinear fractional delay systems by removing the requirement on the permutable matrices:

$$\begin{cases} (^C D_{-\tau^+}^\alpha v)(t) = Av(t) + Bv(t - \tau) + f(t, v(t)) + Cu(t), \ t \in J, \\ v(t) = \psi(t), \ -\tau \leqslant t \leqslant 0, \end{cases} \tag{5.31}$$

where $\alpha \in (0, 1)$, A, B, $C \in \mathbb{R}^{n \times n}$, $AB = BA$, $v \in C^1([-\tau, \tau_1], \mathbb{R}^n)$, $\psi \in C_\tau^1 := C^1([-\tau, 0], \mathbb{R}^n)$, $f \in C(J \times \mathbb{R}^n, \mathbb{R}^n)$, $J = [0, \tau_1]$, $\tau_1 > 0$, and the control function $u(\cdot)$ takes values from $L^2(J, \mathbb{R}^n)$.

From (5.30), we know the solution of (5.31) can be expressed in the following form:

$$v(t) = X_{\tau,\alpha,1}^{A,B}(t + \tau)\psi(-\tau) + \int_{-\tau}^0 X_{\tau,\alpha,\alpha}^{A,B}(t - s)[(^C D_{-\tau^+}^\alpha \psi)(s) - A\psi(s)]ds$$

$$+ \int_0^t X_{\tau,\alpha,\alpha}^{A,B}(t - s)[f(s, v(s)) + Cu(s)]ds. \tag{5.32}$$

By [107, p.1861] and [107, (8)], $\int_0^t (t-s)^{\alpha-1} E_{\alpha,\alpha}(\lambda(t-s)^\alpha)ds = \frac{1}{\lambda}[E_\alpha(\lambda t^\alpha) - 1]$.

Lemma 5.8. *The function* $X_{\tau,\alpha,\beta}^{A,B}(\cdot)$ *is continuous on* $(0, +\infty)$.

Proof. For $(p - 1)\tau < t_* < p\tau$, $p = 1, 2, \cdots$, we have

$$\lim_{t \to t_*} X_{\tau,\alpha,\beta}^{A,B}(t)$$

$$= \lim_{t \to t_*} \left(\sum_{i=0}^\infty A^i \frac{t^{i\alpha+\beta-1}}{\Gamma(i\alpha + \beta)} + \sum_{i=1}^\infty \binom{i}{1} A^{i-1} B \frac{(t - \tau)^{i\alpha+\beta-1}}{\Gamma(i\alpha + \beta)} + \cdots \right.$$

$$\left. + \sum_{i=p-1}^\infty \binom{i}{p-1} A^{i-p+1} B^{p-1} \frac{(t - (p-1)\tau)^{i\alpha+\beta-1}}{\Gamma(i\alpha + \beta)} \right)$$

$$= \sum_{i=0}^\infty A^i \lim_{t \to t_*} \frac{t^{i\alpha+\beta-1}}{\Gamma(i\alpha + \beta)} + \sum_{i=1}^\infty \binom{i}{1} A^{i-1} B \lim_{t \to t_*} \frac{(t - \tau)^{i\alpha+\beta-1}}{\Gamma(i\alpha + \beta)} + \cdots$$

$$+ \sum_{i=p-1}^\infty \binom{i}{p-1} A^{i-p+1} B^{p-1} \lim_{t \to t_*} \frac{(t - (p-1)\tau)^{i\alpha+\beta-1}}{\Gamma(i\alpha + \beta)}$$

$$= \sum_{i=0}^\infty A^i \frac{t_*^{i\alpha+\beta-1}}{\Gamma(i\alpha + \beta)} + \sum_{i=1}^\infty \binom{i}{1} A^{i-1} B \frac{(t_* - \tau)^{i\alpha+\beta-1}}{\Gamma(i\alpha + \beta)} + \cdots$$

$$+ \sum_{i=p-1}^{\infty} \binom{i}{p-1} A^{i-p+1} B^{p-1} \frac{(t_* - (p-1)\tau)^{i\alpha+\beta-1}}{\Gamma(i\alpha+\beta)}$$

$$= X_{\tau,\alpha,\beta}^{A,B}(t_*).$$

For $t_* = p\tau$, $p = 1, 2, \cdots$, we have

$$\lim_{t \to t_*^-} X_{\tau,\alpha,\beta}^{A,B}(t)$$

$$= \lim_{t \to p\tau^-} \left(\sum_{i=0}^{\infty} A^i \frac{t^{i\alpha+\beta-1}}{\Gamma(i\alpha+\beta)} + \sum_{i=1}^{\infty} \binom{i}{1} A^{i-1} B \frac{(t-\tau)^{i\alpha+\beta-1}}{\Gamma(i\alpha+\beta)} + \cdots \right.$$

$$\left. + \sum_{i=p-1}^{\infty} \binom{i}{p-1} A^{i-p+1} B^{p-1} \frac{(t-(p-1)\tau)^{i\alpha+\beta-1}}{\Gamma(i\alpha+\beta)} \right)$$

$$= \sum_{i=0}^{\infty} A^i \frac{p\tau^{i\alpha+\beta-1}}{\Gamma(i\alpha+\beta)} + \sum_{i=1}^{\infty} \binom{i}{1} A^{i-1} B \frac{(p\tau-\tau)^{i\alpha+\beta-1}}{\Gamma(i\alpha+\beta)} + \cdots$$

$$+ \sum_{i=p-1}^{\infty} \binom{i}{p-1} A^{i-p+1} B^{p-1} \frac{(p\tau-(p-1)\tau)^{i\alpha+\beta-1}}{\Gamma(i\alpha+\beta)}$$

$$= X_{\tau,\alpha,\beta}^{A,B}(p\tau) = X_{\tau,\alpha,\beta}^{A,B}(t_*),$$

$$\lim_{t \to t_*^+} X_{\tau,\alpha,\beta}^{A,B}(t)$$

$$= \lim_{t \to p\tau^+} \left(\sum_{i=0}^{\infty} A^i \frac{t^{i\alpha+\beta-1}}{\Gamma(i\alpha+\beta)} + \sum_{i=1}^{\infty} \binom{i}{1} A^{i-1} B \frac{(t-\tau)^{i\alpha+\beta-1}}{\Gamma(i\alpha+\beta)} + \cdots \right.$$

$$+ \sum_{i=p-1}^{\infty} \binom{i}{p-1} A^{i-p+1} B^{p-1} \frac{(t-(p-1)\tau)^{i\alpha+\beta-1}}{\Gamma(i\alpha+\beta)}$$

$$\left. + \sum_{i=p}^{\infty} \binom{i}{p} A^{i-p} B^p \frac{(t-p\tau)^{i\alpha+\beta-1}}{\Gamma(i\alpha+\beta)} \right)$$

$$= \sum_{i=0}^{\infty} A^i \lim_{t \to p\tau^+} \frac{t^{i\alpha+\beta-1}}{\Gamma(i\alpha+\beta)} + \sum_{i=1}^{\infty} \binom{i}{1} A^{i-1} B \lim_{t \to p\tau^+} \frac{(t-\tau)^{i\alpha+\beta-1}}{\Gamma(i\alpha+\beta)} + \cdots$$

$$+ \sum_{i=p-1}^{\infty} \binom{i}{p-1} A^{i-p+1} B^{p-1} \lim_{t \to p\tau^+} \frac{(t-(p-1)\tau)^{i\alpha+\beta-1}}{\Gamma(i\alpha+\beta)}$$

$$+ \sum_{i=p}^{\infty} \binom{i}{p} A^{i-p} B^p \lim_{t \to p\tau^+} \frac{(t-p\tau)^{i\alpha+\beta-1}}{\Gamma(i\alpha+\beta)}$$

$$
= \sum_{i=0}^{\infty} A^i \frac{p\tau^{i\alpha+\beta-1}}{\Gamma(i\alpha+\beta)} + \sum_{i=1}^{\infty} \binom{i}{1} A^{i-1} B \frac{(p\tau-\tau)^{i\alpha+\beta-1}}{\Gamma(i\alpha+\beta)} + \cdots
$$

$$
+ \sum_{i=p-1}^{\infty} \binom{i}{p-1} A^{i-p+1} B^{p-1} \frac{(p\tau-(p-1)\tau)^{i\alpha+\beta-1}}{\Gamma(i\alpha+\beta)}
$$

$$
+ \sum_{i=p}^{\infty} \binom{i}{p} A^{i-p} B^p \lim_{t\to p\tau^+} \frac{(t-p\tau)^{i\alpha+\beta-1}}{\Gamma(i\alpha+\beta)}
$$

$$
= X_{\tau,\alpha,\beta}^{A,B}(p\tau) + \sum_{i=p}^{\infty} \binom{i}{p} A^{i-p} B^p \lim_{t\to p\tau^+} \frac{(t-p\tau)^{i\alpha+\beta-1}}{\Gamma(i\alpha+\beta)}
$$

$$
= X_{\tau,\alpha,\beta}^{A,B}(p\tau) = X_{\tau,\alpha,\beta}^{A,B}(t_*).
$$

To sum up, $X_{\tau,\alpha,\beta}^{A,B}(\cdot)$ is continuous on $(0,+\infty)$. $\qquad\square$

Lemma 5.9. *For all $t \ge 0$, $0 < \alpha < 1$, $0 < \beta \le 1$ satisfying $\alpha + \beta \ge 1$, we have*

$$
\|X_{\tau,\alpha,\beta}^{A,B}(t)\| \le t^{\beta-1} E_{\alpha,\beta}((\|A\|+\|B\|)t^\alpha).
$$

Proof. By (5.27), without loss of generality, for $(p-1)\tau < t \le p\tau$, $p = 1, 2, \cdots$, we have

$$
\|X_{\tau,\alpha,\beta}^{A,B}(t)\|
$$

$$
\le \left\| \sum_{i=0}^{\infty} \sum_{j=0}^{\infty} Q_{i+1}(j\tau) \frac{(t-j\tau)^{i\alpha+\beta-1}}{\Gamma(i\alpha+\beta)} \right\|
$$

$$
\le \sum_{i=0}^{\infty} \|A\|^i \frac{t^{i\alpha+\beta-1}}{\Gamma(i\alpha+\beta)} + \sum_{i=1}^{\infty} \binom{i}{1} \|A\|^{i-1} \|B\| \frac{(t-\tau)^{i\alpha+\beta-1}}{\Gamma(i\alpha+\beta)} + \cdots
$$

$$
+ \sum_{i=p-1}^{\infty} \binom{i}{p-1} \|A\|^{i-p+1} \|B\|^{p-1} \frac{(t-(p-1)\tau)^{i\alpha+\beta-1}}{\Gamma(i\alpha+\beta)}
$$

$$
\le \sum_{i=0}^{\infty} \|A\|^i \frac{t^{i\alpha+\beta-1}}{\Gamma(i\alpha+\beta)} + \sum_{i=1}^{\infty} \binom{i}{1} \|A\|^{i-1} \|B\| \frac{t^{i\alpha+\beta-1}}{\Gamma(i\alpha+\beta)} + \cdots
$$

$$
+ \sum_{i=p-1}^{\infty} \binom{i}{p-1} \|A\|^{i-p+1} \|B\|^{p-1} \frac{t^{i\alpha+\beta-1}}{\Gamma(i\alpha+\beta)}
$$

$$
= \frac{t^{\beta-1}}{\Gamma(\beta)} + (\|A\|+\|B\|) \frac{t^{\alpha+\beta-1}}{\Gamma(\alpha+\beta)}
$$

$$
+ \left[\|A\|^2 + \binom{2}{1} \|A\|\|B\| + \|B\|^2 \right] \frac{t^{2\alpha+\beta-1}}{\Gamma(2\alpha+\beta)}
$$

$$+ \left[\|A\|^3 + \binom{3}{1} \|A\|^2 \|B\| + \binom{3}{2} \|A\| \|B\|^2 + \|B\|^3 \right] \frac{t^{3\alpha+\beta-1}}{\Gamma(3\alpha+\beta)}$$

$$+ \cdots$$

$$+ \left[\|A\|^{p-1} + \binom{p-1}{1} \|A\|^{p-2} \|B\| + \cdots + \|B\|^{p-1} \right] \frac{t^{(p-1)\alpha+\beta-1}}{\Gamma((p-1)\alpha+\beta)}$$

$$+ \sum_{i=p}^{\infty} \left[\|A\|^i + \binom{i}{1} \|A\|^{i-1} \|B\| + \cdots + \binom{i}{p-1} \|A\|^{i-p+1} \|B\|^{p-1} \right]$$

$$\times \frac{t^{i\alpha+\beta-1}}{\Gamma(i\alpha+\beta)}$$

$$\leqslant (\|A\| + \|B\|)^0 \frac{t^{\beta-1}}{\Gamma(\beta)} + (\|A\| + \|B\|)^1 \frac{t^{\alpha+\beta-1}}{\Gamma(\alpha+\beta)}$$

$$+ (\|A\| + \|B\|)^2 \frac{t^{2\alpha+\beta-1}}{\Gamma(2\alpha+\beta)}$$

$$+ \cdots + (\|A\| + \|B\|)^{p-1} \frac{t^{(p-1)\alpha+\beta-1}}{\Gamma((p-1)\alpha+\beta)} + \sum_{i=p}^{\infty} (\|A\| + \|B\|)^i \frac{t^{i\alpha+\beta-1}}{\Gamma(i\alpha+\beta)}$$

$$= \sum_{i=0}^{\infty} (\|A\| + \|B\|)^i \frac{t^{i\alpha+\beta-1}}{\Gamma(i\alpha+\beta)}$$

$$= t^{\beta-1} E_{\alpha,\beta}((\|A\| + \|B\|)t^\alpha).$$

The result is proved. $\qquad\qquad\square$

Remark 5.1. Obviously, for $\beta = 1$, $\|X_{\tau,\alpha,1}^{A,B}(t)\| \leqslant E_{\alpha,1}((\|A\| + \|B\|)t^\alpha) = E_\alpha((\|A\| + \|B\|)t^\alpha)$. If $0 < \alpha = \beta < 1$, then $\|X_{\tau,\alpha,\alpha}^{A,B}(t)\| \leqslant t^{-0.5} E_{\alpha,\alpha}((\|A\| + \|B\|)t^\alpha)$ holds for $\alpha \geq 0.5$.

Definition 5.11. (see [38, Definition 4])System (5.31) is called relatively controllable if for an arbitrary initial vector function $\psi \in C^1([-\tau, 0], \mathbb{R}^n)$, the final state of the vector $v_{\tau_1} \in \mathbb{R}^n$, and time τ_1, there exists a control $u \in L^2(J, \mathbb{R}^n)$ such that system (5.31) has a solution $v \in C^1([-\tau, 0] \cup J, \mathbb{R}^n)$ that satisfies the initial condition ψ and $v(\tau_1) = v_{\tau_1}$.

5.1.2.1 Relative controllability results for linear systems

For $f(t, v(t)) = \mathbf{0}, t \in J$, system (5.31) reduces to the following linear fractional delay controlled system:

$$\begin{cases} (^C D_{-\tau^+}^\alpha v)(t) = Av(t) + Bv(t-\tau) + Cu(t), \ t \in J, \ \tau > 0, \\ v(t) = \psi(t), \ -\tau \leqslant t \leqslant 0, \end{cases} \tag{5.33}$$

with the solution of the form

$$v(t) = X_{\tau,\alpha,1}^{A,B}(t+\tau)\psi(-\tau) + \int_{-\tau}^{0} X_{\tau,\alpha,\alpha}^{A,B}(t-s)[(^{C}D_{-\tau+}^{\alpha}\psi)(s) - A\psi(s)]ds$$

$$+ \int_{0}^{t} X_{\tau,\alpha,\alpha}^{A,B}(t-s)Cu(s)ds.$$

Next, we introduce the following representation of the fractional order delay Grammian matrix, which is an extension of the classical Grammian matrix of linear differential systems:

$$W_{\tau,\alpha}[0,\tau_1] = \int_{0}^{\tau_1} X_{\tau,\alpha,\alpha}^{A,B}(\tau_1 - s)CC^{\top} X_{\tau,\alpha,\alpha}^{A^{\top},B^{\top}}(\tau_1 - s)ds.$$

Theorem 5.13. *System (5.33) is relatively controllable if and only if $W_{\tau,\alpha}[0,\tau_1]$ is nonsingular.*

Proof. If $W_{\tau,\alpha}[0,\tau_1]$ is a nonsingular matrix, its inverse $W_{\tau,\alpha}[0,\tau_1]$ is well defined. One can select a control function as follows:

$$u(t) = C^{\top} X_{\tau,\alpha,\alpha}^{A^{\top},B^{\top}}(\tau_1 - t)W_{\tau,\alpha}^{-1}[0,\tau_1]\eta,$$

where

$$\eta = v_{\tau_1} - X_{\tau,\alpha,1}^{A,B}(\tau_1+\tau)\psi(-\tau) - \int_{-\tau}^{0} X_{\tau,\alpha,\alpha}^{A,B}(\tau_1-s)[(^{C}D_{-\tau+}^{\alpha}\psi)(s) - A\psi(s)]ds.$$

Then

$$v(\tau_1) = X_{\tau,\alpha,1}^{A,B}(\tau_1+\tau)\psi(-\tau) + \int_{-\tau}^{0} X_{\tau,\alpha,\alpha}^{A,B}(\tau_1-s)[(^{C}D_{-\tau+}^{\alpha}\psi)(s) - A\psi(s)]ds$$

$$+ \int_{0}^{\tau_1} X_{\tau,\alpha,\alpha}^{A,B}(\tau_1-s)Cu(s)ds$$

$$= X_{\tau,\alpha,1}^{A,B}(\tau_1+\tau)\psi(-\tau) + \int_{-\tau}^{0} X_{\tau,\alpha,\alpha}^{A,B}(\tau_1-s)[(^{C}D_{-\tau+}^{\alpha}\psi)(s) - A\psi(s)]ds$$

$$+ \int_{0}^{\tau_1} X_{\tau,\alpha,\alpha}^{A,B}(\tau_1-s)CC^{\top} X_{\tau,\alpha,\alpha}^{A^{\top},B^{\top}}(\tau_1-s)W_{\tau,\alpha}^{-1}[0,\tau_1]\eta ds$$

$$= v_{\tau_1}.$$

Next, we prove the necessity result by contradiction. Suppose that $W_{\tau,\alpha}[0,\tau_1]$ is singular, i.e., there exists at least one nonzero state $\tilde{v} \in \mathbb{R}^n$ such that

$$\tilde{v}^{\top} W_{\tau,\alpha}[0,\tau_1]\tilde{v} = 0.$$

Further, one obtains

$$0 = \tilde{v}^{\top} W_{\tau,\alpha}[0,\tau_1]\tilde{v} = \int_{0}^{\tau_1} \tilde{v}^{\top} X_{\tau,\alpha,\alpha}^{A,B}(\tau_1-s)CC^{\top} X_{\tau,\alpha,\alpha}^{A^{\top},B^{\top}}(\tau_1-s)\tilde{v}ds$$

$$= \int_0^{\tau_1} \| \tilde{v}^\top X_{\tau,\alpha,\alpha}^{A,B}(\tau_1 - s)C \|^2 ds,$$

which implies that

$$\tilde{v}^\top X_{\tau,\alpha,\alpha}^{A,B}(\tau_1 - s)C = \mathbf{0}^\top, \ \forall s \in J.$$

Since system (5.33) is relatively controllable, according to Definition 5.11, there exists a control $u_1(t)$ that drives the initial state to zero state at τ_1, namely,

$$v(\tau_1) = X_{\tau,\alpha,1}^{A,B}(\tau_1 + \tau)\psi(-\tau) + \int_{-\tau}^0 X_{\tau,\alpha,\alpha}^{A,B}(\tau_1 - s)[(^C D_{-\tau^+}^\alpha \psi)(s) - A\psi(s)]ds$$

$$+ \int_0^{\tau_1} X_{\tau,\alpha,\alpha}^{A,B}(\tau_1 - s)Cu_1(s)ds = \mathbf{0}. \tag{5.34}$$

Similarly, there also exists a control $u_2(t)$ that drives the initial state to state $\tilde{v}$ at τ_1, i.e.,

$$v(\tau_1) = X_{\tau,\alpha,1}^{A,B}(\tau_1 + \tau)\psi(-\tau) + \int_{-\tau}^0 X_{\tau,\alpha,\alpha}^{A,B}(\tau_1 - s)[(^C D_{-\tau^+}^\alpha \psi)(s) - A\psi(s)]ds$$

$$+ \int_0^{\tau_1} X_{\tau,\alpha,\alpha}^{A,B}(\tau_1 - s)Cu_2(s)ds = \tilde{v}. \tag{5.35}$$

Then by (5.34) and (5.35), we have

$$\tilde{v} = \int_0^{\tau_1} X_{\tau,\alpha,\alpha}^{A,B}(\tau_1 - s)C[u_2(s) - u_1(s)]ds. \tag{5.36}$$

Multiplying both sides of (5.36) by $\tilde{v}^\top$, we obtain

$$\tilde{v}^\top \tilde{v} = \int_0^{\tau_1} \tilde{v}^\top X_{\tau,\alpha,\alpha}^{A,B}(\tau_1 - s)C[u_2(s) - u_1(s)]ds = 0,$$

and sequentially, $\tilde{v} = \mathbf{0}$, which conflicts with $\tilde{v}$ being nonzero. Thus, the fractional delay Grammian matrix $W_{\tau,\alpha}[0, \tau_1]$ is nonsingular. The proof is finished. $\qquad\square$

5.1.2.2 Relative controllability results for semilinear systems

For $f(t, v(t)) \neq \mathbf{0}$, $t \in J$, system (5.31) is a semilinear fractional delay controlled system.

We assume the following hypotheses:

(H_1): The operator $W : L^2(J, \mathbb{R}^n) \to \mathbb{R}^n$ defined by

$$Wu = \int_0^{\tau_1} X_{\tau,\alpha,\alpha}^{A,B}(\tau_1 - s)Cu(s)ds$$

has an inverse operator W^{-1} which takes values in $L^2(J, \mathbb{R}^n)/ker\,W$. Then we set

$$\mathfrak{M} = \|W^{-1}\|_{L_b(\mathbb{R}^n, L^2(J,\mathbb{R}^n)/ker\,W)}.$$

From [87, Remark 3.3], we know

$$\mathfrak{M} = \sqrt{\|W_{\tau,\alpha}^{-1}[0, \tau_1]\|}. \tag{5.37}$$

(H_2): The function $f : J \times \mathbb{R}^n \to \mathbb{R}^n$ is continuous and $L_f(\cdot) \in L^\infty(J, \mathbb{R}^+)$ such that

$$\|f(t, v) - f(t, \mu)\| \leqslant L_f(t)\|v - \mu\|, \ t \in J, \ v, \mu \in \mathbb{R}^n.$$

Theorem 5.14. *Suppose that* $1 > \alpha \geqslant 0.5$, (H_1), *and* (H_2) *are satisfied. Then system (5.31) is relatively controllable provided that*

$$M_2\left(1 + \frac{1}{\lambda}[E_\alpha(\lambda\tau_1^\alpha) - 1]\|C\|\mathfrak{M}\right) < 1, \tag{5.38}$$

where $M_2 = \frac{1}{\lambda}[E_\alpha(\lambda\tau_1^\alpha) - 1]\|L_f\|_{L^\infty(J,\mathbb{R}^+)}$ *and* $\lambda = \|A\| + \|B\|$.

Proof. Using hypothesis (H_1) for an arbitrary function $v(\cdot) \in \mathcal{C}$, it is suitable to define the following control function $u_v(t)$:

$$u_v(t) = W^{-1}\left(v_{\tau_1} - X_{\tau,\alpha,1}^{A,B}(\tau_1 + \tau)\psi(-\tau) - \int_{-\tau}^0 X_{\tau,\alpha,\alpha}^{A,B}(\tau_1 - s)[(^C D_{-\tau^+}^\alpha \psi)(s)\right.$$
$$\left. - A\psi(s)]ds - \int_0^{\tau_1} X_{\tau,\alpha,\alpha}^{A,B}(\tau_1 - s)f(s, v(s))ds\right)(t), \ t \in J.$$

We show that, using this control, the operator $\mathcal{P} : \mathcal{C} \to \mathcal{C}$ defined by

$$(\mathcal{F}v)(t) = X_{\tau,\alpha,1}^{A,B}(t + \tau)\psi(-\tau) + \int_{-\tau}^0 X_{\tau,\alpha,\alpha}^{A,B}(t - s)[(^C D_{-\tau^+}^\alpha \psi)(s) - A\psi(s)]ds$$
$$+ \int_0^t X_{\tau,\alpha,\alpha}^{A,B}(t - s)f(s, v(s))ds + \int_0^t X_{\tau,\alpha,\alpha}^{A,B}(t - s)Cu_v(s)ds$$

has a fixed point v, which is a mild solution of (5.31).

We check that $(\mathcal{F}v)(\tau_1) = v_{\tau_1}$ and $(\mathcal{F}v)(0) = v_0$, which means that u_v steers system (5.31) from v_0 to v_{τ_1} in finite time τ_1. This implies system (5.31) is relatively controllable on J.

For each positive number r, let $\mathcal{B}_r = \{v \in \mathcal{C} : \|v\| \leqslant r\}$. Then, for each r, $\mathcal{B}_r$ is obviously a bound, closed, and convex set of $\mathcal{C}$. Set $R_f = \max_{t \in J} \|f(t, 0)\|$.

In order to make the following process clear we divide it into several steps.

Step 1. We claim that there exists a positive number r such that $\mathcal{F}(\mathcal{B}_r) \subseteq \mathcal{B}_r$. In light of (H_2) and the Hölder inequality, we obtain

$$\int_0^t (t-s)^{\alpha-1} E_{\alpha,\alpha}(\lambda(t-s)^\alpha) L_f(s) ds$$

$$\leqslant \int_0^t (t-s)^{\alpha-1} E_{\alpha,\alpha}(\lambda(t-s)^\alpha) ds \|L_f\|_{L^\infty(J,\mathbb{R}^+)}$$

$$\leqslant \frac{1}{\lambda}[E_\alpha(\lambda t^\alpha) - 1] \|L_f\|_{L^\infty(J,\mathbb{R}^+)}$$

and

$$\int_0^t (t-s)^{\alpha-1} E_{\alpha,\alpha}(\lambda(t-s)^\alpha) \|f(s,0)\| ds$$

$$\leq R_f \int_0^t (t-s)^{\alpha-1} E_{\alpha,\alpha}(\lambda(t-s)^\alpha) ds = \frac{R_f}{\lambda}[E_\alpha(\lambda t^\alpha) - 1].$$

Taking into account Lemma 5.9, using (H_1) and (H_2), we have

$$\|u_v(t)\|$$

$$\leq \|W^{-1}\|_{L_b(\mathbb{R}^n, L^2(J,\mathbb{R}^n)/ker W)} \left(\|v_{\tau_1}\| + \|X_{\tau,\alpha,1}^{A,B}(\tau_1 + \tau)\| \|\psi(-\tau)\| \right.$$

$$+ \int_{-\tau}^0 \|X_{\tau,\alpha,\alpha}^{A,B}(\tau_1 - s)\| \|(^C D_{-\tau^+}^\alpha \psi)(s) - A\psi(s)\| ds$$

$$\left. + \int_0^{\tau_1} \|X_{\tau,\alpha,\alpha}^{A,B}(\tau_1 - s)\| \|f(s, v(s))\| ds \right)$$

$$\leqslant \mathfrak{M} \left[\|v_{\tau_1}\| + E_\alpha(\lambda(\tau_1 + \tau)^\alpha) \|\psi(-\tau)\| \right.$$

$$+ \int_{-\tau}^0 (\tau_1 - s)^{\alpha-1} E_{\alpha,\alpha}(\lambda(\tau_1 - s)^\alpha) \|(^C D_{-\tau^+}^\alpha \psi)(s) - A\psi(s)\| ds$$

$$\left. + \int_0^{\tau_1} (\tau_1 - s)^{\alpha-1} E_{\alpha,\alpha}(\lambda(\tau_1 - s)^\alpha)(\|f(s, v(s)) - f(s,0)\| + \|f(s,0)\|) ds \right]$$

$$\leqslant \mathfrak{M} \left[\|v_{\tau_1}\| + E_\alpha(\lambda(\tau_1 + \tau)^\alpha) \|\psi(-\tau)\| \right.$$

$$+ \int_{-\tau}^0 (\tau_1 - s)^{\alpha-1} E_{\alpha,\alpha}(\lambda(\tau_1 - s)^\alpha) \|(^C D_{-\tau^+}^\alpha \psi)(s) - A\psi(s)\| ds$$

$$\left. + \int_0^{\tau_1} (\tau_1 - s)^{\alpha-1} E_{\alpha,\alpha}(\lambda(\tau_1 - s)^\alpha)(L_f(s)\|v(s)\| + \|f(s,0)\|) ds \right]$$

$$\leqslant \mathfrak{M} \left[\|v_{\tau_1}\| + E_\alpha(\lambda(\tau_1 + \tau)^\alpha) \|\psi(-\tau)\| \right.$$

$$+ \int_{-\tau}^{0} (\tau_1 - s)^{\alpha-1} E_{\alpha,\alpha}(\lambda(\tau_1 - s)^{\alpha}) \|(^C D_{-\tau+}^{\alpha} \psi)(s) - A\psi(s)\| ds$$

$$+ \int_{0}^{\tau_1} (\tau_1 - s)^{\alpha-1} E_{\alpha,\alpha}(\lambda(\tau_1 - s)^{\alpha}) L_f(s) \|v(s)\| ds$$

$$+ \int_{0}^{\tau_1} (\tau_1 - s)^{\alpha-1} E_{\alpha,\alpha}(\lambda(\tau_1 - s)^{\alpha}) \|f(s,0)\| ds \Bigg]$$

$$\leqslant \mathfrak{M} \Bigg[\|v_{\tau_1}\| + E_{\alpha}(\lambda(\tau_1 + \tau)^{\alpha}) \|\psi(-\tau)\|$$

$$+ \int_{-\tau}^{0} (\tau_1 - s)^{\alpha-1} E_{\alpha,\alpha}(\lambda(\tau_1 - s)^{\alpha}) \|(^C D_{-\tau+}^{\alpha} \psi)(s) - A\psi(s)\| ds$$

$$+ \frac{1}{\lambda}[E_{\alpha}(\lambda \tau_1^{\alpha}) - 1]\|L_f\|_{L^{\infty}(J,\mathbb{R}^+)}\|v\| + \frac{R_f}{\lambda}[E_{\alpha}(\lambda \tau_1^{\alpha}) - 1] \Bigg]$$

$$\leqslant \mathfrak{M}\|v_{\tau_1}\| + \mathfrak{M} M_1 + \mathfrak{M} M_2 \|v\|,$$

where

$$M_1 = E_{\alpha}(\lambda(\tau_1 + \tau)^{\alpha}) \|\psi(-\tau)\| + \frac{R_f}{\lambda}[E_{\alpha}(\lambda \tau_1^{\alpha}) - 1]$$

$$+ \int_{-\tau}^{0} (\tau_1 - s)^{\alpha-1} E_{\alpha,\alpha}(\lambda(\tau_1 - s)^{\alpha}) \|(^C D_{-\tau+}^{\alpha} \psi)(s) - A\psi(s)\| ds,$$

and M_2 is defined in the above.

Applying [108, Lemma 2.6], combining conditions (H_1) and (H_2), we obtain

$$\|(\mathcal{F}v)(t)\|$$

$$\leqslant \|X_{\tau,\alpha,1}^{A,B}(t + \tau)\| \|\psi(-\tau)\|$$

$$+ \int_{-\tau}^{0} \|X_{\tau,\alpha,\alpha}^{A,B}(t - s)\| \|(^C D_{-\tau+}^{\alpha} \psi)(s) - A\psi(s)\| ds$$

$$+ \int_{0}^{t} \|X_{\tau,\alpha,\alpha}^{A,B}(t - s)\| \|f(s, v(s))\| ds + \int_{0}^{t} \|X_{\tau,\alpha,\alpha}^{A,B}(t - s)\| \|C\| \|u_v(s)\| ds$$

$$\leqslant E_{\alpha}(\lambda(t + \tau)^{\alpha}) \|\psi(-\tau)\|$$

$$+ \int_{-\tau}^{0} (t - s)^{\alpha-1} E_{\alpha,\alpha}(\lambda(t - s)^{\alpha}) \|(^C D_{-\tau+}^{\alpha} \psi)(s) - A\psi(s)\| ds$$

$$+ \int_{0}^{t} (t - s)^{\alpha-1} E_{\alpha,\alpha}(\lambda(t - s)^{\alpha}) \|f(s, v(s))\| ds$$

$$+ \int_{0}^{t} (t - s)^{\alpha-1} E_{\alpha,\alpha}(\lambda(t - s)^{\alpha}) \|C\| \|u_v(s)\| ds$$

$$\leqslant E_{\alpha}(\lambda(\tau_1 + \tau)^{\alpha}) \|\psi(-\tau)\|$$

$$+ \int_{-\tau}^{0} (\tau_1 - s)^{\alpha-1} E_{\alpha,\alpha}(\lambda(\tau_1 - s)^{\alpha}) \|(^C D_{-\tau^+}^{\alpha} \psi)(s) - A\psi(s)\| ds$$

$$+ \int_{0}^{\tau_1} (\tau_1 - s)^{\alpha-1} E_{\alpha,\alpha}(\lambda(\tau_1 - s)^{\alpha})(L_f(s)\|v(s)\| + \|f(s,0)\|) ds$$

$$+ \int_{0}^{\tau_1} (\tau_1 - s)^{\alpha-1} E_{\alpha,\alpha}(\lambda(\tau_1 - s)^{\alpha}) \|C\|[\mathfrak{M}\|v_{\tau_1}\| + \mathfrak{M} M_1 + \mathfrak{M} M_2 \|v\|] ds$$

$$\leqslant M_1 + M_2\|v\|_C + \frac{1}{\lambda}[E_{\alpha}(\lambda\tau_1^{\alpha}) - 1]\|C\|[\mathfrak{M}\|v_{\tau_1}\| + \mathfrak{M} M_1 + \mathfrak{M} M_2\|v\|]$$

$$= \left(1 + \frac{1}{\lambda}[E_{\alpha}(\lambda\tau_1^{\alpha}) - 1]\|C\|\mathfrak{M}\right) M_1 + \frac{1}{\lambda}[E_{\alpha}(\lambda\tau_1^{\alpha}) - 1]\|C\|\mathfrak{M}\|v_{\tau_1}\|$$

$$+ M_2 \left(1 + \frac{1}{\lambda}[E_{\alpha}(\lambda\tau_1^{\alpha}) - 1]\|C\|\mathfrak{M}\right)\|v\|$$

$$\leqslant \left(1 + \frac{1}{\lambda}[E_{\alpha}(\lambda\tau_1^{\alpha}) - 1]\|C\|\mathfrak{M}\right) M_1 + \frac{1}{\lambda}[E_{\alpha}(\lambda\tau_1^{\alpha}) - 1]\|C\|\mathfrak{M}\|v_{\tau_1}\|$$

$$+ M_2 \left(1 + \frac{1}{\lambda}[E_{\alpha}(\lambda\tau_1^{\alpha}) - 1]\|C\|\mathfrak{M}\right) r = r,$$

for

$$r = \frac{\left(1 + \frac{1}{\lambda}[E_{\alpha}(\lambda\tau_1^{\alpha}) - 1]\|C\|\mathfrak{M}\right) M_1 + \frac{1}{\lambda}[E_{\alpha}(\lambda\tau_1^{\alpha}) - 1]\|C\|\mathfrak{M}\|v_{\tau_1}\|}{1 - M_2\left(1 + \frac{1}{\lambda}[E_{\alpha}(\lambda\tau_1^{\alpha}) - 1]\|C\|\mathfrak{M}\right)}.$$

Hence, we obtain $\mathcal{F}(\mathcal{B}_r) \subseteq \mathcal{B}_r$ for such an r.

Now, we define operators $\mathcal{F}_1$ and $\mathcal{F}_2$ on $\mathcal{B}_r$ as

$$(\mathcal{F}_1 v)(t) = X_{\tau,\alpha,1}^{A,B}(t + \tau)\psi(-\tau) + \int_{-\tau}^{0} X_{\tau,\alpha,\alpha}^{A,B}(t - s)[(^C D_{-\tau^+}^{\alpha} \psi)(s) - A\psi(s)] ds$$

$$+ \int_{0}^{t} X_{\tau,\alpha,\alpha}^{A,B}(t - s) C u_v(s) ds$$

and

$$(\mathcal{F}_2 v)(t) = \int_{0}^{t} X_{\tau,\alpha,\alpha}^{A,B}(t - s) f(s, v(s)) ds,$$

for $t \in J$, respectively.

Step 2. We claim that $\mathcal{F}_1$ is a contraction mapping.

Let $v, \gamma \in \mathcal{B}_r$. In view of (H_1) and (H_2), for each $t \in J$, we have

$$\|u_v(t) - u_\gamma(t)\| \leqslant \mathfrak{M} \int_{0}^{\tau_1} \|X_{\tau,\alpha,\alpha}^{A,B}(\tau_1 - s)\| \|f(s, v(s)) - f(s, \gamma(s))\| ds$$

$$\leqslant \mathfrak{M} \int_0^{\tau_1} (\tau_1 - s)^{\alpha-1} E_{\alpha,\alpha}(\lambda(\tau_1 - s)^\alpha) L_f(s)(\|v(s) - \gamma(s)\|)ds$$

$$\leqslant \mathfrak{M} \int_0^{\tau_1} (\tau_1 - s)^{\alpha-1} E_{\alpha,\alpha}(\lambda(\tau_1 - s)^\alpha) L_f(s)ds \|v - \gamma\|$$

$$\leqslant \frac{\mathfrak{M}}{\lambda}[E_\alpha(\lambda\tau_1^\alpha) - 1]\|L_f\|_{L^\infty(J,\mathbb{R}^+)}\|v - \gamma\|$$

$$\leqslant \mathfrak{M}M_2\|v - \gamma\|.$$

Thus,

$$\|(\mathcal{F}_1 v)(t) - (\mathcal{F}_1 \gamma)(t)\|$$

$$\leqslant \int_0^t \|X_{\tau,\alpha,\alpha}^{A,B}(t - s)\|\|C\|\|u_v(s) - u_\gamma(s)\|ds$$

$$\leqslant \int_0^t (t - s)^{\alpha-1} E_{\alpha,\alpha}(\lambda(t - s)^\alpha)ds \|C\|\mathfrak{M}M_2\|v - \gamma\|$$

$$= \frac{1}{\lambda}[E_\alpha(\lambda t^\alpha) - 1]\|C\|\mathfrak{M}M_2\|v - \gamma\|$$

$$\leqslant \frac{\|C\|\mathfrak{M}M_2}{\lambda}[E_\alpha(\lambda\tau_1^\alpha) - 1]\|v - \gamma\|.$$

So we obtain

$$\|\mathcal{F}_1 v - \mathcal{F}_1 \gamma\| \leqslant P\|v - \gamma\|,$$

where $P = \frac{\|C\|\mathfrak{M}M_2}{\lambda}[E_\alpha(\lambda\tau_1^\alpha) - 1]$.

In view of (5.38), we conclude that $P < 1$, which implies $\mathcal{F}_1$ is a contraction.

Step 3. We claim that $\mathcal{F}_2$ is a compact and continuous operator.

Let $v_n \in \mathcal{B}_r$ with $v_n \to v$ in $\mathcal{B}_r$. Using (H_2), we have $f(s, v_n(s)) \to f(s, v(s))$ in $\mathcal{C}$ and thus, using the dominated convergence theorem,

$$\|(\mathcal{F}_2 v_n)(t) - (\mathcal{F}_2 v)(t)\|$$

$$\leqslant \int_0^t \|X_{\tau,\alpha,\alpha}^{A,B}(t - s)\|\|f(s, v_n(s)) - f(s, v(s))\|ds$$

$$\leqslant \int_0^t (t - s)^{\alpha-1} E_{\alpha,\alpha}(\lambda(t - s)^\alpha)\|f(s, v_n(s)) - f(s, v(s))\|ds$$

$$\to 0 \text{ as } n \to \infty,$$

which implies that $\mathcal{F}_2$ is continuous on $\mathcal{B}_r$.

To check the compactness of $\mathcal{F}_2$, we prove that $\mathcal{F}_2(\mathcal{B}_r) \in \mathcal{C}$ is equicontinuous and uniformly bounded.

In fact, for any $v \in \mathcal{B}_r, 0 < t < t + h \leqslant \tau_1$,

$$(\mathcal{F}_2 v)(t + h) - (\mathcal{F}_2 v)(t)$$

$$
= \int_0^{t+h} X_{\tau,\alpha,\alpha}^{A,B}(t+h-s)f(s,v(s))ds - \int_0^t X_{\tau,\alpha,\alpha}^{A,B}(t-s)f(s,v(s))ds
$$

$$
= \int_t^{t+h} X_{\tau,\alpha,\alpha}^{A,B}(t+h-s)f(s,v(s))ds
$$

$$
+ \int_0^t (X_{\tau,\alpha,\alpha}^{A,B}(t+h-s) - X_{\tau,\alpha,\alpha}^{A,B}(t-s))f(s,v(s))ds.
$$

Denote

$$
W_1 = \int_t^{t+h} X_{\tau,\alpha,\alpha}^{A,B}(t+h-s)f(s,v(s))ds,
$$

$$
W_2 = \int_0^t (X_{\tau,\alpha,\alpha}^{A,B}(t+h-s) - X_{\tau,\alpha,\alpha}^{A,B}(t-s))f(s,v(s))ds.
$$

From the above, we derive

$$
\|(\mathcal{F}_2 v)(t+h) - (\mathcal{F}_2 v)(t)\| \leqslant \|W_1\| + \|W_2\|.
$$

Now, we only need to check $\|W_i\| \to 0$ as $h \to 0$, $i = 1, 2$. Clearly,

$$
\|W_1\| \leqslant \int_t^{t+h} \|X_{\tau,\alpha,\alpha}^{A,B}(t+h-s)\|\|f(s,v(s))\|ds
$$

$$
\leqslant \int_t^{t+h} (t+h-s)^{\alpha-1} E_{\alpha,\alpha}(\lambda(t+h-s)^{\alpha})(L_f(s)\|v(s)\| + \|f(s,0)\|)ds
$$

$$
\leqslant \int_t^{t+h} (t+h-s)^{\alpha-1} E_{\alpha,\alpha}(\lambda(t+h-s)^{\alpha})ds \|L_f\|_{L^{\infty}(J,\mathbb{R}^+)}\|v\|
$$

$$
+ R_f \int_t^{t+h} (t+h-s)^{\alpha-1} E_{\alpha,\alpha}(\lambda(t+h-s)^{\alpha})ds
$$

$$
\leqslant \frac{1}{\lambda}[E_{\alpha}(\lambda h^{\alpha}) - 1]\|L_f\|_{L^{\infty}(J,\mathbb{R}^+)}\|v\| + \frac{R_f}{\lambda}[E_{\alpha}(\lambda h^{\alpha}) - 1]
$$

$$
\to 0 \ \text{as} \ h \to 0
$$

and

$$
\|W_2\| \leqslant \int_0^t \|X_{\tau,\alpha,\alpha}^{A,B}(t+h-s) - X_{\tau,\alpha,\alpha}^{A,B}(t-s)\|\|f(s,v(s))\|ds
$$

$$
\leqslant \int_0^t \|X_{\tau,\alpha,\alpha}^{A,B}(t+h-s) - X_{\tau,\alpha,\alpha}^{A,B}(t-s)\|L_f(s)\|v(s)\|ds
$$

$$
+ \int_0^t \|X_{\tau,\alpha,\alpha}^{A,B}(t+h-s) - X_{\tau,\alpha,\alpha}^{A,B}(t-s)\|\|f(s,0)\|ds
$$

$$
\leqslant \int_0^t \|X_{\tau,\alpha,\alpha}^{A,B}(t+h-s) - X_{\tau,\alpha,\alpha}^{A,B}(t-s)\|ds \|L_f\|_{L^{\infty}(J,\mathbb{R}^+)}\|v\|
$$

$$+ R_f \int_0^t \| X_{\tau,\alpha,\alpha}^{A,B}(t + h - s) - X_{\tau,\alpha,\alpha}^{A,B}(t - s) \| ds$$

$$\to 0 \ \text{as} \ h \to 0.$$

As a result, we immediately obtain

$$\|(\mathcal{F}_2 v)(t + h) - (\mathcal{F}_2 v)(t)\| \to 0 \ \text{as} \ h \to 0,$$

for all $v \in \mathcal{B}_r$. Therefore, $\mathcal{F}_2(\mathcal{B}_r)$ is equicontinuous in $\mathcal{C}$.

Next, repeating the above computations, we have

$$\|(\mathcal{F}_2 v)(t)\| \leqslant \int_0^t \| X_{\tau,\alpha,\alpha}^{A,B}(t - s) \| \| f(s, v(s)) \| ds$$

$$\leqslant \int_0^t (t - s)^{\alpha - 1} E_{\alpha,\alpha}(\lambda (t - s)^{\alpha}) \| f(s, v(s)) \| ds$$

$$\leqslant \int_0^{\tau_1} (\tau_1 - s)^{\alpha - 1} E_{\alpha,\alpha}(\lambda (\tau_1 - s)^{\alpha})(L_f(s) \| v(s) \| + \| f(s, 0) \|) ds$$

$$\leqslant \frac{1}{\lambda} [E_{\alpha}(\lambda \tau_1^{\alpha}) - 1] \| L_f \|_{L^{\infty}(J, \mathbb{R}^+)} \| r + \frac{R_f}{\lambda} [E_{\alpha}(\lambda \tau_1^{\alpha}) - 1].$$

Hence $\mathcal{F}_2(\mathcal{B}_r)$ is bounded. By the Arzela–Ascoli theorem, $\mathcal{F}_2(\mathcal{B}_r) \subset \mathcal{C}$ is relatively compact in $\mathcal{C}$. Thus, $\mathcal{F}_2$ is a compact and continuous operator. Using Krasnoselskii's fixed point theorem (see [82]), $\mathcal{F}$ has a fixed point v on $\mathcal{B}_r$. Obviously, v is a solution of system (5.31) satisfying $v(\tau_1) = v_{\tau_1}$. The boundary condition $v(t) = \psi(t)$, $-\tau \leqslant t \leqslant 0$ holds by (5.32). The proof is completed. $\square$

5.1.2.3 *Numerical examples and discussion*

Example 5.3. Set $\alpha = 0.5$. Consider the following semilinear fractional delay differential controlled system:

$$\begin{cases} (^C D_{-0.5+}^{\alpha} v)(t) = Av(t) + Bv(t - 0.5) + f(t, v(t)) + Cu(t), \ t \in [0, 1.5], \\ v(t) = \begin{pmatrix} 0.4 \\ 0.5 \end{pmatrix}, \ -0.5 \leqslant t \leqslant 0, \end{cases} \tag{5.39}$$

where we set

$$A = \begin{pmatrix} 0.2 & 0 \\ 0 & 0.2 \end{pmatrix}, \ B = \begin{pmatrix} 0.18 & 0 \\ 0 & 0.2 \end{pmatrix},$$

$$C = \begin{pmatrix} 0.4 & 0.1 \\ 0.2 & 0.3 \end{pmatrix}, \ f(t, v(t)) = \begin{pmatrix} 0.2 t v_1(t) \\ 0.15 t v_2(t) \end{pmatrix}.$$

Clearly $\lambda = \|A\| + \|B\| = 0.4$, $\|C\| = 0.5$.

Now we use (5.37) to estimate $\mathfrak{M}$. For this purpose, we need to obtain $W_{\tau,\alpha}[0,\tau_1]$ and then $W_{\tau,\alpha}[0,\tau_1]^{-1}$. The fractional delay Grammian matrix has the following explicit form:

$$W_{\tau,\alpha}[0,\tau_1] = \int_0^{\tau_1} X_{\tau,\alpha,\alpha}^{A,B}(\tau_1-s)CC^\top X_{\tau,\alpha,\alpha}^{A^\top,B^\top}(\tau_1-s)ds$$

$$= \int_0^{1.5} X_{\tau,\alpha,\alpha}^{A,B}(1.5-s)CC^\top X_{\tau,\alpha,\alpha}^{A^\top,B^\top}(1.5-s)ds$$

$$= S_1 + S_2 + S_3,$$

where

$$S_1 = \int_0^{0.5}\left[\sum_{i=0}^\infty A^i\frac{(1.5-s)^{i\alpha+\alpha-1}}{\Gamma(i\alpha+\alpha)} + \sum_{i=1}^\infty\binom{i}{1}A^{i-1}B\frac{(1-s)^{i\alpha+\alpha-1}}{\Gamma(i\alpha+\alpha)}\right.$$
$$+ \sum_{i=2}^\infty\binom{i}{2}A^{i-2}B^2\frac{(0.5-s)^{i\alpha+\alpha-1}}{\Gamma(i\alpha+\alpha)}\Bigg]$$
$$\times CC^\top\left[\sum_{i=0}^\infty(A^\top)^i\frac{(1.5-s)^{i\alpha+\alpha-1}}{\Gamma(i\alpha+\alpha)} + \sum_{i=1}^\infty\binom{i}{1}(A^\top)^{i-1}B^\top\frac{(1-s)^{i\alpha+\alpha-1}}{\Gamma(i\alpha+\alpha)}\right.$$
$$+ \sum_{i=2}^\infty\binom{i}{2}(A^\top)^{i-2}(B^\top)^2\frac{(0.5-s)^{i\alpha+\alpha-1}}{\Gamma(i\alpha+\alpha)}\Bigg]ds,$$

$$S_2 = \int_{0.5}^1\left[\sum_{i=0}^\infty A^i\frac{(1.5-s)^{i\alpha+\alpha-1}}{\Gamma(i\alpha+\alpha)} + \sum_{i=1}^\infty\binom{i}{1}A^{i-1}B\frac{(1-s)^{i\alpha+\alpha-1}}{\Gamma(i\alpha+\alpha)}\right]$$
$$\times C^\top\left[\sum_{i=0}^\infty(A^\top)^i\frac{(1.5-s)^{i\alpha+\alpha-1}}{\Gamma(i\alpha+\alpha)}\right.$$
$$+ \sum_{i=1}^\infty\binom{i}{1}(A^\top)^{i-1}B^\top\frac{(1-s)^{i\alpha+\alpha-1}}{\Gamma(i\alpha+\alpha)}\Bigg]ds,$$

and

$$S_3 = \int_1^{1.5}\left[\sum_{i=0}^\infty A^i\frac{(1.5-s)^{i\alpha+\alpha-1}}{\Gamma(i\alpha+\alpha)}\right]CC^\top\left[\sum_{i=0}^\infty(A^\top)^i\frac{(1.5-s)^{i\alpha+\alpha-1}}{\Gamma(i\alpha+\alpha)}\right]ds.$$

By calculation,

$$S_1 = \begin{pmatrix} 0.0959 & 0.0641 \\ 0.0641 & 0.0783 \end{pmatrix}, \quad S_2 = \begin{pmatrix} 0.1089 & 0.0720 \\ 0.0720 & 0.0870 \end{pmatrix},$$

$$S_3 = \begin{pmatrix} 0.5270 & 0.3410 \\ 0.3410 & 0.4030 \end{pmatrix}.$$

Therefore, we obtain

$$W_{\tau,\alpha}[0, 1.5] = \begin{pmatrix} 0.7318 & 0.4771 \\ 0.4771 & 0.5683 \end{pmatrix},$$

$$W_{\tau,\alpha}^{-1}[0, 1.5] = \begin{pmatrix} 3.0187 & -2.5343 \\ -2.5343 & 3.8872 \end{pmatrix},$$

and

$$\mathfrak{M} = \sqrt{\|W_{\tau,\alpha}^{-1}[0, 1.5]\|} = \sqrt{1.3529} = 1.1631.$$

Further, for any $v, \mu \in \mathbb{R}^n$,

$$\begin{aligned}
\|f(t, v) - f(t, \mu)\| &= \max\{0.2t|v_1 - \mu_1|, 0.15t|v_2 - \mu_2|\} \\
&\le 0.2t \max\{|v_1 - \mu_1|, |v_2 - \mu_2|\} \\
&= 0.2t\|v - \mu\|.
\end{aligned}$$

Note $L_f(t) = 0.2t$ and $\|L_f\|_{L^\infty(J,\mathbb{R}^+)} = 0.2 \times 1.5 = 0.3$, $E_\alpha(\lambda\tau_1^\alpha) = 1.5056$. As a result

$$M_2 = \frac{1}{\lambda}[E_\alpha(\lambda\tau_1^\alpha) - 1]\|L_f\|_{L^\infty(J,\mathbb{R}^+)} = \frac{1}{0.4} \times (1.5056 - 1) \times 0.3 = 0.3792$$

and

$$M_2\left(1 + \frac{1}{\lambda}[E_\alpha(\lambda\tau_1^\alpha) - 1]\|C\|\mathfrak{M}\right)$$

$$= 0.3792 \times (1 + \frac{1}{0.4} \times (1.5056 - 1) \times 0.5 \times 1.1631) = 0.6579 < 1.$$

Thus all the conditions of Theorem 5.14 are satisfied, so (5.39) is relatively controllable on $[0, 1.5]$.

We can find the control function $u = (4t, 5t)^\top$, which makes the solution of system (5.39) satisfy $v(1.5) = (3.3, 4.8)^\top$.

Example 5.4. Consider the relative controllability of system (5.39) (with $f(t, v(t)) \equiv \mathbf{0}$) on J, where A, B, and C are defined in Example 5.3.

Letting $\alpha = 0.7$, by calculation,

$$S_1 = \begin{pmatrix} 0.1378 & 0.0912 \\ 0.0912 & 0.1101 \end{pmatrix}, \quad S_2 = \begin{pmatrix} 0.1232 & 0.0807 \\ 0.0807 & 0.0964 \end{pmatrix},$$

$$S_3 = \begin{pmatrix} 0.2132 & 0.1379 \\ 0.1379 & 0.1630 \end{pmatrix}.$$

Then

$$W_{0.5,0.7}[0, 1.5] = S_1 + S_2 + S_3 = \begin{pmatrix} 0.4742 & 0.3098 \\ 0.3098 & 0.2228 \end{pmatrix},$$

$W_{0.5,0.7}[0, 1.5]$ is a nonsingular matrix, and

$$W_{0.5,0.7}^{-1}[0, 1.5] = \begin{pmatrix} 23.0267 & -32.0183 \\ -32.0183 & 49.0092 \end{pmatrix}.$$

According to Theorem 5.13, we can know that system (5.39) is relative controllability when $f(\cdot, v(\cdot)) = \mathbf{0}$. We can get

$$\eta = v_{\tau_1} - X_{\tau,\alpha,1}^{A,B}(1.5 + 0.5)\psi(-0.5)$$

$$- \int_{-0.5}^{0} X_{\tau,\alpha,\alpha}^{A,B}(1.5 - s)[(^C D_{-\tau+}^{\alpha}\psi)(s) - A\psi(s)]ds$$

$$= v_{\tau_1} - \begin{pmatrix} 0.6802 \\ 0.8713 \end{pmatrix}.$$

By using the form of the control $u(t)$, we get

$$u(t) = C^{\top} X_{\tau,\alpha,\alpha}^{A^{\top},B^{\top}}(\tau_1 - t)W_{\tau,\alpha}^{-1}[0, \tau_1]\eta$$

$$= C^{\top} X_{\tau,\alpha,\alpha}^{A^{\top},B^{\top}}(1.5 - t)W_{0.5,0.7}^{-1}[0, 1.5]\eta$$

$$= \begin{cases} C^{\top}\left[\sum_{i=0}^{\infty}(A^{\top})^{i}\dfrac{(1.5 - t)^{0.7i-0.3}}{\Gamma(0.7i + 0.7)} + \sum_{i=1}^{\infty}\begin{pmatrix} i \\ 1 \end{pmatrix}(A^{\top})^{i-1}B^{\top}\dfrac{(1 - t)^{0.7i-0.3}}{\Gamma(0.7i + 0.7)} \right. \\ \left. + \sum_{i=2}^{\infty}\begin{pmatrix} i \\ 2 \end{pmatrix}(A^{\top})^{i-2}(B^{\top})^{2}\dfrac{(0.5 - t)^{0.7i-0.3}}{\Gamma(0.7i + 0.7)}\right] W_{0.5,0.7}^{-1}[0, 1.5]\eta, \\ \quad 0 \leqslant t < 0.5, \\ C^{\top}\left[\sum_{i=0}^{\infty}(A^{\top})^{i}\dfrac{(1.5 - t)^{0.7i-0.3}}{\Gamma(0.7i + 0.7)} + \sum_{i=1}^{\infty}\begin{pmatrix} i \\ 1 \end{pmatrix}(A^{\top})^{i-1}B^{\top}\dfrac{(1 - t)^{0.7i-0.3}}{\Gamma(0.7i + 0.7)}\right] \\ \quad \times W_{0.5,0.7}^{-1}[0, 1.5]\eta, \quad 0.5 \leqslant t < 1, \\ C^{\top}\left[\sum_{i=0}^{\infty}(A^{\top})^{i}\dfrac{(1.5 - t)^{0.7i-0.3}}{\Gamma(0.7i + 0.7)}\right] W_{\tau,\alpha}^{-1}[0, 1.5]\eta, \quad 1 \leqslant t < 1.5, \\ C^{\top} W_{0.5,0.7}^{-1}[0, 1.5]\eta, \quad t = 1.5. \end{cases}$$

5.1.2.4 Conclusion

Based on the fractional delay Grammian matrix defined by a two-parameter Mittag-Leffler-type delay matrix function, we show sufficient and necessary

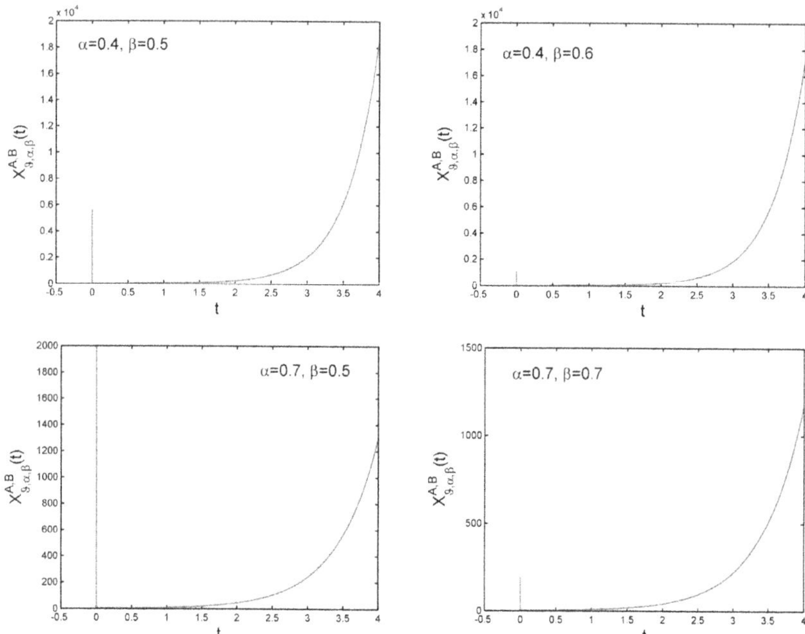

FIGURE 5.1 The trajectories of $X_{\vartheta,\alpha,\beta}^{A,B}(t)$ for $A = 0.8$, $B = 0.9$, and $\vartheta = 0.2$.

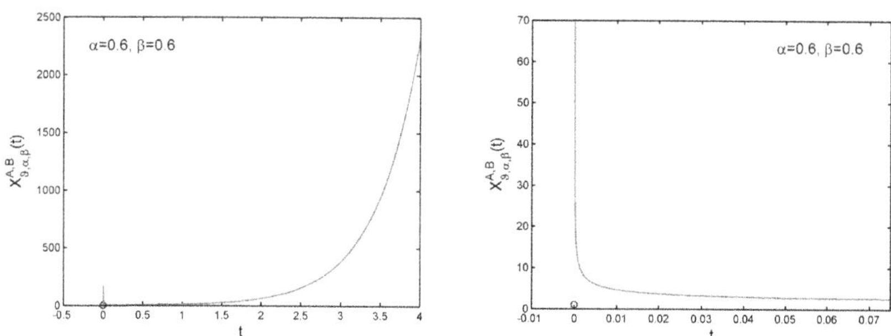

FIGURE 5.2 The trajectories of $X_{\vartheta,\alpha,\beta}^{A,B}(t)$ for $A = 0.8$, $B = 0.9$, $\vartheta = 0.2$, $\alpha = 0.6$, and $\beta = 0.6$. The right is an enlarged view of the left, and $\circ$ is the coordinate $(0, 1)$, that is to say, the value of $X_{0.2,0.6,0.6}^{0.8,0.9}(t)$ at 0 is 1.

conditions to guarantee a linear controlled system is relatively controllable. Next, we present sufficient conditions for controllability results by Krasnoselskii's fixed point theorem. Finally, we use two numerical examples to illustrate our theories (see Figs. 5.1 and 5.2). The results in this part are motivated from [109].

5.2   Finite time stability and controllability for Riemann–Liouville type

5.2.1   Finite time stability for Riemann–Liouville type

In this section, we study the finite time stability of linear fractional delay differential systems given by

$$
\begin{cases}
(^{RL}D^{\alpha}_{-\tau+}y)(x) = By(x - \tau) + f(x), \ \ B \in \mathbb{R}^{n \times n}, \ x \in (0, T], \ \tau > 0, \\
y(x) = \omega(x), \ \omega(x) \in \mathbb{R}^{n}, \ \ -\tau \leq x \leq 0, \\
(\mathbb{I}^{1-\alpha}_{-\tau+}y)(-\tau^{+}) = \omega(-\tau), \ \omega(-\tau) \in \mathbb{R}^{n},
\end{cases}
$$

$$(5.40)$$

where $^{RL}D^{\alpha}_{-\tau+}y$ denotes the Riemann–Liouville derivative of order $\alpha \in (0, 1)$ (see Definition 5.3), $\mathbb{I}^{1-\alpha}_{-\tau+}y$ denotes the Riemann–Liouville fractional integral of order $1 - \alpha$ (see Definition 5.2), $f \in C([-\tau, T], \mathbb{R}^{n})$, and ω is an arbitrary Riemann–Liouville differentiable vector function, i.e., $^{RL}D^{\alpha}_{-\tau+}\omega$ exists.

5.2.1.1   *Representation of solutions for linear systems*

Lemma 5.10. *Letting* $(k - 1)\tau < x \leq k\tau$, $-\tau \leq s \leq t$, *and* $k \in \mathbb{N}^{+}$ *being a fixed number, we have*

$$
\int_{s}^{x} (x - t)^{-\alpha} \mathbb{E}^{B(t-\tau-s)^{\alpha}}_{\tau,\alpha} \, dt = \sum_{i=0}^{k} \int_{i\tau+s}^{x} (x - t)^{-\alpha} B^{i} \frac{(t - i\tau - s)^{(i+1)\alpha-1}}{\Gamma(i\alpha + \alpha)} \, dt.
$$

Proof. The proof is similar to that of Lemma 5.3, so we omit it here.   □

Firstly, we use (5.4) to construct the explicit formula of solutions of

$$
\begin{cases}
(^{RL}D^{\alpha}_{-\tau+}y)(x) = By(x - \tau), \ y(x) \in \mathbb{R}^{n}, \ x \in (0, T], \ \tau > 0, \\
y(x) = \omega(x), \ \omega(x) \in \mathbb{R}^{n}, \ \ -\tau \leq x \leq 0, \\
(\mathbb{I}^{1-\alpha}_{-\tau+}y)(-\tau^{+}) = \omega(-\tau), \ \omega(-\tau) \in \mathbb{R}^{n}.
\end{cases}
$$

$$(5.41)$$

Theorem 5.15. *For the delayed Mittag-Leffler-type matrix* $\mathbb{E}^{B\cdot\alpha}_{\tau,\alpha} : \mathbb{R} \to \mathbb{R}^{n \times n}$, *one has* $(^{RL}D^{\alpha}_{-\tau+}\mathbb{E}^{Bt^{\alpha}}_{\tau,\alpha})(x) = B\mathbb{E}^{B(x-\tau)^{\alpha}}_{\tau,\alpha}$, *i.e.,* $\mathbb{E}^{Bx^{\alpha}}_{\tau,\alpha}$ *is a solution of* $(^{RL}D^{\alpha}_{-\tau+}y)(x) = By(x - \tau)$ *that satisfies initial conditions* $\mathbb{E}^{Bx^{\alpha}}_{\tau,\alpha} = I\frac{(\tau+x)^{\alpha-1}}{\Gamma(\alpha)}$, $-\tau \leq x \leq 0$.

Proof. For $x \in ((k - 1)\tau, k\tau]$ and $k \in \mathbb{N}^{+}$, we adopt mathematical induction to prove our result.

(i) For $k = 1, 0 < x \leq \tau$, we have

$$
y(x) = \mathbb{E}^{Bx^{\alpha}}_{\tau,\alpha} = I\frac{(\tau + x)^{\alpha-1}}{\Gamma(\alpha)} + B\frac{x^{2\alpha-1}}{\Gamma(\alpha + \alpha)}.
$$

$$(5.42)$$

Using Definition 5.3 for $\mathbb{E}_{\tau,\alpha}^{B,\alpha}$, via (5.42) and Lemma 5.1 we obtain

$$
(^{RL}D_{-\tau+}^{\alpha}\mathbb{E}_{\tau,\alpha}^{Bt^{\alpha}})(x)
$$
$$
= \frac{1}{\Gamma(1-\alpha)}\frac{d}{dx}\left(\int_{-\tau}^{0}(x-t)^{-\alpha}y(t)dt + \int_{0}^{x}(x-t)^{-\alpha}y(t)dt\right)
$$
$$
= \frac{1}{\Gamma(1-\alpha)}
$$
$$
\times \frac{d}{dx}\left(\int_{-\tau}^{x}(x-t)^{-\alpha}I\frac{(\tau+t)^{\alpha-1}}{\Gamma(\alpha)}dt + \int_{0}^{x}(x-t)^{-\alpha}B\frac{t^{2\alpha-1}}{\Gamma(\alpha+\alpha)}dt\right)
$$
$$
= \frac{1}{\Gamma(1-\alpha)}\frac{d}{dx}\left(I\frac{\mathbb{B}[1-\alpha,\alpha]}{\Gamma(\alpha)} + \frac{B}{\Gamma(2\alpha)}x^{\alpha}\mathbb{B}[1-\alpha,2\alpha]\right) = B\frac{x^{\alpha-1}}{\Gamma(\alpha)}.
$$

(ii) For $k=2$, $\tau < x \le 2\tau$, we have

$$
y(x) = \mathbb{E}_{\tau,\alpha}^{Bx^{\alpha}} = I\frac{(\tau+x)^{\alpha-1}}{\Gamma(\alpha)} + B\frac{x^{2\alpha-1}}{\Gamma(\alpha+\alpha)} + B^2\frac{(x-\tau)^{3\alpha-1}}{\Gamma(2\alpha+\alpha)}. \tag{5.43}
$$

Using Definition 5.3 for $\mathbb{E}_{\tau,\alpha}^{B,\alpha}$, via (5.43) and Lemma 5.1 we obtain

$$
(^{RL}D_{-\tau+}^{\alpha}\mathbb{E}_{\tau,\alpha}^{Bt^{\alpha}})(x)
$$
$$
= \frac{1}{\Gamma(1-\alpha)}\frac{d}{dx}\left(\int_{-\tau}^{\tau}(x-t)^{-\alpha}y(t)dt + \int_{\tau}^{x}(x-t)^{-\alpha}y(t)dt\right)
$$
$$
= B\frac{x^{\alpha-1}}{\Gamma(\alpha)} + \frac{1}{\Gamma(1-\alpha)}\frac{d}{dx}\int_{\tau}^{x}(x-t)^{-\alpha}B^2\frac{(t-\tau)^{3\alpha-1}}{\Gamma(2\alpha+\alpha)}dt
$$
$$
= B\frac{x^{\alpha-1}}{\Gamma(\alpha)} + B^2\frac{(x-\tau)^{2\alpha-1}}{\Gamma(2\alpha)}.
$$

(iii) Supposing $k=n$, $(n-1)\tau < x \le n\tau$ and $n \in \mathbb{N}^+$, the following relation holds:

$$
(^{RL}D_{-\tau+}^{\alpha}\mathbb{E}_{\tau,\alpha}^{Bt^{\alpha}})(x) = B\frac{x^{\alpha-1}}{\Gamma(\alpha)} + B^2\frac{(x-\tau)^{2\alpha-1}}{\Gamma(2\alpha)} + \cdots + B^n\frac{(x-(n-1)\tau)^{n\alpha-1}}{\Gamma(n\alpha)}.
$$

Next, for $k=n+1$, $n\tau < x \le (n+1)\tau$, we have

$$
y(x) = \mathbb{E}_{\tau,\alpha}^{Bx^{\alpha}}
$$
$$
= I\frac{(\tau+x)^{\alpha-1}}{\Gamma(\alpha)} + B\frac{x^{2\alpha-1}}{\Gamma(\alpha+\alpha)} + \cdots + B^{n+1}\frac{(x-n\tau)^{(n+2)\alpha-1}}{\Gamma((n+1)\alpha+\alpha)}. \tag{5.44}
$$

By using Definition 5.3 for $\mathbb{E}_{\tau,\alpha}^{B,\alpha}$, via (5.44) and Lemma 5.1 we obtain

$$(^{RL}D^{\alpha}_{-\tau^+}\mathbb{E}^{Bt^{\alpha}}_{\tau,\alpha})(x)$$

$$= \frac{1}{\Gamma(1-\alpha)}\frac{d}{dx}\left(\int_{-\tau}^{0}(x-t)^{-\alpha}y(t)dt + \cdots + \int_{n\tau}^{x}(x-t)^{-\alpha}y(t)dt\right)$$

$$= B\frac{x^{\alpha-1}}{\Gamma(\alpha)} + B^2\frac{(x-\tau)^{2\alpha-1}}{\Gamma(2\alpha)} + \cdots + B^n\frac{(x-(n-1)\tau)^{n\alpha-1}}{\Gamma(n\alpha)}$$

$$+ \frac{1}{\Gamma(1-\alpha)}\frac{d}{dx}\left(\frac{B^{n+1}}{\Gamma((n+1)\alpha+\alpha)}\int_{n\tau}^{x}(x-t)^{-\alpha}(t-n\tau)^{(n+2)\alpha-1}dt\right)$$

$$= B\frac{x^{\alpha-1}}{\Gamma(\alpha)} + B^2\frac{(x-\tau)^{2\alpha-1}}{\Gamma(2\alpha)} + \cdots + B^{n+1}\frac{(x-n\tau)^{(n+1)\alpha-1}}{\Gamma((n+1)\alpha)}.$$

Therefore, for any $(k-1)\tau < x \leq k\tau$ and $k \in \mathbb{N}^+$, by mathematical induction, we have

$$(^{RL}D^{\alpha}_{-\tau^+}\mathbb{E}^{Bt^{\alpha}}_{\tau,\alpha})(x)$$

$$= B\left(\frac{x^{\alpha-1}}{\Gamma(\alpha)} + B\frac{(x-\tau)^{2\alpha-1}}{\Gamma(2\alpha)} + \cdots + B^{k-1}\frac{(x-(k-1)\tau)^{k\alpha-1}}{\Gamma(k\alpha)}\right)$$

$$= B\mathbb{E}^{B(x-\tau)^{\alpha}}_{\tau,\alpha}.$$

The proof is completed. $\square$

Theorem 5.16. *Let $(n-1)\tau < x \leq n\tau$ for all $n \in \{0,1,2,\cdots,k^*\}$. A solution $y \in C(X,\mathbb{R}^n)$ of (5.41) can be expressed by the following formula:*

$$y(x) = \mathbb{E}^{Bx^{\alpha}}_{\tau,\alpha}\,\omega(-\tau) + \int_{-\tau}^{0}\mathbb{E}^{B(x-\tau-s)^{\alpha}}_{\tau,\alpha}\,(^{RL}D^{\alpha}_{-\tau^+}\omega)(s)ds,$$

where either $X = ((n-1)\tau, n\tau]$ for $0 < \alpha < 1/(n+1)$ or $X = [(n-1)\tau, n\tau]$ for $\alpha \geq 1/(n+1)$.

Proof. Let matrix $Y_0(x) = \mathbb{E}^{Bx^{\alpha}}_{\tau,\alpha}$ satisfy Theorem 5.15. We should search any solution of (5.41) satisfying initial conditions $y(x) = \omega(x)$, $-\tau \leq x \leq 0$, in the form

$$y(x) = Y_0(x)C + \int_{-\tau}^{0}\mathbb{E}^{B(x-\tau-s)^{\alpha}}_{\tau,\alpha}z(s)ds, \tag{5.45}$$

where C is an unknown vector constant and $z(\cdot)$ is an unknown Riemann–Liouville differentiable vector function. According to the matrix $Y_0(x)$ is a solution of (5.41). Therefore, we choose C satisfying $(\mathbb{I}^{1-\alpha}_{-\tau^+}y)(-\tau^+) = \omega(-\tau)$.

Let us assume $x = -\tau$. Using (5.4), we obtain $\mathbb{E}^{B(-2\tau-s)^{\alpha}}_{\tau,\alpha} = \Theta$ with $-\tau \leq s \leq 0$. For $-\tau < x \leq 0$, we have

$$\omega(-\tau) = (\mathbb{I}^{1-\alpha}_{-\tau^+}y)(-\tau^+) = \lim_{x\to-\tau^+}(\mathbb{I}^{1-\alpha}_{-\tau^+}y)(x)$$

$$
\begin{aligned}
&= \lim_{x \to -\tau^+} \left(\frac{1}{\Gamma(1-\alpha)} \int_{-\tau}^{x} (x-t)^{-\alpha} Y_0(t) C \, dt \right) \\
&= \lim_{x \to -\tau^+} \frac{C}{\Gamma(1-\alpha)} \left(\int_{-\tau}^{x} (x-t)^{-\alpha} (\tau+t)^{\alpha-1} dt \right) = \lim_{x \to -\tau^+} C = C,
\end{aligned}
$$

which implies that formula (5.45) takes the form

$$
y(x) = \mathbb{E}_{\tau,\alpha}^{Bx^\alpha} \omega(-\tau) + \int_{-\tau}^{0} \mathbb{E}_{\tau,\alpha}^{B(x-\tau-s)^\alpha} z(s) ds.
$$

Since $-\tau \leq x \leq 0$, one should divide the interval into two subintervals:

(i) For $-\tau \leq s \leq x$ and $-\tau \leq x - \tau - s \leq x$, the delayed Mittag-Leffler-type matrix $\mathbb{E}_{\tau,\alpha}^{B(x-\tau-s)^\alpha} = I \frac{(x-s)^{\alpha-1}}{\Gamma(\alpha)}$.

(ii) For $x \leq s \leq 0$ and $x - \tau \leq x - \tau - s \leq -\tau$, the delayed Mittag-Leffler-type matrix $\mathbb{E}_{\tau,\alpha}^{B(x-\tau-s)^\alpha} = \Theta$. Thus on the interval $-\tau \leq x \leq 0$, we have

$$
\omega(x) = I \frac{(\tau+x)^{\alpha-1}}{\Gamma(\alpha)} \omega(-\tau) + \int_{-\tau}^{x} I \frac{(x-s)^{\alpha-1}}{\Gamma(\alpha)} z(s) ds. \tag{5.46}
$$

By Riemann–Liouville fractional differentiation on both sides of (5.46), we obtain

$$
\begin{aligned}
&(^{RL} D_{-\tau^+}^\alpha \omega)(x) \\
&= \frac{1}{\Gamma(1-\alpha)} \\
&\quad \times \frac{d}{dx} \int_{-\tau}^{x} (x-t)^{-\alpha} \left(I \frac{(\tau+t)^{\alpha-1}}{\Gamma(\alpha)} \omega(-\tau) + \int_{-\tau}^{x} I \frac{(t-s)^{\alpha-1}}{\Gamma(\alpha)} z(s) ds \right) dt \\
&= \frac{I}{\Gamma(1-\alpha)} \frac{d}{dx} \int_{-\tau}^{x} \frac{z(s)}{\Gamma(\alpha)} \left(\int_{s}^{x} (x-t)^{-\alpha} (t-s)^{\alpha-1} dt \right) ds \\
&= \frac{d}{dx} \int_{-\tau}^{x} z(s) ds = z(x).
\end{aligned}
$$

The proof is completed. $\qquad\square$

Now we are ready to derive a special solution of (5.40) with zero initial conditions.

Theorem 5.17. *A solution* $\tilde{y} \in C([-\tau, T], \mathbb{R}^n)$ *of* (5.40) *satisfying initial conditions* $y(x) = 0$, $x \in [-\tau, 0]$ *has the form*

$$
\tilde{y}(x) = \int_{-\tau}^{x} \mathbb{E}_{\tau,\alpha}^{B(x-\tau-t)^\alpha} f(t) dt, \, x \in [0, T].
$$

Proof. Using the method of variation of constants, we will search the solution of the nonhomogeneous system $\widetilde{y}(x)$ in the form

$$\widetilde{y}(x) = \int_{-\tau}^{x} \mathbb{E}_{\tau,\alpha}^{B(x-\tau-t)^{\alpha}} c(t)dt, \qquad (5.47)$$

where $c(\cdot)$, $-\tau \leq t \leq x$, is an unknown vector function and $\widetilde{y}(0) = 0$.

Taking the Riemann–Liouville fractional differentiate for formula (5.47), we obtain:

(i) For $k = 1$ and $0 < x \leq \tau$, according to (5.40), we have

$$(^{RL}D_{-\tau+}^{\alpha}\widetilde{y})(x) = B\widetilde{y}(x-\tau) + f(x) = B\int_{-\tau}^{x-\tau} \mathbb{E}_{\tau,\alpha}^{B(x-2\tau-t)^{\alpha}} c(t)dt + f(x)$$

$$= B\int_{-\tau}^{x-\tau} I\frac{(x-\tau-t)^{\alpha-1}}{\Gamma(\alpha)} c(t)dt + f(x).$$

However, according to Definition 5.3 and Lemma 5.10, we have

$$(^{RL}D_{-\tau+}^{\alpha}\widetilde{y})(x)$$

$$= \frac{1}{\Gamma(1-\alpha)}\frac{d}{dx}\int_{-\tau}^{x}(x-t)^{-\alpha}\left(\int_{-\tau}^{t}\mathbb{E}_{\tau,\alpha}^{B(t-\tau-s)^{\alpha}}c(s)ds\right)dt$$

$$= \frac{1}{\Gamma(1-\alpha)}\frac{d}{dx}\int_{-\tau}^{x}c(s)\left(\int_{s}^{x}(x-t)^{-\alpha}\mathbb{E}_{\tau,\alpha}^{B(t-\tau-s)^{\alpha}}dt\right)ds$$

$$= \frac{1}{\Gamma(1-\alpha)}\frac{d}{dx}\int_{-\tau}^{x}c(s)\left(\int_{s}^{x}(x-t)^{-\alpha}I\frac{(t-s)^{\alpha-1}}{\Gamma(\alpha)}dt\right.$$

$$\left. + \int_{\tau+s}^{x}(x-t)^{-\alpha}B\frac{(t-\tau-s)^{2\alpha-1}}{\Gamma(2\alpha)}dt\right)ds$$

$$= \frac{d}{dx}\int_{-\tau}^{x}c(s)ds$$

$$+ \frac{1}{\Gamma(1-\alpha)}\frac{d}{dx}\int_{-\tau}^{x-\tau}c(s)\left(\int_{\tau+s}^{x}(x-t)^{-\alpha}B\frac{(t-\tau-s)^{2\alpha-1}}{\Gamma(2\alpha)}dt\right)ds$$

$$= c(x) + \frac{B}{\Gamma(\alpha+1)}\frac{d}{dx}\int_{-\tau}^{x-\tau}c(s)(x-\tau-s)^{\alpha}ds$$

$$= c(x) + B\int_{-\tau}^{x-\tau}\frac{(x-\tau-s)^{\alpha-1}}{\Gamma(\alpha)}c(s)ds.$$

Hence, we obtain $c(x) = f(x)$.

(ii) For $(k-1)\tau < x \leq k\tau$ and $k \in \mathbb{N}^{+}$, according to (5.40), we have

$$(^{RL}D_{-\tau+}^{\alpha}\widetilde{y})(x) = B\widetilde{y}(x-\tau) + f(x) = B\int_{-\tau}^{x-\tau} \mathbb{E}_{\tau,\alpha}^{B(x-2\tau-t)^{\alpha}} c(t)dt + f(x)$$

$$= \sum_{i=1}^{k} \int_{-\tau}^{x-i\tau} B^i \frac{(x - i\tau - t)^{i\alpha-1}}{\Gamma(i\alpha)} c(t)dt + f(x).$$

However, according to Definition 5.3, we have

$$({}^{RL}D_{-\tau+}^{\alpha}\tilde{y})(x) = \frac{1}{\Gamma(1-\alpha)} \frac{d}{dx} \int_{-\tau}^{x} (x-t)^{-\alpha} \left(\int_{-\tau}^{t} \mathbb{E}_{\tau,\alpha}^{B(t-\tau-s)^{\alpha}} c(s)ds \right) dt$$

$$= \frac{1}{\Gamma(1-\alpha)} \frac{d}{dx} \int_{-\tau}^{x} c(s) \left(\int_{s}^{x} (x-t)^{-\alpha} \mathbb{E}_{\tau,\alpha}^{B(t-\tau-s)^{\alpha}} dt \right) ds.$$

According to Lemma 5.1 and (5.40), we obtain

$$({}^{RL}D_{-\tau+}^{\alpha}\tilde{y})(x)$$

$$= \frac{d}{dx} \int_{-\tau}^{x} c(s)ds + \sum_{i=1}^{k} \frac{1}{\Gamma(1-\alpha)}$$

$$\times \frac{d}{dx} \int_{-\tau}^{x-i\tau} c(s) \left(\int_{i\tau+s}^{x} (x-t)^{-\alpha} B^i \frac{(t - i\tau - s)^{(i+1)\alpha-1}}{\Gamma(i\alpha + \alpha)} dt \right) ds$$

$$= c(x) + \sum_{i=1}^{k} \frac{B^i}{\Gamma(i\alpha + 1)} \frac{d}{dx} \int_{-\tau}^{x-i\tau} c(s)(x - i\tau - s)^{i\alpha} ds$$

$$= c(x) + \sum_{i=1}^{k} \int_{-\tau}^{x-i\tau} B^i \frac{(x - i\tau - t)^{i\alpha-1}}{\Gamma(i\alpha)} c(t)dt.$$

Hence, we obtain $c(x) = f(x)$. The proof is completed. $\qquad\square$

Linking Theorems 5.16 and 5.17 via the superposition principle, we have the following result.

Theorem 5.18. *A solution* $y \in C(X, \mathbb{R}^n) \cap C([-\tau, T], \mathbb{R}^n)$ *of (5.40) satisfying initial conditions* $y(x) = \omega(x)$, $-\tau \leq x \leq 0$, *has the form*

$$y(x) = \mathbb{E}_{\tau,\alpha}^{Bx^{\alpha}} \omega(-\tau) + \int_{-\tau}^{0} \mathbb{E}_{\tau,\alpha}^{B(x-\tau-s)^{\alpha}} ({}^{RL}D_{-\tau+}^{\alpha}\omega)(s)ds$$

$$+ \int_{-\tau}^{x} \mathbb{E}_{\tau,\alpha}^{B(x-\tau-t)^{\alpha}} f(t)dt.$$

Remark 5.2. A discontinuity effect is displayed in (5.40) with $x = 0$.

I. Since the variable of the equation $({}^{RL}D_{-\tau+}^{\alpha}y)(x) = Ay(x - \tau) + f(x)$ takes value in the interval $(0, \infty)$, the time point $x = 0$ is a discontinuity point of the first kind, where ${}^{RL}D_{-\tau+}^{\alpha}y$ denotes a Riemann–Liouville fractional derivative.

II. The equation $({}^{RL}D^{\alpha}_{-\tau^+}y)(x) = Ay(x-\tau) + f(x)$, $x > 0$, at point $x = 0$ is continuous if and only if $\lim\limits_{x\to 0^+} y(x) = \lim\limits_{x\to 0^-} \omega(x) = \omega(0)$, where

$$\lim_{x\to 0^+} y(x)$$

$$= I\frac{\tau^{\alpha-1}}{\Gamma(\alpha)}\omega(-\tau) + \int_{-\tau}^{0} \mathbb{E}^{A(-\tau-s)^{\alpha}}_{\tau,\alpha}\,({}^{RL}D^{\alpha}_{-\tau^+}\omega)(s)ds + \int_{-\tau}^{0}\mathbb{E}^{A(-\tau-t)^{\alpha}}_{\tau,\alpha}f(t)dt$$

$$= I\frac{\tau^{\alpha-1}}{\Gamma(\alpha)}\omega(-\tau) + \int_{-\tau}^{0}\frac{(-s)^{\alpha-1}}{\Gamma(\alpha)}\,({}^{RL}D^{\alpha}_{-\tau^+}\omega)(s)ds + \int_{-\tau}^{0}\frac{(-t)^{\alpha-1}}{\Gamma(\alpha)}f(t)dt.$$

Remark 5.3. The Mittag-Leffler-type matrix $\mathbb{E}^{A\cdot\alpha}_{\tau,\alpha}: \mathbb{R} \to \mathbb{R}^{n\times n}$ in (5.4) is nonnegative. However, the monotonicity of the Mittag-Leffler-type matrix $\mathbb{E}^{A\cdot\alpha}_{\tau,\alpha}: \mathbb{R} \to \mathbb{R}^{n\times n}$ depends on constants $0 < \alpha < 1$ and $n \in \mathbb{N}$.

5.2.1.2 Finite time stability for linear systems

In this section, we study the finite time stability of

$$\begin{cases} ({}^{RL}D^{\alpha}_{-\tau^+}y)(x) = Ay(x-\tau) + f(x),\ 0 < \alpha < 1,\ A \in \mathbb{R}^{n\times n}, \\ x \in (0, T],\ \tau > 0, \\ y(x) = \mu(x),\ \mu(x) \in \mathbb{R}^n,\ -\tau < x \le 0, \\ (\mathbb{I}^{1-\alpha}_{-\tau^+}y)(-\tau^+) = \mathbf{b},\ \mathbf{b} \in \mathbb{R}^n. \end{cases} \tag{5.48}$$

For $0 < \gamma < 1$, we denote $C_{\gamma}([a,b],\mathbb{R}^n) = \left\{y(x) \in C((a,b],\mathbb{R}^n)\colon (x-a)^{\gamma}y(x) \in C([a,b],\mathbb{R}^n)\right\}$. Then, $C_{\gamma}([a,b],\mathbb{R}^n)$ is a Banach space endowed with $\|y\|_{C_{\gamma}} = \max\limits_{a\le x\le b} \|(x-a)^{\gamma}y(x)\|$.

Definition 5.12. (see [24]) Eq. (5.48) is finite time stable with respect to $\{0, \bar{J}, \tau, \delta, \eta\}$ if and only if $\|\mu\|_C < \delta$ and $\|\mathbf{b}\| < \delta$ implies a solution y of (5.48) satisfying $\|y\|_{C_{\gamma}} < \eta$ and $\delta < \eta$, $\delta, \eta > 0$, where $\bar{J} := ((n-1)\tau, n\tau]$.

Lemma 5.11. *For any $x \in ((n-1)\tau, n\tau]$, $n \in \{1, 2, \cdots, k^*\}$ and $0 < \gamma < 1$, we obtain:*

(I) Letting $0 < \alpha < \frac{1}{n+1}$, we have

$$\|(x-(n-1)\tau)^{\gamma}\mathbb{E}^{Ax^{\alpha}}_{\tau,\alpha}\| \le (x-(n-1)\tau)^{\alpha+\gamma-1}E_{\alpha,\alpha}(\|A\|(x-(n-1)\tau)^{\alpha}).$$

(II) Letting $\frac{1}{n+1} \le \alpha \le \frac{1}{2}$, we have

$$\|(x-(n-1)\tau)^{\gamma}\mathbb{E}^{Ax^{\alpha}}_{\tau,\alpha}\| \le (x-(n-1)\tau)^{\gamma}\sum_{j=0}^{n}\|A\|^j\frac{(x-(j-1)\tau)^{(j+1)\alpha-1}}{\Gamma(j\alpha+\alpha)}.$$

(III) Letting $\alpha > \frac{1}{2}$, we have

$$\|(x - (n-1)\tau)^\gamma \mathbb{E}_{\tau,\alpha}^{Ax^\alpha}\| \le x^{\alpha+\gamma-1} E_{\alpha,\alpha}(\|A\|x^\alpha).$$

Proof. According to the definition of $\mathbb{E}_{\tau,\alpha}^{A.\alpha}$, one has the following three cases:

(I) Letting $x \in ((n-1)\tau, n\tau]$ and $0 < \alpha < \frac{1}{n+1}$, we have

$$\|(x - (n-1)\tau)^\gamma \mathbb{E}_{\tau,\alpha}^{Ax^\alpha}\|$$

$$\le (x - (n-1)\tau)^\gamma \left(\frac{(\tau+x)^{\alpha-1}}{\Gamma(\alpha)} + \|A\| \frac{x^{2\alpha-1}}{\Gamma(\alpha+\alpha)} + \|A\|^2 \frac{(x-\tau)^{3\alpha-1}}{\Gamma(2\alpha+\alpha)} \right.$$

$$\left. + \cdots + \|A\|^n \frac{(x-(n-1)\tau)^{(n+1)\alpha-1}}{\Gamma(n\alpha+\alpha)} \right)$$

$$\le (x - (n-1)\tau)^\gamma \left(\frac{(x-(n-1)\tau)^{\alpha-1}}{\Gamma(\alpha)} + \|A\| \frac{(x-(n-1)\tau)^{2\alpha-1}}{\Gamma(\alpha+\alpha)} \right.$$

$$\left. + \|A\|^2 \frac{(x-(n-1)\tau)^{3\alpha-1}}{\Gamma(2\alpha+\alpha)} + \cdots + \|A\|^n \frac{(x-(n-1)\tau)^{(n+1)\alpha-1}}{\Gamma(n\alpha+\alpha)} \right)$$

$$\le (x - (n-1)\tau)^\gamma \sum_{j=0}^{n} \|A\|^j \frac{(x-(n-1)\tau)^{(j+1)\alpha-1}}{\Gamma(j\alpha+\alpha)}$$

$$\le (x - (n-1)\tau)^{\alpha+\gamma-1} E_{\alpha,\alpha}(\|A\|(x-(n-1)\tau)^\alpha).$$

(II) Letting $x \in ((n-1)\tau, n\tau]$ and $\frac{1}{n+1} \le \alpha \le \frac{1}{2}$, we have

$$\|(x - (n-1)\tau)^\gamma \mathbb{E}_{\tau,\alpha}^{Ax^\alpha}\|$$

$$\le (x - (n-1)\tau)^\gamma \left(\frac{(\tau+x)^{\alpha-1}}{\Gamma(\alpha)} + \|A\| \frac{x^{2\alpha-1}}{\Gamma(\alpha+\alpha)} + \|A\|^2 \frac{(x-\tau)^{3\alpha-1}}{\Gamma(2\alpha+\alpha)} \right.$$

$$\left. + \cdots + \|A\|^n \frac{(x-(n-1)\tau)^{(n+1)\alpha-1}}{\Gamma(n\alpha+\alpha)} \right)$$

$$\le (x - (n-1)\tau)^\gamma \sum_{j=0}^{n} \|A\|^j \frac{(x-(j-1)\tau)^{(j+1)\alpha-1}}{\Gamma(j\alpha+\alpha)}.$$

(III) Letting $x \in ((n-1)\tau, n\tau]$ and $\alpha > \frac{1}{2}$, we have

$$\|(x - (n-1)\tau)^\gamma \mathbb{E}_{\tau,\alpha}^{Ax^\alpha}\|$$

$$\le x^\gamma \left(\frac{(\tau+x)^{\alpha-1}}{\Gamma(\alpha)} + \|A\| \frac{x^{2\alpha-1}}{\Gamma(\alpha+\alpha)} + \|A\|^2 \frac{(x-\tau)^{3\alpha-1}}{\Gamma(2\alpha+\alpha)} \right.$$

$$\left. + \cdots + \|A\|^n \frac{(x-(n-1)\tau)^{(n+1)\alpha-1}}{\Gamma(n\alpha+\alpha)} \right)$$

$$\leq x^{\gamma}\left(\frac{x^{\alpha-1}}{\Gamma(\alpha)}+\|A\|\frac{x^{2\alpha-1}}{\Gamma(\alpha+\alpha)}+\|A\|^{2}\frac{x^{3\alpha-1}}{\Gamma(2\alpha+\alpha)}+\cdots+\|A\|^{n}\frac{x^{(n+1)\alpha-1}}{\Gamma(n\alpha+\alpha)}\right)$$

$$\leq x^{\alpha+\gamma-1}E_{\alpha,\alpha}(\|A\|x^{\alpha}).$$

This proof is finished. $\qquad\square$

Lemma 5.12. *For any $x \in ((n-1)\tau, n\tau]$, $n \in \{1, 2, \cdots, k^{*}\}$, we have*

$$\int_{-\tau}^{0}\|\mathbb{E}_{\tau,\alpha}^{A(x-\tau-s)^{\alpha}}\|ds \leq \sum_{j=0}^{n}\frac{\|A\|^{j}}{\Gamma((j+1)\alpha+1)}(x-(j-1)\tau)^{(j+1)\alpha}$$

$$-\sum_{j=1}^{n}\frac{\|A\|^{j-1}}{\Gamma(j\alpha+1)}(x-(j-1)\tau)^{j\alpha}.$$

Proof. According to the definition of $\mathbb{E}_{\tau,\alpha}^{A,\alpha}$, we obtain

$$\int_{-\tau}^{0}\|\mathbb{E}_{\tau,\alpha}^{A(x-\tau-s)^{\alpha}}\|ds$$

$$=\int_{-\tau}^{x-n\tau}\|\mathbb{E}_{\tau,\alpha}^{A(x-\tau-s)^{\alpha}}\|ds+\int_{x-n\tau}^{0}\|\mathbb{E}_{\tau,\alpha}^{A(x-\tau-s)^{\alpha}}\|ds$$

$$\leq\int_{-\tau}^{x-n\tau}\left(\frac{(x-s)^{\alpha-1}}{\Gamma(\alpha)}+\|A\|\frac{(x-\tau-s)^{2\alpha-1}}{\Gamma(\alpha+\alpha)}+\|A\|^{2}\frac{(x-2\tau-s)^{3\alpha-1}}{\Gamma(2\alpha+\alpha)}\right.$$

$$\left.+\cdots+\|A\|^{n-1}\frac{(x-(n-1)\tau-s)^{n\alpha-1}}{\Gamma((n-1)\alpha+\alpha)}\right)ds$$

$$+\int_{-\tau}^{x-n\tau}\|A\|^{n}\frac{(x-n\tau-s)^{(n+1)\alpha-1}}{\Gamma(n\alpha+\alpha)}ds$$

$$+\int_{x-n\tau}^{0}\left(\frac{(x-s)^{\alpha-1}}{\Gamma(\alpha)}+\|A\|\frac{(x-\tau-s)^{2\alpha-1}}{\Gamma(\alpha+\alpha)}+\|A\|^{2}\frac{(x-2\tau-s)^{3\alpha-1}}{\Gamma(2\alpha+\alpha)}\right.$$

$$\left.+\cdots+\|A\|^{n-1}\frac{(x-(n-1)\tau-s)^{n\alpha-1}}{\Gamma((n-1)\alpha+\alpha)}\right)ds$$

$$\leq\int_{-\tau}^{x-n\tau}\|A\|^{n}\frac{(x-n\tau-s)^{(n+1)\alpha-1}}{\Gamma(n\alpha+\alpha)}ds$$

$$+\int_{-\tau}^{0}\left(\frac{(x-s)^{\alpha-1}}{\Gamma(\alpha)}+\|A\|\frac{(x-\tau-s)^{2\alpha-1}}{\Gamma(\alpha+\alpha)}\right.$$

$$\left.+\|A\|^{2}\frac{(x-2\tau-s)^{3\alpha-1}}{\Gamma(2\alpha+\alpha)}+\cdots+\|A\|^{n-1}\frac{(x-(n-1)\tau-s)^{n\alpha-1}}{\Gamma((n-1)\alpha+\alpha)}\right)ds$$

$$\leq\int_{-\tau}^{x-n\tau}\|A\|^{n}\frac{(x-n\tau-s)^{(n+1)\alpha-1}}{\Gamma(n\alpha+\alpha)}ds$$

$$+ \sum_{j=1}^{n} \frac{\|A\|^{j-1}}{\Gamma(j\alpha)} \int_{-\tau}^{0} (x - (j-1)\tau - s)^{j\alpha-1} ds$$

$$\leq \frac{\|A\|^{n}}{\Gamma((n+1)\alpha + 1)} (x - (n-1)\tau)^{(n+1)\alpha}$$

$$+ \sum_{j=1}^{n} \frac{\|A\|^{j-1}}{\Gamma(j\alpha + 1)} \left((x - (j-2)\tau)^{j\alpha} - (x - (j-1)\tau)^{j\alpha} \right)$$

$$\leq \sum_{j=0}^{n} \frac{\|A\|^{j}}{\Gamma((j+1)\alpha + 1)} (x - (j-1)\tau)^{(j+1)\alpha}$$

$$- \sum_{j=1}^{n} \frac{\|A\|^{j-1}}{\Gamma(j\alpha + 1)} (x - (j-1)\tau)^{j\alpha}.$$

This proof is completed. $\qquad\qquad\qquad\qquad\qquad\qquad\qquad\qquad$ $\square$

Lemma 5.13. *For any $\alpha > 1 - \frac{1}{p}$ ($p > 1$), $x \in ((n-1)\tau, n\tau]$ and $n \in \{1, 2, \cdots, k^*\}$, we have*

$$\int_{-\tau}^{0} \|\mathbb{E}_{\tau,\alpha}^{A(x-\tau-s)^{\alpha}}\| \|(^{RL}D_{-\tau^+}^{\alpha}\mu)(s)\| ds$$

$$\leq \sum_{j=0}^{n} \left(\frac{\|A\|^{j}}{\Gamma(j\alpha + \alpha)} \frac{(x - (j-1)\tau)^{(j+1)\alpha-1+\frac{1}{p}}}{(p(j+1)\alpha - p + 1)^{\frac{1}{p}}} \right)$$

$$\times \left(\int_{-\tau}^{0} \|(^{RL}D_{-\tau^+}^{\alpha}\mu)(s)\|^{q} ds \right)^{\frac{1}{q}}.$$

Proof. According to the definition of $\mathbb{E}_{\tau,\alpha}^{A,\alpha}$, using Hölder's inequality, we derive

$$\int_{-\tau}^{0} \|\mathbb{E}_{\tau,\alpha}^{A(x-\tau-s)^{\alpha}}\| \|(^{RL}D_{-\tau^+}^{\alpha}\mu)(s)\| ds$$

$$= \int_{-\tau}^{x-n\tau} \|\mathbb{E}_{\tau,\alpha}^{A(x-\tau-s)^{\alpha}}\| \|(^{RL}D_{-\tau^+}^{\alpha}\mu)(s)\| ds$$

$$+ \int_{x-n\tau}^{0} \|\mathbb{E}_{\tau,\alpha}^{A(x-\tau-s)^{\alpha}}\| \|(^{RL}D_{-\tau^+}^{\alpha}\mu)(s)\| ds$$

$$\leq \int_{-\tau}^{x-n\tau} \|A\|^{n} \frac{(x - n\tau - s)^{(n+1)\alpha-1}}{\Gamma(n\alpha + \alpha)} \|(^{RL}D_{-\tau^+}^{\alpha}\mu)(s)\| ds$$

$$+ \sum_{j=1}^{n} \frac{\|A\|^{j-1}}{\Gamma(j\alpha)} \int_{-\tau}^{0} (x - (j-1)\tau - s)^{j\alpha-1} \|(^{RL}D_{-\tau^+}^{\alpha}\mu)(s)\| ds$$

$$\leq \frac{\|A\|^n}{\Gamma(n\alpha+\alpha)}\left(\int_{-\tau}^{x-n\tau}(x-n\tau-s)^{p((n+1)\alpha-1)}ds\right)^{\frac{1}{p}}$$

$$\times\left(\int_{-\tau}^{x-n\tau}\|(^{RL}D_{-\tau+}^{\alpha}\mu)(s)\|^q ds\right)^{\frac{1}{q}}$$

$$+\sum_{j=0}^{n-1}\frac{\|A\|^j}{\Gamma(j\alpha+\alpha)}\left(\int_{-\tau}^{0}(x-j\tau-s)^{p((j+1)\alpha-1)}ds\right)^{\frac{1}{p}}$$

$$\times\left(\int_{-\tau}^{0}\|(^{RL}D_{-\tau+}^{\alpha}\mu)(s)\|^q ds\right)^{\frac{1}{q}}$$

$$\leq\sum_{j=0}^{n}\left(\frac{\|A\|^j}{\Gamma(j\alpha+\alpha)}\frac{(x-(j-1)\tau)^{(j+1)\alpha-1+\frac{1}{p}}}{(p(j+1)\alpha-p+1)^{\frac{1}{p}}}\right)$$

$$\times\left(\int_{-\tau}^{0}\|(^{RL}D_{-\tau+}^{\alpha}\mu)(s)\|^q ds\right)^{\frac{1}{q}},$$

where $\frac{1}{p}+\frac{1}{q}=1$ for every $p>1$. This proof is completed. $\qquad\square$

We now try to give finite time stable results. We impose the following assumptions:

$[A_1]$ $0<M=\sup_{-\tau\leq s\leq 0}\|(^{RL}D_{-\tau+}^{\alpha}\mu)(s)\|<\infty$;

$[A_2]$ $0<M_1=\int_{-\tau}^{0}\|(^{RL}D_{-\tau+}^{\alpha}\mu)(s)\|ds<\infty$;

$[A_3]$ $0<M_2=\left(\int_{-\tau}^{0}\|(^{RL}D_{-\tau+}^{\alpha}\mu)(s)\|^q ds\right)^{\frac{1}{q}}<\infty$, $\frac{1}{q}=1-\frac{1}{p}$, $p>1$;

$[A_4]$ for every constant $0<\gamma<1$ and $0<\alpha<1$, we have $\alpha+\gamma-1\geq 0$;

$[A_5]$ we suppose $f(\cdot)\in C([-\tau,T],\mathbb{R}^n)$ and $\|f\|=\max_{x\in[-\tau,T]}\|f(x)\|<\infty$;

$[A_6]$ $\exists\,v(\cdot)\in L^q([-\tau,T],\mathbb{R}^+)$, $\frac{1}{q}=1-\frac{1}{p}$, $p>1$, such that $\|f(x)\|\leq v(x)$

for $x\in[-\tau,T]$ and $\phi(x):=\left(\int_{-\tau}^{x}v(t)^q dt\right)^{\frac{1}{q}}<\infty$.

For every $(n-1)\tau<x\leq n\tau$ and $n\in\{1,2,\cdots,k^*\}$, define

$$\Phi_1(x)=\begin{cases}(x-(n-1)\tau)^{\alpha+\gamma-1}E_{\alpha,\alpha}(\|A\|(x-(n-1)\tau)^{\alpha}),\ 0<\alpha<\frac{1}{n+1},\\[2mm](x-(n-1)\tau)^{\gamma}\sum_{j=0}^{n}\|A\|^j\frac{(x-(j-1)\tau)^{(j+1)\alpha-1}}{\Gamma(j\alpha+\alpha)},\ \frac{1}{n+1}\leq\alpha\leq\frac{1}{2},\\[2mm]x^{\alpha+\gamma-1}E_{\alpha,\alpha}(\|A\|x^{\alpha}),\ \alpha>\frac{1}{2},\end{cases}$$

and

$$\Phi_2(x)=\sum_{j=0}^{n}\frac{\|A\|^j}{\Gamma((j+1)\alpha+1)}(x-(j-1)\tau)^{(j+1)\alpha}$$

$$-\sum_{j=1}^{n}\frac{\|A\|^{j-1}}{\Gamma(j\alpha+1)}(x-(j-1)\tau)^{j\alpha},$$

$$\Phi_3(x)=\sum_{j=0}^{n}\frac{\|A\|^{j}}{\Gamma(j\alpha+\alpha)}\frac{(x-(j-1)\tau)^{(j+1)\alpha}}{(j+1)\alpha},$$

$$\Phi_4(x)=\sum_{j=0}^{n}\left(\frac{\|A\|^{j}}{\Gamma(j\alpha+\alpha)}\frac{(x-(j-1)\tau)^{(j+1)\alpha-1+\frac{1}{p}}}{(p(j+1)\alpha-p+1)^{\frac{1}{p}}}\right).$$

Theorem 5.19. *Assume that $[A_1]$, $[A_4]$, and $[A_5]$ hold. Eq. (5.48) is finite time stable with respect to $\{0,\bar{J},\tau,\delta,\eta\}$ provided that*

$$\delta\Phi_1(x)+(x-(n-1)\tau)^{\gamma}(M\Phi_2(x)+\|f\|\Phi_3(x))<\eta,\ \forall\, x\in\bar{J}.\qquad(5.49)$$

Proof. By Theorem 5.18, the solution of (5.48) can be given by

$$y(x)=\mathbb{E}^{Ax^{\alpha}}_{\tau,\alpha}\mathbf{b}+\int_{-\tau}^{0}\mathbb{E}^{A(x-\tau-s)^{\alpha}}_{\tau,\alpha}(^{RL}D^{\alpha}_{-\tau+}\mu)(s)ds+\int_{-\tau}^{x}\mathbb{E}^{A(x-\tau-t)^{\alpha}}_{\tau,\alpha}f(t)dt.$$

Using Lemmas 5.11 and 5.12, via (5.49) we obtain

$$\|(x-(n-1)\tau)^{\gamma}y(x)\|$$
$$\leq\|(x-(n-1)\tau)^{\gamma}\mathbb{E}^{Ax^{\alpha}}_{\tau,\alpha}\mathbf{b}\|$$
$$+\int_{-\tau}^{0}\|(x-(n-1)\tau)^{\gamma}\mathbb{E}^{A(x-\tau-s)^{\alpha}}_{\tau,\alpha}(^{RL}D^{\alpha}_{-\tau+}\mu)(s)\|ds$$
$$+\int_{-\tau}^{x}(x-(n-1)\tau)^{\gamma}\|\mathbb{E}^{A(x-\tau-t)^{\alpha}}_{\tau,\alpha}\|\|f(t)\|dt$$
$$\leq\delta\Phi_1(x)+(x-(n-1)\tau)^{\gamma}\left(M\int_{-\tau}^{0}\|\mathbb{E}^{A(x-\tau-s)^{\alpha}}_{\tau,\alpha}\|ds\right.$$
$$\left.+\|f\|\int_{-\tau}^{x}\|\mathbb{E}^{A(x-\tau-t)^{\alpha}}_{\tau,\alpha}\|dt\right)$$
$$\leq\delta\Phi_1(x)+(x-(n-1)\tau)^{\gamma}\left(M\Phi_2(x)\right.$$
$$\left.+\|f\|\sum_{j=0}^{n}\frac{\|A\|^{j}}{\Gamma(j\alpha+\alpha)}\int_{-\tau}^{x-j\tau}(x-j\tau-t)^{(j+1)\alpha-1}dt\right)$$
$$\leq\delta\Phi_1(x)+(x-(n-1)\tau)^{\gamma}(M\Phi_2(x)+\|f\|\Phi_3(x))<\eta,\ \forall\, x\in\bar{J}.$$

This proof is completed. $\qquad\square$

Theorem 5.20. *Assume that* $[A_2]$, $[A_4]$, *and* $[A_5]$ *hold. For all* $\alpha > \frac{1}{2}$, (5.48) *is finite time stable with respect to* $\{0, \bar{J}, \tau, \delta, \eta\}$ *provided that*

$$\delta \Phi_1(x) + x^{\alpha+\gamma-1} E_{\alpha,\alpha}(\|A\| x^{\alpha}) M_1 + x^{\gamma} \|f\| \Phi_3(x) < \eta, \ \forall x \in \bar{J}. \qquad (5.50)$$

Proof. By using Lemma 5.11 via (5.50), we obtain

$$\|(x - (n-1)\tau)^{\gamma} y(x)\|$$

$$\leq \|(x - (n-1)\tau)^{\gamma} \mathbb{E}_{\tau,\alpha}^{A x^{\alpha}} \mathbf{b}\|$$

$$\quad + \int_{-\tau}^{0} \|(x - (n-1)\tau)^{\gamma} \mathbb{E}_{\tau,\alpha}^{A(x-\tau-s)^{\alpha}} (^{RL} D_{-\tau+}^{\alpha} \mu)(s)\| ds$$

$$\quad + \int_{-\tau}^{x} (x - (n-1)\tau)^{\gamma} \|\mathbb{E}_{\tau,\alpha}^{A(x-\tau-t)^{\alpha}}\| \|f(t)\| dt$$

$$\leq \delta \Phi_1(x) + (x - (n-1)\tau)^{\gamma} \left(\int_{-\tau}^{0} \|\mathbb{E}_{\tau,\alpha}^{A(x-\tau-s)^{\alpha}} (^{RL} D_{-\tau+}^{\alpha} \mu)(s)\| ds \right.$$

$$\quad \left. + \|f\| \int_{-\tau}^{x} \|\mathbb{E}_{\tau,\alpha}^{A(x-\tau-t)^{\alpha}}\| dt \right)$$

$$\leq \delta \Phi_1(x) + x^{\gamma} \left(x^{\alpha-1} E_{\alpha,\alpha}(\|A\| x^{\alpha}) M_1 + \|f\| \Phi_3(x) \right)$$

$$\leq \delta \Phi_1(x) + x^{\alpha+\gamma-1} E_{\alpha,\alpha}(\|A\| x^{\alpha}) M_1 + x^{\gamma} \|f\| \Phi_3(x) < \eta, \ \forall x \in \bar{J}.$$

This proof is completed. $\qquad\qquad \square$

Theorem 5.21. *Assume that* $[A_3]$, $[A_4]$, $[A_6]$, *and* $\alpha > 1 - \frac{1}{p}$ $(p > 1)$ *hold. Eq.* (5.48) *is finite time stable with respect to* $\{0, \bar{J}, \tau, \delta, \eta\}$ *provided that*

$$\delta \Phi_1(x) + (x - (n-1)\tau)^{\gamma} \Phi_4(x) \left(M_2 + \phi(x) \right) < \eta, \ \forall x \in \bar{J}. \qquad (5.51)$$

Proof. Using Lemmas 5.11 and 5.12, via (5.51) we obtain

$$\|(x - (n-1)\tau)^{\gamma} y(x)\|$$

$$\leq \delta \Phi_1(x) + (x - (n-1)\tau)^{\gamma} \left(\int_{-\tau}^{0} \|\mathbb{E}_{\tau,\alpha}^{A(x-\tau-s)^{\alpha}} (^{RL} D_{-\tau+}^{\alpha} \mu)(s)\| ds \right.$$

$$\quad \left. + \int_{-\tau}^{x} \|\mathbb{E}_{\tau,\alpha}^{A(x-\tau-t)^{\alpha}}\| \|f(t)\| dt \right)$$

$$\leq \delta \Phi_1(x) + (x - (n-1)\tau)^{\gamma} \left(M_2 \Phi_4(x) \right.$$

$$\quad \left. + \sum_{j=0}^{n} \frac{\|A\|^{j}}{\Gamma(j\alpha+\alpha)} \int_{-\tau}^{x-j\tau} (x - j\tau - t)^{(j+1)\alpha-1} v(t) dt \right)$$

$$\leq \delta\Phi_1(x) + (x - (n-1)\tau)^\gamma \left(M_2\Phi_4(x) + \sum_{j=0}^{n} \frac{\|A\|^j}{\Gamma(j\alpha + \alpha)} \right.$$

$$\times \left. \left(\int_{-\tau}^{x-j\tau} (x - j\tau - t)^{p((j+1)\alpha - 1)} dt \right)^{\frac{1}{p}} \left(\int_{-\tau}^{x} v(t)^q dt \right)^{\frac{1}{q}} \right)$$

$$\leq \delta\Phi_1(x) + (x - (n-1)\tau)^\gamma \Phi_4(x) \left(M_2 + \phi(x) \right) < \eta, \ \forall\, x \in \bar{J}.$$

This proof is completed. $\qquad\square$

Theorem 5.22. *Assume that* $[A_2]$, $[A_4]$, *and* $[A_6]$ *hold. For all* $\alpha > \max\{\frac{1}{2}, 1 - \frac{1}{p}\}$ $(p > 1)$, *(5.48) is finite time stable with respect to* $\{0, \bar{J}, \tau, \delta, \eta\}$ *provided that*

$$\delta\Phi_1(x) + \left(x^{\alpha+\gamma-1} M_1 + x^\gamma \phi(x) \frac{(x+\tau)^{\alpha-1+\frac{1}{p}}}{(p(\alpha-1)+1)^{\frac{1}{p}}} \right) E_{\alpha,\alpha}(\|A\|x^\alpha)$$

$$< \eta, \ \forall\, x \in \bar{J}. \tag{5.52}$$

Proof. Using Lemma 5.11, via (5.52) we obtain

$$\|(x - (n-1)\tau)^\gamma y(x)\|$$

$$\leq \delta\Phi_1(x) + (x - (n-1)\tau)^\gamma \left(\int_{-\tau}^{0} \|\mathbb{E}_{\tau,\alpha}^{A(x-\tau-s)^\alpha}({}^{RL}D_{-\tau+}^\alpha \mu)(s)\| ds \right.$$

$$\left. + \int_{-\tau}^{x} \|\mathbb{E}_{\tau,\alpha}^{A(x-\tau-t)^\alpha}\| \|f(t)\| dt \right)$$

$$\leq \delta\Phi_1(x) + x^\gamma \left(x^{\alpha-1} E_{\alpha,\alpha}(\|A\|x^\alpha) M_1 + \int_{-\tau}^{x} \|\mathbb{E}_{\tau,\alpha}^{A(x-\tau-t)^\alpha}\| \|f(t)\| dt \right)$$

$$\leq \delta\Phi_1(x) + x^{\alpha+\gamma-1} E_{\alpha,\alpha}(\|A\|x^\alpha) M_1$$

$$+ x^\gamma \sum_{j=0}^{n} \frac{\|A\|^j}{\Gamma(j\alpha + \alpha)} \int_{-\tau}^{x-j\tau} (x - j\tau - t)^{(j+1)\alpha - 1} v(t) dt$$

$$\leq \delta\Phi_1(x) + x^{\alpha+\gamma-1} E_{\alpha,\alpha}(\|A\|x^\alpha) M_1$$

$$+ x^\gamma \sum_{j=0}^{n} \frac{(\|A\|x^\alpha)^j}{\Gamma(j\alpha + \alpha)} \int_{-\tau}^{x} (x - t)^{\alpha - 1} v(t) dt$$

$$\leq \delta\Phi_1(x) + \left(x^{\alpha+\gamma-1} M_1 + x^\gamma \phi(x) \frac{(x+\tau)^{\alpha-1+\frac{1}{p}}}{(p(\alpha-1)+1)^{\frac{1}{p}}} \right) E_{\alpha,\alpha}(\|A\|x^\alpha)$$

$$< \eta, \ \forall\, x \in \bar{J}.$$

This proof is completed. $\qquad\square$

5.2.1.3 Numerical example and discussion

Example 5.5. Set $\alpha = 0.6$, $\tau = 0.2$, $k^* = 3$, $T = 0.6$. We consider

$$
\begin{cases}
{}^{RL}D_{-0.2^+}^{0.6}\, y(x) = Ay(x - 0.2) + f(x), & x \in (0, 0.6], \\
\mu(x) = (x + 0.2, 2(x + 0.2)^2)^\top, & -0.2 < x \le 0, \\
(\mathbb{I}_{-0.2^+}^{0.4}\, y)(-0.2^+) = \mathbf{b} = (0.05, 0.05)^\top,
\end{cases}
\tag{5.53}
$$

where

$$
y(x) = \begin{pmatrix} y_1(x) \\ y_2(x) \end{pmatrix}, \quad
A = \begin{pmatrix} 0.2 & 0.1 \\ 0.3 & 0.5 \end{pmatrix}, \quad
f(x) = \begin{pmatrix} x \\ x^2 \end{pmatrix}.
$$

By Theorem 5.18, for $x \in (0.2(n-1), 0.2n]$, $n \in \{1, 2, 3\}$, any solution of (5.53) has the following form:

$$
y(x) = \mathbb{E}_{0.2,0.6}^{Ax^{0.6}}\mathbf{b} + \int_{-0.2}^{0} \mathbb{E}_{0.2,0.6}^{A(x-0.2-s)^{0.6}} ({}^{RL}D_{-0.2^+}^{0.6}\mu)(s)ds
$$

$$
+ \int_{-0.2}^{x} \mathbb{E}_{0.2,0.6}^{A(x-0.2-t)^{0.6}} f(t)dt,
$$

where

$$
\mathbb{E}_{0.2,0.6}^{Ax^{0.6}} =
\begin{cases}
I\dfrac{(x+0.2)^{-0.4}}{\Gamma(0.6)} + A\dfrac{x^{0.2}}{\Gamma(1.2)}, & 0 < x \le 0.2, \\[2mm]
I\dfrac{(x+0.2)^{-0.4}}{\Gamma(0.6)} + A\dfrac{x^{0.2}}{\Gamma(1.2)} + A^2\dfrac{(x-0.2)^{0.8}}{\Gamma(1.8)}, & 0.2 < x \le 0.4, \\[2mm]
I\dfrac{(x+0.2)^{-0.4}}{\Gamma(0.6)} + A\dfrac{x^{0.2}}{\Gamma(1.2)} + A^2\dfrac{(x-0.2)^{0.8}}{\Gamma(1.8)} \\[1mm]
\quad + A^3\dfrac{(x-0.4)^{1.4}}{\Gamma(2.4)}, & 0.4 < x \le 0.6,
\end{cases}
$$

$$
\int_{-0.2}^{0} \mathbb{E}_{0.2,0.6}^{A(x-0.2-s)^{0.6}} ({}^{RL}D_{-0.2^+}^{0.6}\mu)(s)ds
$$

$$
= \int_{-0.2}^{0} \mathbb{E}_{0.2,0.6}^{A(x-0.2-s)^{0.6}}
\begin{pmatrix}
\dfrac{1.4}{\Gamma(0.4)}\mathbb{B}[2, 0.4](s + 0.2)^{0.4} \\[2mm]
\dfrac{4.8}{\Gamma(0.4)}\mathbb{B}[3, 0.4](s + 0.2)^{1.4}
\end{pmatrix} ds,
$$

and

$$
\int_{-0.2}^{x} \mathbb{E}_{0.2,0.6}^{A(x-0.2-t)^{0.6}} f(t)dt
$$

$$= \begin{cases} \int_{-0.2}^{x} I \frac{(x-t)^{-0.4}}{\Gamma(0.6)} \begin{pmatrix} t \\ t^2 \end{pmatrix} dt + \int_{-0.2}^{x-0.2} A \frac{(x-0.2-t)^{0.2}}{\Gamma(1.2)} \begin{pmatrix} t \\ t^2 \end{pmatrix} dt, \ 0 < x \le 0.2, \\[1.5em] \int_{-0.2}^{x} I \frac{(x-t)^{-0.4}}{\Gamma(0.6)} \begin{pmatrix} t \\ t^2 \end{pmatrix} dt + \int_{-0.2}^{x-0.2} A \frac{(x-0.2-t)^{0.2}}{\Gamma(1.2)} \begin{pmatrix} t \\ t^2 \end{pmatrix} dt \\[1.5em] \quad + \int_{-0.2}^{x-0.4} A^2 \frac{(x-0.4-t)^{0.8}}{\Gamma(1.8)} \begin{pmatrix} t \\ t^2 \end{pmatrix} dt, \qquad\qquad\qquad\quad 0.2 < x \le 0.4, \\[1.5em] \int_{-0.2}^{x} I \frac{(x-t)^{-0.4}}{\Gamma(0.6)} \begin{pmatrix} t \\ t^2 \end{pmatrix} dt + \int_{-0.2}^{x-0.2} A \frac{(x-0.2-t)^{0.2}}{\Gamma(1.2)} \begin{pmatrix} t \\ t^2 \end{pmatrix} dt \\[1.5em] \quad + \int_{-0.2}^{x-0.4} A^2 \frac{(x-0.4-t)^{0.8}}{\Gamma(1.8)} \begin{pmatrix} t \\ t^2 \end{pmatrix} dt + \int_{-0.2}^{x-0.6} A^3 \frac{(x-0.6-t)^{1.4}}{\Gamma(2.4)} \begin{pmatrix} t \\ t^2 \end{pmatrix} dt, \\[1.5em] 0.4 < x \le 0.6. \end{cases}$$

Now put $n = 3$, $p = 2$, $q = 2$, and $\gamma = 0.5$. By calculation, one has $\|\mu\| = 0.28$, $\|f\| = 0.96$, $M = 0.9304$, $M_1 = 0.1128$, $M_2 = 0.2713$, $\phi(0.6) = 0.1621$. Next, $\Phi_1(0.6) = 0.6^{0.1} E_{0.6,0.6}(0.6^{0.7}) = 2.2993$, $\Phi_2(0.6) = \sum_{i=0}^{3} \frac{0.6^i}{\Gamma(0.6i+1.6)} \times (0.8 - 0.2i)^{(i+1)0.6} - \sum_{i=1}^{3} \frac{0.6^{i-1}}{\Gamma(0.6i+1)} (0.8 - 0.2i)^{0.6i} = 0.2998$, $\Phi_3(0.6) = \sum_{i=0}^{3} \frac{0.6^i (0.8-0.2i)^{(i+1)0.6}}{0.6(i+1)\Gamma(0.6(i+1))} = 1.3167$, and $\Phi_4(0.6) = \sum_{i=0}^{3} \frac{0.6^i (0.8-0.2i)^{0.6i+0.1}}{\Gamma(0.6(i+1))(1.2i+0.2)^{\frac{1}{2}}} = 1.9317$.

By Definition 5.12, we seek a suitable η making $\|x\|_{C_\gamma}$ of (5.48) not exceed η on $\bar{J}$. On the one hand, we can use the explicit formula of solution to (5.53) via numerical simulation to find a corresponding $\eta = 0.875$ for a fixed $T = 0.6$. On the other hand, by checking conditions in Theorems 5.19, 5.20, 5.21, and 5.22 for $[-0.2, 0.6]$, one can choose the better value $\eta = 1.10$.

5.2.1.4 Conclusions

We construct the explicit formulas of solutions for Caputo- and Riemann–Liouville-type fractional systems and give sufficient conditions to guarantee finite time stability and controllability for these fractional systems. The results in this part are motivated from [110,111].

5.2.2 Relative controllability for Riemann–Liouville type

In this section, we study the relative controllability of

$$\begin{cases} (^{RL}D_{-\tau^+}^{\alpha} y)(x) = Ay(x - \tau) + Bu(x), \ x \in (0, x_1], \ \tau > 0, \\ y(x) = \mu(x), \ \mu(x) \in \mathbb{R}^n, \ -\tau < x \le 0, \\ (\mathbb{I}_{-\tau^+}^{1-\alpha} y)(-\tau^+) = \mathbf{b}, \ \mathbf{b} \in \mathbb{R}^n, \end{cases} \tag{5.54}$$

where $y : J = (-\tau, x_1] \to \mathbb{R}^n$ is Riemann–Liouville differentiable on $(-\tau, x_1]$ with $(n-1)\tau < x_1 \leq n\tau$, $A, B \in \mathbb{R}^{n \times n}$, and the control function $u \in L^p(J, \mathbb{R}^n)$, $p \in (1, \infty)$.

Definition 5.13. Eq. (5.54) is called relatively controllable if for $\forall x_1 \in \mathbb{R}^n$, $\exists$ $u^* \in L^p(J, \mathbb{R}^n)$ such that

$$
\begin{cases}
(^{RL}D^{\alpha}_{-\tau+} y)(x) = Ay(x - \tau) + Bu^*(x), \ x \in (0, x_1], \ x_1 > 0, \ \tau > 0, \\
y(x) = \mu(x), \ \mu(x) \in \mathbb{R}^n, \ -\tau < x \leq 0, \\
(\mathbb{I}^{1-\alpha}_{-\tau+} y)(-\tau^+) = \mathbf{b}, \ \mathbf{b} \in \mathbb{R}^n,
\end{cases}
$$

$$(5.55)$$

has a solution $y(x, u^*) := y^*(x)$ satisfying $y^*(x) = \mu(x)$, $-\tau < x \leq 0$, via $y^*(x_1) = y_1$.

Lemma 5.14. *A solution $y \in C(\Omega, \mathbb{R}^n) \cap C([-\tau, T], \mathbb{R}^n)$ of (5.54) can be given by*

$$
y(x) = \mathbb{E}^{Ax^\alpha}_{\tau,\alpha} \mathbf{b} + \int_{-\tau}^{0} \mathbb{E}^{A(x-\tau-s)^\alpha}_{\tau,\alpha} (^{RL}D^{\alpha}_{-\tau+}\mu)(s)ds + \int_{-\tau}^{x} \mathbb{E}^{A(x-\tau-t)^\alpha}_{\tau,\alpha} Bu(t)dt.
$$

Consider a linear mapping $\Psi_\tau(u(\cdot)) : L^p(J, \mathbb{R}^n) \to \mathbb{R}^n$ given by

$$
\Psi_\tau(u(x)) = \int_{-\tau}^{x} \mathbb{E}^{A(x-\tau-t)^\alpha}_{\tau,\alpha} Bu(t)dt.
$$

Now we prove that $\Psi_\tau(u(\cdot))$ is bounded.

Lemma 5.15. *Let $x \in ((n-1)\tau, n\tau]$ and $n \in \{1, 2, \cdots, k^*\}$. If $\alpha > \max\{\frac{1}{2}, \frac{1}{p}\}$ with $1 < p < \infty$, then $\Psi_\tau(u(\cdot))$ is bounded.*

Proof. Letting $x \in ((n-1)\tau, n\tau]$ and $n \in \{1, 2, \cdots, k^*\}$, we have

$$
\int_{-\tau}^{x} \|\mathbb{E}^{A(x-\tau-t)^\alpha}_{\tau,\alpha} Bu(t)\|dt
$$

$$
= \|B\| \left(\int_{-\tau}^{x-n\tau} \|\mathbb{E}^{A(x-\tau-t)^\alpha}_{\tau,\alpha}\| \|u(t)\|dt + \int_{x-n\tau}^{x-(n-1)\tau} \|\mathbb{E}^{A(x-\tau-t)^\alpha}_{\tau,\alpha}\| \|u(t)\|dt \right.
$$

$$
\left. + \cdots + \int_{x-\tau}^{x} \|\mathbb{E}^{A(x-\tau-t)^\alpha}_{\tau,\alpha}\| \|u(t)\|dt \right)
$$

$$
\leq \|B\| \left(\int_{-\tau}^{x-n\tau} \left(\frac{(x-t)^{\alpha-1}}{\Gamma(\alpha)} + \|A\| \frac{(x-\tau-t)^{2\alpha-1}}{\Gamma(\alpha+\alpha)} \right. \right.
$$

$$
+ \|A\|^2 \frac{(x-2\tau-t)^{3\alpha-1}}{\Gamma(2\alpha+\alpha)} + \cdots + \|A\|^{n-1} \frac{(x-(n-1)\tau-t)^{n\alpha-1}}{\Gamma((n-1)\alpha+\alpha)}
$$

$$
\left. \left. + \|A\|^n \frac{(x-n\tau-t)^{(n+1)\alpha-1}}{\Gamma(n\alpha+\alpha)} \right) \|u(t)\|dt \right.
$$

$$
+ \int_{x-n\tau}^{x-(n-1)\tau} \left(\frac{(x-t)^{\alpha-1}}{\Gamma(\alpha)} + \|A\| \frac{(x-\tau-t)^{2\alpha-1}}{\Gamma(\alpha+\alpha)} \right.
$$

$$
\left. + \|A\|^2 \frac{(x-2\tau-t)^{3\alpha-1}}{\Gamma(2\alpha+\alpha)} + \cdots + \|A\|^{n-1} \frac{(x-(n-1)\tau-t)^{n\alpha-1}}{\Gamma((n-1)\alpha+\alpha)} \right)
$$

$$
\times \|u(t)\| dt + \cdots
$$

$$
+ \int_{x-2\tau}^{x-\tau} \left(\frac{(x-t)^{\alpha-1}}{\Gamma(\alpha)} + \|A\| \frac{(x-\tau-t)^{2\alpha-1}}{\Gamma(\alpha+\alpha)} \right) \|u(t)\| dt
$$

$$
+ \int_{x-\tau}^{x} \frac{(x-t)^{\alpha-1}}{\Gamma(\alpha)} \|u(t)\| dt \Big)
$$

$$
\leq \|B\| \left(\sum_{j=0}^{n} \frac{\|A\|^j}{\Gamma(j\alpha+\alpha)} \int_{-\tau}^{x-j\tau} (x-j\tau-t)^{(j+1)\alpha-1} \|u(t)\| dt \right)
$$

$$
\leq \|B\| \left(\sum_{j=0}^{n} \frac{(\|A\|x^\alpha)^j}{\Gamma(i\alpha+\alpha)} \int_{-\tau}^{x} (x-t)^{\alpha-1} \|u(t)\| dt \right)
$$

$$
\leq \|B\| E_{\alpha,\alpha}(\|A\|x^\alpha) \frac{(x+\tau)^{\alpha-\frac{1}{p}}}{(1+q(\alpha-1))^{\frac{1}{q}}} \|u\|_{L^p}, \quad \frac{1}{p} + \frac{1}{q} = 1.
$$

This proof is completed. $\qquad\qquad\qquad\qquad\qquad\qquad\qquad\qquad\square$

Lemma 5.16. *(see [112, Lemma 3.2])Let $\alpha > \max\{\frac{1}{2}, \frac{1}{p}\}$ with $1 < p < \infty$. If a pair (A, B) of (5.54) is controllable, i.e., rank $S_k = k$, where $S_k = \{B, AB, A^2B, \cdots, A^{k-1}B\}$, then the matrix function $\omega_{\tau,\alpha}(x) = \mathbb{E}_{\tau,\alpha}^{Ax^\alpha} B$ consists of elements that are row linearly independent on the interval $-\tau < x \leq x_1$ with $x_1 > (k-1)\tau$, i.e., there exists no nonzero vector $\bar{b}^\top = (b_1, b_2, \cdots, b_n)$ such that $\bar{b}^\top \omega_{\tau,\alpha}(x) = 0$.*

5.2.2.1 Relative controllability for linear systems

We study the relative controllability of (5.54) by imposing $\alpha > \max\{\frac{1}{2}, \frac{1}{p}\}$ with $1 < p < \infty$. Consider

$$
W_{\tau,\alpha}[-\tau, x_1] = \int_{-\tau}^{x_1} \mathbb{E}_{\tau,\alpha}^{A(x_1-\tau-t)^\alpha} B B^\top \mathbb{E}_{\tau,\alpha}^{A^\top(x_1-\tau-t)^\alpha} dt. \tag{5.56}
$$

Now we will give the delayed fractional Grammian matrix criterion result.

Theorem 5.23. $W_{\tau,\alpha}[-\tau, x_1]$ *defined in (5.56) is nonsingular if and only if system (5.54) is relatively controllable.*

Proof. Note that $W_{\tau,\alpha}[-\tau, x_1]$ is a nonsingular matrix, which guarantees $W_{\tau,\alpha}^{-1}[-\tau, x_1]$ exists. For any $y_1 \in \mathbb{R}^n$, one can choose $u(\cdot) \in L^p(J, \mathbb{R}^n)$ such that

$$
u(t) = B^\top \mathbb{E}_{\tau,\alpha}^{A^\top(x_1-\tau-t)^\alpha} W_{\tau,\alpha}^{-1}[-\tau, x_1]\zeta, \tag{5.57}
$$

where

$$\zeta = y_1 - \mathbb{E}_{\tau,\alpha}^{A x_1^\alpha} \mathbf{b} - \int_{-\tau}^{0} \mathbb{E}_{\tau,\alpha}^{A(x_1-\tau-s)^\alpha} (^{RL}D_{-\tau+}^\alpha \mu)(s)ds. \tag{5.58}$$

By using Lemma 5.14 and (5.57), we have

$$y(x_1) = \mathbb{E}_{\tau,\alpha}^{A x_1^\alpha} \mathbf{b} + \int_{-\tau}^{0} \mathbb{E}_{\tau,\alpha}^{A(x_1-\tau-s)^\alpha} (^{RL}D_{-\tau+}^\alpha \mu)(s)ds$$

$$+ \int_{-\tau}^{x_1} \mathbb{E}_{\tau,\alpha}^{A(x_1-\tau-t)^\alpha} B B^\top \mathbb{E}_{\tau,\alpha}^{A^\top(x_1-\tau-t)^\alpha} dt \, W_{\tau,\alpha}^{-1}[-\tau, x_1]\zeta. \tag{5.59}$$

Linking (5.56) and (5.59), via (5.58) we have

$$y(x_1) = \mathbb{E}_{\tau,\alpha}^{A x_1^\alpha} \mathbf{b} + \int_{-\tau}^{0} \mathbb{E}_{\tau,\alpha}^{A(x_1-\tau-s)^\alpha} (^{RL}D_{-\tau+}^\alpha \mu)(s)ds + \zeta = y_1.$$

By Definition 5.13 and Lemma 5.14, $y(x) = \mu(x)$ with $-\tau < x \le 0$ holds. Therefore, (5.54) is relatively controllable.

Suppose $W_{\tau,\alpha}[-\tau, x_1]$ is a singular matrix and one has at least one nonzero state $\widetilde{y} \in \mathbb{R}^n$ satisfying $\widetilde{y}^\top W_{\tau,\alpha}[-\tau, x_1]\widetilde{y} = 0$. Therefore, we have

$$0 = \widetilde{y}^\top W_{\tau,\alpha}[-\tau, x_1]\widetilde{y}$$

$$= \int_{-\tau}^{x_1} \widetilde{y}^\top \mathbb{E}_{\tau,\alpha}^{A(x_1-\tau-t)^\alpha} B B^\top \mathbb{E}_{\tau,\alpha}^{A^\top(x_1-\tau-t)^\alpha} \widetilde{y} dt$$

$$= \int_{-\tau}^{x_1} \left\| \widetilde{y}^\top \mathbb{E}_{\tau,\alpha}^{A(x_1-\tau-t)^\alpha} B \right\|^2 dt,$$

which implies

$$\widetilde{y}^\top \mathbb{E}_{\tau,\alpha}^{A(x_1-\tau-t)^\alpha} B = \mathbf{0}^\top, \ \forall \, t \in J. \tag{5.60}$$

Note that (5.54) is relatively controllable. Therefore, $\exists \, \bar{u}(\cdot)$ such that $y(x_1) = \mathbf{0}$, i.e.,

$$y(x_1) = \mathbb{E}_{\tau,\alpha}^{A x_1^\alpha} \mathbf{b} + \int_{-\tau}^{0} \mathbb{E}_{\tau,\alpha}^{A(x_1-\tau-s)^\alpha} (^{RL}D_{-\tau+}^\alpha \mu)(s)ds$$

$$+ \int_{-\tau}^{x_1} \mathbb{E}_{\tau,\alpha}^{A(x_1-\tau-t)^\alpha} B\bar{u}(t)dt = \mathbf{0}. \tag{5.61}$$

Analogically, $\exists \, \widetilde{u}(\cdot)$ such that $y(x_1) = \widetilde{y}$, i.e.,

$$y(x_1) = \mathbb{E}_{\tau,\alpha}^{A x_1^\alpha} \mathbf{b} + \int_{-\tau}^{0} \mathbb{E}_{\tau,\alpha}^{A(x_1-\tau-s)^\alpha} (^{RL}D_{-\tau+}^\alpha \mu)(s)ds$$

$$+ \int_{-\tau}^{x_1} \mathbb{E}_{\tau,\alpha}^{A(x_1-\tau-t)^\alpha} B\widetilde{u}(t)dt = \widetilde{y}. \tag{5.62}$$

Then, by (5.61) and (5.62), we have

$$\widetilde{y} = \int_{-\tau}^{x_1} \mathbb{E}_{\tau,\alpha}^{A(x_1-\tau-t)^\alpha} B\triangle u(t)dt, \quad \triangle u = \widetilde{u} - \bar{u}. \tag{5.63}$$

According to (5.63), we have

$$\widetilde{y}^\top \widetilde{y} = \int_{-\tau}^{x_1} \widetilde{y}^\top \mathbb{E}_{\tau,\alpha}^{A(x_1-\tau-t)^\alpha} B\triangle u(t)dt.$$

Note that by (5.60), one can get $\widetilde{y}^\top \widetilde{y} = 0$. This contradicts the hypothesis that $\widetilde{y} \neq \mathbf{0}$. Therefore, $W_{\tau,\alpha}[-\tau, x_1]$ is nonsingular. $\qquad\square$

Next, we will give the rank criterion result.

Theorem 5.24. *System (5.54) is relatively controllable if and only if $x_1 > (k - 1)\tau$ and rank $S_k = k$.*

Proof. Assume that (5.54) is relatively controllable. Then, $\exists\ u^*(\cdot)$ such that (5.55) has a solution $y^*(\cdot)$ satisfying $y^*(x_1) = y_1$.

By Lemma 5.14, a solution $y \in C(\Omega, \mathbb{R}^n) \cap C([-\tau, T], \mathbb{R}^n)$ of (5.54) can be given by

$$y(x) = \mathbb{E}_{\tau,\alpha}^{Ax^\alpha} \mathbf{b} + \int_{-\tau}^{0} \mathbb{E}_{\tau,\alpha}^{A(x-\tau-s)^\alpha} (^{RL}D_{-\tau+}^\alpha \mu)(s)ds + \int_{-\tau}^{x} \mathbb{E}_{\tau,\alpha}^{A(x-\tau-t)^\alpha} Bu^*(t)dt.$$

By Definition 5.13, we have

$$y(x_1) = \mathbb{E}_{\tau,\alpha}^{Ax_1^\alpha} \mathbf{b} + \int_{-\tau}^{0} \mathbb{E}_{\tau,\alpha}^{A(x_1-\tau-s)^\alpha} (^{RL}D_{-\tau+}^\alpha \mu)(s)ds$$

$$+ \int_{-\tau}^{x_1} \mathbb{E}_{\tau,\alpha}^{A(x_1-\tau-t)^\alpha} Bu^*(t)dt. \tag{5.64}$$

Consider an arbitrary vector ξ satisfying the follow equation:

$$y_1 - \mathbb{E}_{\tau,\alpha}^{Ax_1^\alpha} \mathbf{b} - \int_{-\tau}^{0} \mathbb{E}_{\tau,\alpha}^{A(x_1-\tau-s)^\alpha} (^{RL}D_{-\tau+}^\alpha \mu)(s)ds = \xi. \tag{5.65}$$

Without loss of generality, let $x_1 \in ((n - 1)\tau, n\tau]$ and note that with the representation of $\mathbb{E}_{\tau,\alpha}^{A(x_1-\tau-s)^\alpha}$ in (5.4), we get

$$\int_{-\tau}^{x_1} \mathbb{E}_{\tau,\alpha}^{A(x_1-\tau-s)^\alpha} Bu^*(t)dt$$

$$= B \int_{-\tau}^{x_1} \frac{(x_1 - t)^{\alpha-1}}{\Gamma(\alpha)} u^*(t)dt + AB \int_{-\tau}^{x_1-\tau} \frac{(x_1 - \tau - t)^{2\alpha-1}}{\Gamma(2\alpha)} u^*(t)dt + \cdots +$$

$$+ A^{(n-1)}B \int_{-\tau}^{x_1-(n-1)\tau} \frac{(x_1 - (n-1)\tau - t)^{n\alpha-1}}{\Gamma(n\alpha)} u^*(t)dt.$$

Denote

$$\varphi_1(x_1) = \int_{-\tau}^{x_1} \frac{(x_1 - t)^{\alpha-1}}{\Gamma(\alpha)} u^*(t)dt,$$

$$\varphi_2(x_1) = \int_{-\tau}^{x_1-\tau} \frac{(x_1 - \tau - t)^{2\alpha-1}}{\Gamma(2\alpha)} u^*(t)dt, \quad \cdots,$$

$$\varphi_n(x_1) = \int_{-\tau}^{x_1-(n-1)\tau} \frac{(x_1 - (n-1)\tau - t)^{n\alpha-1}}{\Gamma(n\alpha)} u^*(t)dt.$$

Note that by (5.64) and (5.65), we have

$$B\varphi_1(x_1) + AB\varphi_2(x_1) + \cdots + A^{(n-1)}B\varphi_n(x_1) = \xi. \tag{5.66}$$

By the fact that (5.54) is relatively controllable, (5.66) has a solution for $\forall \xi$. Note that for $n \geq k$, (5.54) is relatively controllable such that $x_1 > (n-1)\tau \geq (k-1)\tau$.

The matrix A can be expressed by a linear combination of $I, A, A^2, \cdots, A^{k-1}$ for $\forall A^i$, $i \geq k$ [89,113]. For $n \geq k$ Eq. (5.66) can be replaced by

$$B\bar{\varphi}_1(x_1) + AB\bar{\varphi}_2(x_1) + \cdots + A^{(k-1)}B\bar{\varphi}_k(x_1) = \xi, \tag{5.67}$$

where $\bar{\varphi}_j(x_1)$ with $j = 1, 2, \cdots, k$ are some functions at x_1.

If (5.67) has the solution $\bar{\varphi}_j(x_1)$ for arbitrary ξ, then $\operatorname{rank} S_k = k$.

We prove our results by contradiction. Suppose that (5.54) is not relatively controllable and $\operatorname{rank} S_k = k$. By Theorem 5.23, the matrix $W_{\tau,\alpha}[-\tau, x_1]$ is singular. Therefore, there exists at least one nonzero state $\bar{y} \in \mathbb{R}^n$ such that

$$0 = \bar{y}^\top W_{\tau,\alpha}[-\tau, x_1]\bar{y} = \int_{-\tau}^{x_1} \bar{y}^\top \mathbb{E}_{\tau,\alpha}^{A(x_1-\tau-t)^\alpha} BB^\top \mathbb{E}_{\tau,\alpha}^{A^\top(x_1-\tau-t)^\alpha} \bar{y} dt$$

$$= \int_{-\tau}^{x_1} \left(\bar{y}^\top \mathbb{E}_{\tau,\alpha}^{A(x_1-\tau-t)^\alpha} B\right)\left(\bar{y}^\top \mathbb{E}_{\tau,\alpha}^{A(x_1-\tau-t)^\alpha} B\right)^\top dt,$$

which implies

$$\bar{y}^\top \mathbb{E}_{\tau,\alpha}^{A(x_1-\tau-t)^\alpha} B = \mathbf{0}^\top, \quad -\tau < t \leq x_1.$$

Changing variables $x_1 - \tau - t = x$, we have

$$\bar{y}^\top \mathbb{E}_{\tau,\alpha}^{Ax^\alpha} B = \mathbf{0}^\top, \quad -\tau < x \leq x_1.$$

This contradicts Lemma 5.16. This proof is completed. $\qquad\square$

5.2.2.2 *Numerical example and discussion*

Example 5.6. Set $\alpha = 0.6$, $p = 2$, $\tau = 0.4$, $x_1 = 0.8$, and $k = 2$. We study

$$
\begin{cases}
{}^{RL}D^{0.6}_{-0.4+}\,y(x) = Ay(x - 0.4) + Bu(x), \quad x \in (0, x_1], \\
\mu(x) = (x, 2x^2)^\top, \quad -0.4 < x \le 0, \\
(\mathbb{I}^{0.4}_{-0.4+}\,y)(-0.4^+) = \mathbf{b} = (-0.4, 0.32)^\top,
\end{cases}
\tag{5.68}
$$

where $u \in L^2([-0.4, x_1], \mathbb{R}^2)$ and

$$
y(x) = \begin{pmatrix} y_1(x) \\ y_2(x) \end{pmatrix}, \quad
A = \begin{pmatrix} 1 & 0.2 \\ 0.2 & 0.3 \end{pmatrix}, \quad
B = \begin{pmatrix} 1 & 0.5 \\ 0.2 & 1 \end{pmatrix}.
$$

One computes that the fractional delayed Grammian matrix of system (5.68) via (5.56) can be achieved:

$$
W_{0.4,0.6}[-0.4, 0.8] = \int_{-0.4}^{0.8} \mathbb{E}^{A(0.4-t)^{0.6}}_{0.4,0.6}\, BB^\top \mathbb{E}^{A^\top(0.4-t)^{0.6}}_{0.4,0.6}\, dt = W_1 + W_2 + W_3,
$$

where

$$
W_1 = \int_{-0.4}^{0} \left(I\frac{(0.8 - t)^{-0.4}}{\Gamma(0.6)} + A\frac{(0.4 - t)^{0.2}}{\Gamma(1.2)} + A^2\frac{(-t)^{0.8}}{\Gamma(1.8)} \right) BB^\top
$$
$$
\times \left(I\frac{(0.8 - t)^{-0.4}}{\Gamma(0.6)} + A^\top\frac{(0.4 - t)^{0.2}}{\Gamma(1.2)} + (A^\top)^2\frac{(-t)^{0.8}}{\Gamma(1.8)} \right) dt,
$$
$$
W_2 = \int_{0}^{0.4} \left(I\frac{(0.8 - t)^{-0.4}}{\Gamma(0.6)} + A\frac{(0.4 - t)^{0.2}}{\Gamma(1.2)} \right) BB^\top \left(I\frac{(0.8 - t)^{-0.4}}{\Gamma(0.6)} \right.
$$
$$
\left. + A^\top\frac{(0.4 - t)^{0.2}}{\Gamma(1.2)} \right) dt,
$$
$$
W_3 = \int_{0.4}^{0.8} \left(I\frac{(0.8 - t)^{-0.4}}{\Gamma(0.6)} \right) BB^\top \left(I\frac{(0.8 - t)^{-0.4}}{\Gamma(0.6)} \right) dt.
$$

By a direct computation, one can get

$$
W_1 = \begin{pmatrix} 2.47 & 1.04 \\ 1.04 & 0.67 \end{pmatrix}, \quad
W_2 = \begin{pmatrix} 1.48 & 0.7 \\ 0.7 & 0.607 \end{pmatrix}, \quad
W_3 = \begin{pmatrix} 2.35 & 1.31 \\ 1.31 & 1.95 \end{pmatrix}.
$$

Therefore, we have

$$
W_{0.4,0.6}[-0.4, 0.8] = \begin{pmatrix} 6.3 & 3.05 \\ 3.05 & 3.22 \end{pmatrix},
$$

$$
W^{-1}_{0.4,0.6}[-0.4, 0.8] = \begin{pmatrix} 0.28 & -0.29 \\ -0.29 & 0.56 \end{pmatrix}.
$$

Set $y(x_1) = (y_1, y_2)$. By (5.57), one can construct $u \in L^2([-0.4, x_1], \mathbb{R}^2)$ as

$$u(t) = B^\top \mathbb{Z}_{0.4,0.6}^{A^\top (0.4-t)^{0.6}} W_{0.4,0.6}^{-1}[-0.4, 0.8]\zeta$$

$$= \begin{cases} B^\top \left(I \frac{(0.8-t)^{-0.4}}{\Gamma(0.6)} + A^\top \frac{(0.4-t)^{0.2}}{\Gamma(1.2)} + (A^\top)^2 \frac{(-t)^{0.8}}{\Gamma(1.8)} \right) W_{0.4,0.6}^{-1}[-0.4, 0.8]\zeta, \\ \qquad\qquad\qquad\qquad\qquad\qquad\qquad\qquad\qquad\qquad t \in [-0.4, 0), \\[2mm] B^\top \left(I \frac{(0.8-t)^{-0.4}}{\Gamma(0.6)} + A^\top \frac{(0.4-t)^{0.2}}{\Gamma(1.2)} \right) W_{0.4,0.6}^{-1}[-0.4, 0.8]\zeta, \ t \in [0, 0.4), \\[2mm] B^\top \left(I \frac{(0.8-t)^{-0.4}}{\Gamma(0.6)} \right) W_{0.4,0.6}^{-1}[-0.4, 0.8]\zeta, \qquad\qquad\qquad t \in [0.4, 0.8), \end{cases}$$

$$(5.69)$$

where

$$\zeta = x(x_1) - \mathbb{E}_{0.4,0.6}^{A0.8^{0.6}} \mathbf{b} - \int_{-0.4}^{0} \mathbb{E}_{0.4,0.6}^{A(0.4-s)^{0.6}} (^{RL} D_{-0.4+}^{0.6} \mu)(s) ds$$

$$= \begin{pmatrix} y_1 + 0.449 \\ y_2 - 0.092 \end{pmatrix},$$

where

$$(^{RL} D_{-0.4+}^{0.6} \mu)(s)$$

$$= \frac{1}{\Gamma(0.4)} \frac{d}{ds} \int_{-0.4}^{s} (s-t)^{-0.6} \begin{pmatrix} t \\ 2t^2 \end{pmatrix} dt$$

$$= \frac{1}{\Gamma(0.4)} \begin{pmatrix} \frac{25}{14}(s+0.4)^{0.4} + \frac{1}{35}(25s-4)(s+0.4)^{-0.6} \\ \frac{5}{42}(50s-4)(s+0.4)^{0.4} + \frac{1}{21}(25s^2 - 4s + \frac{28}{25})(s+0.4)^{0.6} \end{pmatrix}.$$

Consider

$$S_k = \{B, AB\} = \begin{pmatrix} 1 & 0.5 & 1.04 & 0.7 \\ 0.2 & 1 & 0.26 & 0.4 \end{pmatrix}.$$

Obviously, rank$S_k = 2$.

From the above, system (5.68) is relatively controllable via Theorems 5.23 and 5.24.

From Lemma 5.14 and (5.69), the solution of system (5.68) has the following form:

$$y(x) = \mathbb{E}_{0.4,0.6}^{Ax^{0.6}} \mathbf{b} + \int_{-0.4}^{0} \mathbb{E}_{0.4,0.6}^{A(x-0.4-s)^{0.6}} (^{RL} D_{-0.4+}^{0.6} \mu)(s) ds$$

$$+ \int_{-0.4}^{x} \mathbb{E}_{0.4,0.6}^{A(x-0.4-t)^{0.6}} Bu(t) dt.$$

Now we consider the integral term $\int_{-0.4}^{x} \mathbb{Z}_{0.4,0.6}^{A(x-0.4-t)^{0.6}} Bu(t)dt$ in (5.69).

Letting $-0.4 \le t \le x$ and $0 < x \le 0.4$, we can obtain $-0.4 < x - 0.4 - t \le x - 0.4$ and $x - 0.4 < x - 0.4 - t \le x$. Therefore, the integral term $\int_{-0.4}^{x} \mathbb{E}_{0.4,0.6}^{A(x-0.4-t)^{0.6}} Bu(t)dt$ in (5.69) can be given by

$$
\int_{-0.4}^{x} \mathbb{E}_{0.4,0.6}^{A(x-0.4-t)^{0.6}} Bu(t)dt
$$

$$
= \left(\int_{-0.4}^{x} \frac{(x-t)^{-0.4}}{\Gamma(0.6)} BB^{\top} \frac{(0.8-t)^{-0.4}}{\Gamma(0.6)} dt \right.
$$

$$
+ \int_{-0.4}^{x} \frac{(x-t)^{-0.4}}{\Gamma(0.6)} BB^{\top} A^{\top} \frac{(0.4-t)^{0.2}}{\Gamma(1.2)} dt
$$

$$
+ \int_{-0.4}^{0} \frac{(x-t)^{-0.4}}{\Gamma(0.6)} BB^{\top} (A^{\top})^2 \frac{(-t)^{0.8}}{\Gamma(1.8)} dt
$$

$$
+ \int_{-0.4}^{x-0.4} A \frac{(x-0.4-t)^{0.2}}{\Gamma(1.2)} BB^{\top}
$$

$$
\left. \times \left(I \frac{(0.8-t)^{-0.4}}{\Gamma(0.6)} + A^{\top} \frac{(0.4-t)^{0.2}}{\Gamma(1.2)} + (A^{\top})^2 \frac{(-t)^{0.8}}{\Gamma(1.8)} \right) dt \right)
$$

$$
\times W_{0.4,0.6}^{-1}[-0.4, 0.8]\zeta.
$$

Letting $-0.4 \le t \le x$ and $0.4 < x \le 0.8$, we can obtain $-0.4 < x - 0.4 - t \le x - 0.8$, $x - 0.8 < x - 0.4 - t \le x - 0.4$, and $x - 0.4 < x - 0.4 - t \le x$. Therefore, the integral term $\int_{-0.4}^{x} \mathbb{E}_{0.4,0.6}^{A(x-0.4-t)^{0.6}} Bu(t)dt$ in $\int_{-0.4}^{x} \mathbb{E}_{0.4,0.6}^{A(x-0.4-t)^{0.6}} Bu(t)dt$ in (5.69) can be given by

$$
\int_{-0.4}^{x} \mathbb{E}_{0.4,0.6}^{A(x-0.4-t)^{0.6}} Bu(t)dt
$$

$$
= \left(\int_{-0.4}^{x} \frac{(x-t)^{-0.4}}{\Gamma(0.6)} BB^{\top} \frac{(0.8-t)^{-0.4}}{\Gamma(0.6)} dt \right.
$$

$$
+ \int_{-0.4}^{0.4} \frac{(x-t)^{-0.4}}{\Gamma(0.6)} BB^{\top} A^{\top} \frac{(0.4-t)^{0.2}}{\Gamma(1.2)} dt
$$

$$
+ \int_{-0.4}^{0} \frac{(x-t)^{-0.4}}{\Gamma(0.6)} BB^{\top} (A^{\top})^2 \frac{(-t)^{0.8}}{\Gamma(1.8)} dt
$$

$$
+ \int_{-0.4}^{0} A \frac{(x-0.4-t)^{0.2}}{\Gamma(1.2)} BB^{\top} (A^{\top})^2 \frac{(-t)^{0.8}}{\Gamma(1.8)} dt
$$

$$
+ \int_{-0.4}^{x-0.4} A \frac{(x-0.4-t)^{0.2}}{\Gamma(1.2)} BB^{\top} \left(I \frac{(0.8-t)^{-0.4}}{\Gamma(0.6)} + A^{\top} \frac{(0.4-t)^{0.2}}{\Gamma(1.2)} \right) dt
$$

$$
+ \int_{-0.4}^{x-0.8} A^2 \frac{(x-0.8-t)^{0.8}}{\Gamma(1.8)} BB^{\top} \left(I \frac{(0.8-t)^{-0.4}}{\Gamma(0.6)} + A^{\top} \frac{(0.4-t)^{0.2}}{\Gamma(1.2)} \right.
$$

$$+ (A^\top)^2 \frac{(-t)^{0.8}}{\Gamma(1.8)}\bigg)dt\bigg) W^{-1}_{0.4,0.6}[-0.4, 0.8]\zeta.$$

5.2.2.3 Conclusions

We give the delayed Grammian matrix and rank criterion to examine whether a linear delay system is relatively controllable. The results in this part are motivated from [111].

Chapter 6

Difference delay systems

6.1 Controllability

6.1.1 Controllability for linear discrete delay systems

In this part, we find other control functions for the problem from [4]. Moreover, we state an equivalent condition for the relative controllability assuming the final point to be from a specific linear subspace, i.e., for the restricted relative controllability. We introduce the notation $\mathbb{Z}_p^q := \{p, p+1, \ldots, q-1, q\}$, where $p, q \in \mathbb{Z}$, $\mathbb{Z} := \{0, \pm 1, \pm 2, \ldots\}$ with $p \le q$.

The definition of discrete delayed matrix exponential $e_m^{B\cdot}$ was given in [2,3].

Definition 6.1. For an $n \times n$ constant matrix B, we define the discrete matrix delayed exponential $e_m^{B\cdot} : \mathbb{R} \to \mathbb{R}^{n \times n}$ by the following discrete matrix function:

$$
e_m^{Bk} := \begin{cases}
\Theta, & \text{if } k \in \mathbb{Z}_{-\infty}^{-m-1}, \\
I, & \text{if } k \in \mathbb{Z}_{-m}^{0}, \\
I + B\frac{k!}{1!(k-1)!} + B^2\frac{(k-m)!}{2!(k-m-2)!} + \ldots + B^s\frac{(k-(s-1)m)!}{s!(k-(s-1)m-s)!}, \\
\quad \text{if } k \in \mathbb{Z}_{(s-1)(m+1)+1}^{s(m+1)}, \ s = 1, 2, \ldots,
\end{cases}
\tag{6.1}
$$

where $m \ge 1$.

The main property of e_m^{Bk} proved in [3, Theorem 2.1] is

$$
\Delta e_m^{Bk} := e_m^{B(k+1)} - e_m^{Bt} = B e_m^{B(k-m)}, \quad k \in \mathbb{Z}_{-m}^{\infty}.
$$

We shall always consider the forward difference $\Delta x(k) := x(k+1) - x(k)$ and investigate the relative controllability in the sense of the following definition.

Definition 6.2. Let $m \in \mathbb{N}$, $b \in \mathbb{R}^n$ and let B be an $n \times n$ matrix. System

$$
\Delta x(k) = Bx(k-m) + bu(k), \quad k \in \mathbb{Z}_0^{\infty},
\tag{6.2}
$$

is relatively controllable if for any initial function $\varphi : \mathbb{Z}_{-m}^{0} \to \mathbb{R}^n$, any finite terminal state $x^* \in \mathbb{R}^n$, and any finite terminal point $k_1 \ge k^* \in \mathbb{N}$ with a fixed k^* there exists a discrete function $u^* : \mathbb{Z}_0^{k_1-1} \to \mathbb{R}$ such that the system

$$
\Delta x(k) = Bx(k-m) + bu^*(k), \quad k \in \mathbb{Z}_0^{k_1-1},
$$

Stability and Controls Analysis for Delay Systems. https://doi.org/10.1016/B978-0-32-399792-8.00012-8

Copyright © 2023 Elsevier Inc. All rights reserved.

has a solution $x : \mathbb{Z}^{k_1}_{-m} \to \mathbb{R}^n$ such that $x(k_1) = x^*$ and

$$x(k) = \varphi(k), \quad k \in \mathbb{Z}^0_{-m}. \tag{6.3}$$

We recall that a solution $x : \mathbb{Z}^{\infty}_{-m} \to \mathbb{R}^n$ of Cauchy problem (6.2)–(6.3) can be represented in the form [3]

$$x(k) = e_m^{Bk}\varphi(-m) + \sum_{j=-m+1}^{0} e_m^{B(k-m-j)}\Delta\varphi(j-1) + \sum_{j=1}^{k} e_m^{B(k-m-j)}bu(j-1). \tag{6.4}$$

Now, we recall two results on relative controllability from [4] which we shall extend in this section.

Theorem 6.1. *Problem*

$$\Delta x(k) = Bx(k-m) + bu(k), \quad k \in \mathbb{Z}^{k_1-1}_0, \tag{6.5}$$

$$x(k) = \varphi(k), \quad k \in \mathbb{Z}^0_{-m}, \tag{6.6}$$

$$x(k_1) = x^* \tag{6.7}$$

is relatively controllable if and only if assumptions

$$rank(b, Bb, B^2b, \ldots, B^{n-1}b) = n \tag{6.8}$$

and

$$k_1 \geq k^* := (n-1)(m+1) + 1 \tag{6.9}$$

hold simultaneously.

Theorem 6.2. *Let conditions (6.8) and (6.9) be valid. Then a control function* $u = u^*$ *for the problem (6.5)–(6.7) can be expressed in the form*

$$u^*(k) = b^\top (e_m^{B(k_1-m-k-1)})^\top G^{-1}\xi \tag{6.10}$$

for $k \in \mathbb{Z}^{k_1-1}_0$, *where*

$$G = \sum_{j=1}^{k_1} e_m^{B(k_1-m-j)}bb^\top e_m^{B^\top(k_1-m-j)}, \tag{6.11}$$

$$\xi = x^* - e_m^{Bk_1}\varphi(-m) - \sum_{j=-m+1}^{0} e_m^{B(k_1-m-j)}\Delta\varphi(j-1). \tag{6.12}$$

Note that the authors of [4] followed a classical construction of a control function (6.10) for linear difference equations (see, e.g., [114]), although they used the discrete delayed matrix exponential and obtained results for delay difference equations. In this section, we generalize this approach by constructing new control functions.

Throughout this section, $[u, v]$ shall denote the linear span of vectors u, v and θ_p shall denote a p-dimensional zero vector (we omit the index if $p = n$).

We present a simple corollary for the relative controllability of weakly nonlinear systems.

Corollary 6.1. *Let $f : \mathbb{Z}_0^\infty \times \mathbb{R}^n \times \mathbb{R}^n \times \mathbb{R} \times \mathbb{R} \to \mathbb{R}^n$ be a given C^1-smooth function in all arguments and let $\varepsilon \in \mathbb{R}$ be close to 0. The problem consisting of equation*

$$\Delta x(k) = Bx(k - m) + bu(k) + \varepsilon f(k, x(k), x(k - m), u(k), \varepsilon) \tag{6.13}$$

for $k \in \mathbb{Z}_0^{k_1 - 1}$ and conditions (6.6)–(6.7) is relatively controllable if conditions (6.8)–(6.9) hold simultaneously.

Proof. Consider the Banach spaces $X = \mathbb{R}^{k_1}$, $Y = \mathbb{R}$, $Z = \mathbb{R}^n$ with usual maximum norms. Problem (6.13), (6.6), (6.7) is relatively controllable if there is a solution u of equation $\mathcal{F}(u, \varepsilon) = \theta$ with $\mathcal{F} : X \times Y \to Z$ given by

$$\mathcal{F}(u, \varepsilon) := \sum_{j=1}^{k_1} e_m^{B(k_1 - m - j)} (bu(j - 1)$$

$$+ \varepsilon f(j - 1, x(j - 1), x(j - m - 1), u(j - 1), \varepsilon)) - \xi$$

for ξ of (6.12) and $u = (u(0), u(1), \ldots, u(k_1 - 1))$. By Theorem 6.1 we get the existence of $\hat{u} \in \mathbb{R}^{k_1}$ satisfying $\mathcal{F}(\hat{u}, 0) = \theta$, and moreover, we find that the linear mapping

$$K = D_u \mathcal{F}(\hat{u}, 0) \colon \mathbb{R}^{k_1} \to \mathbb{R}^n : v \mapsto \sum_{j=1}^{k_1} e_m^{B(k_1 - m - j)} bv(j - 1) \tag{6.14}$$

is surjective. So, the surjective function theorem can be applied, and one obtains the existence of neighborhoods $U_1 \times U_2 \subset \mathbb{R}^{k_1}$ of $\hat{u}$ in $\mathbb{R}^{k_1}$ for $U_1 \subset \ker K$ with $\dim \ker K = k_1 - n$ and $U_2 \subset X_2$ with a complement X_2 of $\ker K$ in $\mathbb{R}^{k_1}$, so $\dim X_2 = n$, and $V \subset \mathbb{R}$ of 0 in $\mathbb{R}$, and a unique C^1-function $u_2(u_1, \varepsilon)$ such that $\mathcal{F}(u, \varepsilon) = \theta$ for $(u, \varepsilon) \in U_1 \times U_2 \times V$ if and only if $u = (u_1, u_2(u_1, \varepsilon))$. Moreover, $u_2(u_1, 0) = \hat{u}_2$ for $\hat{u} = (\hat{u}_1, \hat{u}_2) \in U_1 \times U_2$. The proof is complete. $\square$

Remark 6.1. We note that neighborhoods U_1, U_2, V and the smallness of ε depend on the inputs in (6.6)–(6.7). On the other hand, when f is globally Lipschitz continuous in the second, third, and fourth variables with a constant L, then writing

$$\mathcal{F}(u, \varepsilon) = Ku + \varepsilon \mathcal{G}(u, \varepsilon) - \xi,$$

we see that $\mathcal{G}\colon \mathbb{R}^{k_1} \times \mathbb{R} \to \mathbb{R}^n$ is also globally Lipschitz continuous with a constant $L_{\mathcal{G}}$, where, of course, $L_{\mathcal{G}}$ can be computed by L but this formula is awkward so we omit it. Since $K\colon \mathbb{R}^{k_1} \to \mathbb{R}^n$ is surjective, it has a right inverse, one is given by the Moore–Penrose pseudoinverse $K^\top (K K^\top)^{-1}$ from Theorem 6.2 determined by (6.10) and (6.11). To see this, we take the matrix representation of linear mapping K given by (6.14),

$$
Ku = \left(e_m^{B(k_1-m-1)}b \quad e_m^{B(k_1-m-2)}b \quad \cdots \quad e_m^{B(-m)}b \right) u. \tag{6.15}
$$

Denoting $s_j := e_m^{B(k_1-m-j)}b = (s_{j1}, s_{j2}, \ldots, s_{jn})^\top$ for $j \in \mathbb{Z}_1^{k_1}$ the columns of K we get

$$
K K^\top = (s_1, s_2, \ldots, s_{k_1})
\begin{pmatrix}
s_1^\top \\
s_2^\top \\
\vdots \\
s_{k_1}^\top
\end{pmatrix}
$$

$$
= (s_1 s_{11} + s_2 s_{21} + \cdots + s_{k_1} s_{k_1 1}, s_1 s_{12} + s_2 s_{22} + \cdots + s_{k_1} s_{k_1 2}, \ldots,
$$
$$
s_1 s_{1n} + s_2 s_{2n} + \cdots + s_{k_1} s_{k_1 n})
$$
$$
= (s_1 s_{11}, s_1 s_{12}, \ldots, s_1 s_{1n}) + (s_2 s_{21}, s_2 s_{22}, \ldots, s_2 s_{2n}) + \cdots
$$
$$
+ (s_{k_1} s_{k_1 1}, s_{k_1} s_{k_1 2}, \ldots, s_{k_1} s_{k_1 n})
$$
$$
= s_1 s_1^\top + s_2 s_2^\top + \cdots + s_{k_1} s_{k_1}^\top = \sum_{j=1}^{k_1} s_j s_j^\top,
$$

which is exactly matrix G of (6.11). Moreover, $(K^\top u)_j = s_j^\top u \; \forall j \in \mathbb{Z}_1^{k_1}$ for the j-th coordinate of the vector $K^\top u$, and formula (6.10) follows.

Clearly, any right inverse $K^{-1}\colon \mathbb{R}^n \to \mathbb{R}^{k_1}$ gives X_2 and vice versa. Then we split $u = u_1 + u_2$, $u_1 \in \ker K$, $u_2 \in X_2$, and $\mathcal{F}(u, \varepsilon) = \theta$ is equivalent to

$$
u_2 = K^{-1} (\xi - \varepsilon \mathcal{G}(u_1 + u_2, \varepsilon)). \tag{6.16}
$$

Certainly, the map

$$
X_2 \ni u_2 \mapsto K^{-1} (\xi - \varepsilon \mathcal{G}(u_1 + u_2, \varepsilon)) \in X_2
$$

is globally Lipschitz with a constant $|\varepsilon| \|K^{-1}\| L_{\mathcal{G}}$, so if $|\varepsilon| \|K^{-1}\| L_{\mathcal{G}} < 1$, then the Banach fixed point theorem gives a unique solution $u_2 = u_2(\xi, u_1, \varepsilon)$ of (6.16) for any $\xi \in \mathbb{R}^n$, $u_1 \in \ker K$, and $u_2(\xi, u_1, \varepsilon)$ is globally Lipschitz in ξ and u_1.

6.1.1.1 New control functions

In this section, we provide control functions for the boundary problem (6.5)–(6.7) which are, in general, different from u^* given by (6.10). Later, we state conditions for relative controllability restricted to a special subspace of $\mathbb{R}^n$. It will be shown that if the right-hand condition x^* in (6.7) lies in this subspace, it can be achieved sooner by the choice of u^*, i.e., k_1 can be lower than in (6.9).

First, we state the results on the control functions.

Theorem 6.3. *Let* $v\colon \mathbb{Z}_1^{k_1} \to \mathbb{R}^n$ *be such that the matrix*

$$W := \sum_{j=1}^{k_1} e_m^{B(k_1-m-j)} b v^\top(j) \tag{6.17}$$

is nonsingular and the conditions for relative controllability (6.8)–(6.9) are valid. Then a control function $u = u^*$ for the problem (6.5)–(6.7) can be expressed in the form

$$u^*(k) = v^\top(k+1)W^{-1}\xi, \quad k \in \mathbb{Z}_0^{k_1-1}. \tag{6.18}$$

Proof. We look for the control function in the form $u(j-1) = v^\top(j)D$ with constant vector $D = (D_1, \dots, D_n)^\top \in \mathbb{R}^n$. Since $u = u^*$ has to satisfy (6.4) at k_1 and (6.7) at once, we get

$$x^* = e_m^{Bk_1}\varphi(-m) + \sum_{j=-m+1}^{0} e_m^{B(k_1-m-1)}\Delta\varphi(j-1) + \sum_{j=1}^{k_1} e_m^{B(k_1-m-j)}bu(j-1)$$

or, in the notation (6.12),

$$\sum_{j=1}^{k_1} e_m^{B(k_1-m-j)} b v^\top(j)D = \xi.$$

Using the property of matrix W we obtain $D = W^{-1}\xi$. Hence $u(j-1) = v^\top(j)W^{-1}\xi$. $\quad\square$

Remark 6.2. When one sets $v(k) = e_m^{B(k_1-m-k)}b$ in the latter theorem, one gets

$$u^*(k) = b^\top e_m^{B^\top(k_1-m-k-1)} \left[\sum_{j=1}^{k_1} e_m^{B(k_1-m-j)}bb^\top e_m^{B^\top(k_1-m-j)} \right]^{-1} \xi.$$

Thus, in this case, the theorem coincides with Theorem 6.2. We note that in [4, Lemma 3.4] it is also proved that the matrix G of (6.11) is nonsingular.

The next lemma provides a necessary condition for function $v(j)$ to satisfy the assumption of Theorem 6.3.

Lemma 6.1. *If matrix*

$$\widetilde{W} := \sum_{j=1}^{k_1} w(j)v^\top(j) \tag{6.19}$$

with some $w \colon \mathbb{Z}_1^{k_1} \to \mathbb{R}^n$ *is nonsingular, then the elements* $(v(j))_i$, $i = 1, \ldots, n$, *are linearly independent on* $\mathbb{Z}_1^{k_1}$, *i.e., there is no nontrivial constant vector* $\mu = (\mu_1, \ldots, \mu_n)^\top$ *such that* $\mu^\top v(j) = 0$ *for all* $j \in \mathbb{Z}_1^{k_1}$.

Proof. Let there exist $\mu = (\mu_1, \ldots, \mu_n)^\top \neq \theta$ such that $\mu^\top v(j) = 0$ on $\mathbb{Z}_1^{k_1}$. Then

$$\left(\sum_{j=1}^{k_1} w(j)v^\top(j)\right)\mu = \sum_{j=1}^{k_1} w(j)(v^\top(j)\mu) = 0,$$

which is a contradiction with the nonsingularity of the matrix $\widetilde{W}$. $\qquad\square$

The next example compares Theorem 6.2 and Theorem 6.3.

Example 6.1. Consider the following boundary value problem:

$$
\begin{aligned}
\Delta x(k) &= x(k-2) + 2y(k-2) + u(k), & k &\in \mathbb{Z}_0^3, \\
\Delta y(k) &= x(k-2), & k &\in \mathbb{Z}_0^3, \\
x(k) &= y(k) = 0, & k &\in \mathbb{Z}_{-2}^0, \\
(x(4), y(4)) &= (\xi_1, \xi_2) \in \mathbb{R}^2.
\end{aligned}
$$

It is easy to see that $n = m = 2$, $k_1 = 4$, $\varphi(k) = (0,0)^\top$ for $k \in \mathbb{Z}_{-2}^0$, $B = \left(\begin{smallmatrix} 1 & 2 \\ 1 & 0 \end{smallmatrix}\right)$, and $b = (1,0)^\top$. Hence $Bb = (1,1)^\top$, $k^* = 4$, and conditions (6.8)–(6.9) are immediately satisfied. Theorem 6.1 yields the existence of $u \colon \mathbb{Z}_0^3 \to \mathbb{R}$ such that there is a solution of this problem. From the definition of the discrete delayed matrix exponential we get

$$
e_m^{B(k_1 - m - j)} b = \begin{cases} (2,1)^\top, & j = 1, \\ (1,0)^\top, & j = 2, 3, 4. \end{cases} \tag{6.20}
$$

Following Theorem 6.2, one constructs

$$
G = \begin{pmatrix} 7 & 2 \\ 2 & 1 \end{pmatrix}, \quad G^{-1} = \frac{1}{3}\begin{pmatrix} 1 & -2 \\ -2 & 7 \end{pmatrix}, \quad \xi = (x(k^*), y(k^*))^\top = (\xi_1, \xi_2)^\top
$$

and control function $u = u^*$, which is concluded in Table 6.1 together with the values of solution (x, y) given by (6.4). On the other side, using Theorem 6.3

TABLE 6.1 Control function and the corresponding solution of Example 6.1.

k	-2	-1	0	1	2	3	4
$u^*(k)$			ξ_2	$\frac{\xi_1-2\xi_2}{3}$	$\frac{\xi_1-2\xi_2}{3}$	$\frac{\xi_1-2\xi_2}{3}$	
$x(k)$	0	0	0	ξ_2	$\frac{\xi_1+\xi_2}{3}$	$\frac{2\xi_1-\xi_2}{3}$	ξ_1
$y(k)$	0	0	0	0	0	0	ξ_2

with

$$
v(j) = \begin{cases} (1,0)^\top, & j=1, \\ (0,1)^\top, & j=2, \\ (0,0)^\top, & j=3,4 \end{cases}
$$

(which does not violate the necessary condition of Lemma 6.1) leads to

$$
W = \begin{pmatrix} 2 \\ 1 \end{pmatrix} \begin{pmatrix} 1 & 0 \end{pmatrix} + \begin{pmatrix} 1 \\ 0 \end{pmatrix} \begin{pmatrix} 0 & 1 \end{pmatrix} = \begin{pmatrix} 2 & 1 \\ 1 & 0 \end{pmatrix}, \quad W^{-1} = \begin{pmatrix} 0 & 1 \\ 1 & -2 \end{pmatrix}
$$

by (6.17) and, consequently, to Table 6.2 by (6.18) and (6.4).

TABLE 6.2 Another control function and the corresponding solution of Example 6.1.

k	-2	-1	0	1	2	3	4
$u^*(k)$			ξ_2	$\xi_1 - 2\xi_2$	0	0	
$x(k)$	0	0	0	ξ_2	$\xi_1 - \xi_2$	$\xi_1 - \xi_2$	ξ_1
$y(k)$	0	0	0	0	0	0	ξ_2

From the previous example, one can see how to find a nonzero control function implying a nonzero solution that satisfies the zero initial condition and even the zero final condition $x(k_1) = \theta$. It suffices to set $u^* = u_1 - u_2$, where u_1, u_2 are two different control functions of one problem with a nonzero final condition. Such a nonzero control function u^* leading to a nonzero solution satisfying zero initial and final conditions can be found neither by Theorem 6.2 nor by Theorem 6.3, due to multiplication by ξ in (6.10) and (6.18).

For instance, subtracting control functions of Table 6.1 and Table 6.2 leads to a family of control functions $\{u_\zeta^*\}_{\zeta \in \mathbb{R}}$ parametrized by $\zeta = \frac{\xi_1 - 2\xi_2}{3} \in \mathbb{R}$, and each u_ζ^* results in the zero final condition of Example 6.2 (see Table 6.3).

Another way is to use the next result providing a complete characterization of control functions for problem (6.5)–(6.7). In what follows, we denote by $|_Z$ the restriction onto a linear space Z and by $(\ker K)^\perp$ an orthogonal complement to a null space of a linear operator K.

TABLE 6.3 Control function and the corresponding solution of Example 6.1 with $\xi = (0, 0)$.

k	-2	-1	0	1	2	3	4
$u_\zeta^*(k)$			0	-2ζ	ζ	ζ	
$x_\zeta(k)$	0	0	0	0	-2ζ	$-\zeta$	0
$y_\zeta(k)$	0	0	0	0	0	0	0

Theorem 6.4. *Let system (6.5)–(6.7) be relatively controllable. Then each control function* $u^* : \mathbb{Z}_0^{k_1-1} \to \mathbb{R}^n$ *fulfills*

$$\overline{u}^* = K|_{(\ker K)^\perp}^{-1} \xi + \overline{U}$$

for some $\overline{U} \in \ker K$, *where* $\overline{u}^* = (u^*(0), \dots, u^*(k_1 - 1))^\top$ *is a vector representation of function* u^* *and* $K : \mathbb{R}^{k_1} \to \mathbb{R}^n$ *is a linear operator defined by (6.14) and represented by the* $n \times k_1$ *matrix (6.15).*

Proof. Let u^* be a control function for (6.5)–(6.7). Then by (6.4) and (6.7), u^* satisfies

$$\sum_{j=1}^{k_1} e_m^{B(k_1-m-j)} b u^*(j-1) = \xi.$$

Equivalently, we can use matrix notation and write $K\overline{u}^* = \xi$. Simultaneously, there is a unique decomposition $\overline{u}^* = \overline{u}_1^* + \overline{u}_2^*$, $\overline{u}_1^* \in (\ker K)^\perp$, $\overline{u}_2^* \in \ker K$. Therefore, $\overline{u}_1^* = K|_{(\ker K)^\perp}^{-1} \xi$ and $\overline{u}_2^* = \overline{U}$. $\square$

Remark 6.3. Of course we can take instead of $K|_{(\ker K)^\perp}^{-1}$ the Moore–Penrose pseudoinverse $K^T (K K^T)^{-1}$.

Now, we apply the above theorem on Example 6.1 to find all control functions leading to zero final conditions, i.e., we look for the controls for Example 6.1 with $\xi = (0, 0)^\top$. By (6.20) one gets $K = \left(\begin{smallmatrix} 2 & 1 & 1 & 1 \\ 1 & 0 & 0 & 0 \end{smallmatrix}\right)$, $\ker K = [(0, 1, -1, 0)^\top, (0, 1, 0, -1)^\top]$, $(\ker K)^\perp = [(1, 0, 0, 0)^\top, (0, 1, 1, 1)^\top]$. So in fact, there is a family of two-parameter control functions $\{u_{\zeta_1, \zeta_2}^*\}_{\zeta_1, \zeta_2 \in \mathbb{R}}$ such that the corresponding vector representation is $\overline{u}_{\zeta_1, \zeta_2}^* = (0, \zeta_1 + \zeta_2, -\zeta_1, -\zeta_2)^\top$. One can see that the control u_ζ^* of Table 6.3 is just a special case when $\zeta_1 = \zeta_2 = -\zeta$.

6.1.1.2 Relative controllability

Theorem 6.5. *Let* $M \subset \mathbb{R}^n$ *be B an invariant linear subspace of dimension* p, $x^* \in M$, $\varphi(k) \in M$ *for each* $k \in \mathbb{Z}_{-m}^0$, *and* $b \in M$. *Problem (6.5)–(6.7) is relatively controllable if and only if assumptions*

$$rank(b, Bb, B^2b, \dots, B^{p-1}b) = p \tag{6.21}$$

and

$$k_1 \geq k_p^* := (p-1)(m+1) + 1 \tag{6.22}$$

hold simultaneously.

Proof. Let $\{v_1, v_2, \ldots, v_p\}$ be a basis of M and let

$$\{v_1, v_2, \ldots, v_p, w_{p+1}, w_{p+2}, \ldots, w_n\}$$

be a basis of $\mathbb{R}^n$. From [115, Section 39] we know that the matrix representation of the linear transformation B in coordinate system

$$\{v_1, v_2, \ldots, v_p, w_{p+1}, w_{p+2}, \ldots, w_n\}$$

has the form $\left(\begin{smallmatrix} \widetilde{B} & C_1 \\ \Theta_{n-p,p} & C_2 \end{smallmatrix}\right)$, where $\widetilde{B}$ is the $p \times p$ matrix of B considered as linear transformation on the space M with respect to the coordinate system $\{v_1, v_2, \ldots, v_p\}$, $\Theta_{n-p,p}$ is an $(n-p) \times p$ zero matrix, and C_1, C_2 are some arrays of scalars of dimensions $p \times (n-p)$, $(n-p) \times (n-p)$, respectively. In other words, $S^{-1}BS = \left(\begin{smallmatrix} \widetilde{B} & C_1 \\ \Theta_{n-p,p} & C_2 \end{smallmatrix}\right)$ with a regular matrix

$$S := (v_1, v_2, \ldots, v_p, w_{p+1}, w_{p+2}, \ldots, w_n). \tag{6.23}$$

Since $S^{-1}S = E$, we have $S^{-1}v_i = e_i$ for each $i = 1, \ldots, p$, where

$$e_i = (\underbrace{0, \ldots, 0}_{i-1}, 1, \underbrace{0, \ldots, 0}_{n-i})^\top$$

are canonical vectors in $\mathbb{R}^n$. Thus if $a = \sum_{i=1}^{p} a_i v_i \subset M$, then

$$S^{-1}a = \sum_{i=1}^{p} a_i S^{-1} v_i = \sum_{i=1}^{p} a_i e_i = (a_1, \ldots, a_p, 0, \ldots, 0)^\top,$$

i.e., $S^{-1}M = \mathbb{R}^p \times \{0\}^{n-p}$, in particular $S^{-1}b =: \left(\begin{smallmatrix} \widetilde{b} \\ \theta_{n-p} \end{smallmatrix}\right)$. We introduce the change of coordinates $x = Sy$. Accordingly,

$$\Delta y(k) = \Delta S^{-1}x(k) = S^{-1}\Delta x(k) = S^{-1}Bx(k-m) + S^{-1}bu(k)$$
$$= S^{-1}BSy(k-m) + S^{-1}bu(k).$$

So we have the next problem for $y = \left(\begin{smallmatrix} y_1 \\ y_2 \end{smallmatrix}\right) \in \mathbb{R}^p \times \mathbb{R}^{n-p}$:

$$\Delta y_1(k) = \widetilde{B} y_1(k-m) + C_1 y_2(k-m) + \widetilde{b}u(k), \qquad k \in \mathbb{Z}_0^{k_1-1},$$
$$\Delta y_2(k) = C_2 y_2(k-m), \qquad k \in \mathbb{Z}_0^{k_1-1},$$
$$y(k) = \widetilde{\varphi}(k) := \begin{pmatrix} \widetilde{\varphi}_1(k) \\ \widetilde{\varphi}_2(k) \end{pmatrix}, \qquad k \in \mathbb{Z}_{-m}^{0},$$

$$y(k_1) = y^* = \begin{pmatrix} y_1^* \\ y_2^* \end{pmatrix},$$

with $\widetilde{\varphi} := S^{-1}\varphi$, $y^* := S^{-1}x^*$. Note that by the assumptions of the theorem, $y_2^* = \theta_{n-p}$, $\widetilde{\varphi}_2(k) = \theta_{n-p}$ for each $k \in \mathbb{Z}_{-m}^0$. Hence we get immediately the solution $y_2(k) = \theta_{n-p}$ $\forall k \in \mathbb{Z}_{-m}^{k_1}$. Now, it is easy to see that the problem (6.5)–(6.7) is relatively controllable if and only if the problem

$$\Delta y_1(k) = \widetilde{B} y_1(k - m) + \widetilde{b}u(k), \qquad\qquad k \in \mathbb{Z}_0^{k_1-1}, \qquad (6.24)$$

$$y_1(k) = \widetilde{\varphi}_1(k), \qquad\qquad k \in \mathbb{Z}_{-m}^0, \qquad (6.25)$$

$$y_1(k_1) = y_1^* \qquad\qquad (6.26)$$

is relatively controllable.

If problem (6.24)–(6.26) is relatively controllable, then (by Definition 6.2) there is a $k_p^* \in \mathbb{N}$ such that for each $\widetilde{\varphi}_1 : \mathbb{Z}_{-m}^0 \to \mathbb{R}^p$, $y_1^* \in \mathbb{R}^p$, $k_1 \geq k_p^*$ there exists a control function $u^* : \mathbb{Z}_0^{k_1-1} \to \mathbb{R}$ such that there is a solution y_1 of the system (6.24)–(6.26) with $u = u^*$. By (6.4), at the point k_1 we have

$$y_1^* = y_1(k_1) = e_m^{\widetilde{B}k_1}\widetilde{\varphi}_1(-m) + \sum_{j=-m+1}^{0} e_m^{\widetilde{B}(k_1-m-j)}\Delta\widetilde{\varphi}_1(j-1)$$
$$+ \sum_{j=1}^{k_1} e_m^{\widetilde{B}(k_1-m-j)}\widetilde{b}u^*(j-1),$$

i.e.,

$$\sum_{j=1}^{k_1} e_m^{\widetilde{B}(k_1-m-j)}\widetilde{b}u^*(j-1) = \xi_1 \qquad (6.27)$$

with

$$\xi_1 := y_1^* - e_m^{\widetilde{B}k_1}\widetilde{\varphi}_1(-m) - \sum_{j=-m+1}^{0} e_m^{\widetilde{B}(k_1-m-j)}\Delta\widetilde{\varphi}_1(j-1).$$

Definition 6.1 of discrete delayed matrix exponential yields that the left-hand side of (6.27) is a linear combination of $q + 1$ vectors $\widetilde{b}, \widetilde{B}\widetilde{b}, \dots, \widetilde{B}^q\widetilde{b}$, where

$$q := \left\lceil \frac{k_1 - m - 1}{m + 1} \right\rceil = \left\lceil \frac{k_1}{m + 1} \right\rceil - 1 = \left\lfloor \frac{k_1 - 1}{m + 1} \right\rfloor$$

and $\lceil \cdot \rceil$ and $\lfloor \cdot \rfloor$ are the ceiling function and floor function, respectively. By the Caley–Hamilton theorem, matrix $\widetilde{B}^p$ can be written as a linear combination of matrices $E, \widetilde{B}, \ldots, \widetilde{B}^{p-1}$. Thus, we can consider

$$q = \min\left\{ \left\lfloor \frac{k_1 - 1}{m + 1} \right\rfloor, p - 1 \right\}. \tag{6.28}$$

So Eq. (6.27) has the form of a linear equation,

$$\sum_{j=0}^{q} \widetilde{B}^j \widetilde{b} \alpha_j = \xi_1,$$

with constants $\alpha_0, \alpha_1, \ldots, \alpha_q$ depending on $u^*(0), u^*(1), \ldots, u^*(k_1 - 1)$. A solution $(\alpha_0, \alpha_1, \ldots, \alpha_q)$ of this system exists for any right-hand side $\xi_1 \in \mathbb{R}^p$ if and only if two conditions are fulfilled simultaneously:

1. the number of equations is not greater than the number of variables,
2. matrix $(\widetilde{b}, \widetilde{B}\widetilde{b}, \ldots, \widetilde{B}^{p-1}\widetilde{b})$ has full rank.

The first condition means that $q \geq p - 1$, i.e.,

$$\frac{k_1 - 1}{m + 1} \geq \left\lfloor \frac{k_1 - 1}{m + 1} \right\rfloor \geq p - 1,$$

$$k_1 \geq (p - 1)(m + 1) + 1.$$

By (6.28), it is equivalent to $q = p - 1$. Now, the second condition is equivalent to $rank(\widetilde{b}, \widetilde{B}\widetilde{b}, \ldots, \widetilde{B}^{p-1}\widetilde{b}) = p$. Clearly, the *rank* of a matrix will not change if we add $n - p$ zero rows or multiply this matrix by a nonsingular matrix. Hence,

$$rank(\widetilde{b}, \widetilde{B}\widetilde{b}, \ldots, \widetilde{B}^{p-1}\widetilde{b}) = rank(\begin{pmatrix} \widetilde{b} \\ \theta_{n-p} \end{pmatrix}, \begin{pmatrix} \widetilde{B}\widetilde{b} \\ \theta_{n-p} \end{pmatrix}, \ldots, \begin{pmatrix} \widetilde{B}^{p-1}\widetilde{b} \\ \theta_{n-p} \end{pmatrix})$$

$$= rank(S^{-1}b, S^{-1}Bb, \ldots, S^{-1}B^{p-1}b) = rank(b, Bb, \ldots, B^{p-1}b).$$

So the second condition is equivalent to (6.21).

Now, assume the conditions (6.21) and (6.22) are valid. We show that the problem (6.24)–(6.26) is relatively controllable. Denote by

$$Q_\psi := \{ y_1(k_1) \mid y_1 \text{ satisfies } (6.24)–(6.25) \text{ with } \widetilde{\varphi}_1 = \psi \}$$

the domain of reachability for a fixed initial function ψ with the values in $\mathbb{R}^p$. If $\widetilde{\varphi}_1$ does not vanish on $\mathbb{Z}^0_{-m}$, one can split problem (6.24)–(6.26) to a homogeneous problem with nonhomogeneous initial function and a nonhomogeneous problem with homogeneous initial function. More precisely, we write $y_1 = u_{\widetilde{\varphi}_1} + v$, where $u_{\widetilde{\varphi}_1}$ solves

$$\Delta u_{\widetilde{\varphi}_1}(k) = \widetilde{B} u_{\widetilde{\varphi}_1}(k - m), \qquad k \in \mathbb{Z}_0^{k_1 - 1},$$

$$u_{\widetilde{\varphi}_1}(k) = \widetilde{\varphi}_1(k), \qquad k \in \mathbb{Z}^0_{-m},$$

and v is a solution of (6.24)–(6.26) with $\widetilde{\varphi}_1(k) = \theta_p \ \forall k \in \mathbb{Z}^0_{-m}$. Obviously, (6.24)–(6.26) is relatively controllable if and only if the corresponding problem for v is relatively controllable. Therefore, it suffices to investigate only the case $\widetilde{\varphi}_1(k) = \theta_p \ \forall k \in \mathbb{Z}^0_{-m}$ with the corresponding domain of reachability Q_0. First, we prove that $\dim Q_0 = p$, i.e., there are p linearly independent final vectors $y^*_1 \in Q_0 \subset \mathbb{R}^p$ that can be reached at k_1 by the choice of a convenient control u^*. Let us suppose, on the contrary, that there is a nontrivial constant vector $z \in \mathbb{R}^p$ such that for any control function $u : \mathbb{Z}^{k_1-1}_0 \to \mathbb{R}$ and the corresponding solution y_1 of (6.24)–(6.25) we have $z^\top y_1(k_1) = 0$. Then, by (6.4) this equation becomes

$$z^\top \sum_{j=1}^{k_1} e_m^{\widetilde{B}(k_1-m-j)} \widetilde{b} u(j-1) = 0$$

for any control function u. Accordingly, $z^\top e_m^{\widetilde{B}(k_1-m-j)} \widetilde{b} = 0 \ \forall j \in \mathbb{Z}^{k_1}_1$. On setting $j = k_1$, we get $z^\top \widetilde{b} = 0$. Next, if $j = k_1 - (m+1)$, then

$$z^\top e_m^{\widetilde{B}(k_1-m-j)} \widetilde{b} = z^\top e_m^{\widetilde{B} \cdot 1} \widetilde{b} = z^\top \widetilde{b} + z^\top \widetilde{B} \widetilde{b} = z^\top \widetilde{B} \widetilde{b} = 0.$$

Subsequently, we obtain

$$z^\top \widetilde{b} = z^\top \widetilde{B} \widetilde{b} = \cdots = z^\top \widetilde{B}^{p-1} \widetilde{b} = 0$$

(in the last term $j = k_1 - (p-1)(m+1) \geq 1$ by (6.22)) which is a contradiction with nonzero z due to condition (6.21).

Clearly, if Q_0 contains point $y_1(k_1)$, then also $-y_1(k_1) \in Q_0$ (one simply takes $-u$ and applies formula (6.4)). Similarly, it can be shown that Q_0 is linear, i.e., $Q_0 = \mathbb{R}^p$. This says that for the zero initial condition (and also for any $\varphi \not\equiv \theta_p$ on $\mathbb{Z}^0_{-m}$) and any final point y^*_1 there exists a control u^* and the corresponding solution of (6.24)–(6.26). That is exactly the relative controllability of this system. The proof is finished. $\qquad\square$

The following statement for weakly nonlinear systems immediately results from the latter theorem.

Corollary 6.2. *Let M, x^*, φ, b be as in Theorem 6.5, let $f : \mathbb{R} \times \mathbb{R}^n \times \mathbb{R}^n \times \mathbb{R} \times \mathbb{R} \to M$ be a given function C^1-smooth in all arguments, and let $\varepsilon \in \mathbb{R}$ be close to 0. The problem (6.13), (6.6), (6.7) is relatively controllable if conditions (6.21)–(6.22) hold simultaneously.*

Proof. In the notation of the proof of Theorem 6.5, we write $x = Sy$ to get the following problem for $y = \left(\begin{smallmatrix} y_1 \\ y_2 \end{smallmatrix}\right) \in \mathbb{R}^p \times \mathbb{R}^{n-p}$:

$$\Delta y_1(k) = \widetilde{B} y_1(k-m) + C_1 y_2(k-m) + \widetilde{b} u(k)$$
$$+ \varepsilon \widetilde{f}(k, y(k), y(k-m), u(k), \varepsilon), \qquad k \in \mathbb{Z}^{k_1-1}_0,$$

$$\Delta y_2(k) = C_2 y_2(k - m), \qquad\qquad k \in \mathbb{Z}_0^{k_1-1},$$

$$y(k) = \widetilde{\varphi}(k) := \begin{pmatrix} \widetilde{\varphi}_1(k) \\ \widetilde{\varphi}_2(k) \end{pmatrix}, \qquad\qquad k \in \mathbb{Z}_{-m}^0,$$

$$y(k_1) = \begin{pmatrix} y_1^* \\ y_2^* \end{pmatrix},$$

where

$$S^{-1} f(k, Sy(k), Sy(k-m), u(k), \varepsilon) = \begin{pmatrix} \widetilde{f}(k, y(k), y(k-m), u(k), \varepsilon) \\ \theta_{n-p} \end{pmatrix}.$$

As in the mentioned proof $y_2^* = \theta_{n-p}$, $\widetilde{\varphi}_2(k) = \theta_{n-p}$ $\forall k \in \mathbb{Z}_{-m}^0$, and the problem (6.13), (6.6), (6.7) is relatively controllable if and only if the problem consisting of equation

$$\Delta y_1(k) = \widetilde{B} y_1(k - m) + \widetilde{b} u(k) + \varepsilon \widetilde{f}(k, (\begin{smallmatrix} y_1(k) \\ \theta_{n-p} \end{smallmatrix}), (\begin{smallmatrix} y_1(k-m) \\ \theta_{n-p} \end{smallmatrix}), u(k), \varepsilon)$$

for $k \in \mathbb{Z}_0^{k_1-1}$ and conditions (6.25), (6.26) is relatively controllable. Here one applies Corollary 6.1 with p instead of n. The statement is proved. $\qquad\square$

Remark 6.1 can be extended to Corollary 6.2.

Motivated by Theorem 6.3, we state the next result on the control function for the restricted relative controllability from Theorem 6.5.

Theorem 6.6. *Let M, x^*, φ, b be as in Theorem 6.5, let $v \colon \mathbb{Z}_1^{k_1} \to \mathbb{R}^n$ be such that $rank\, W = p$ for*

$$W := \sum_{j=1}^{k_1} e_m^{B(k_1-m-j)} b v^\top(j), \tag{6.29}$$

and let conditions (6.21)–(6.22) be valid. Then a control function $u = u^*$ for the problem (6.5)–(6.7) can be expressed in the form

$$u^*(k) = v^\top(k+1)(W|_{(\ker W)^\perp}^{-1} \xi + F), \quad k \in \mathbb{Z}_0^{k_1-1}, \tag{6.30}$$

for an arbitrary fixed $F \in \ker W$.

Proof. As in the proof of Theorem 6.3, we look for a control function in the form $u(j - 1) = v^T(j)D$ with a constant vector $D \in \mathbb{R}^n$. In this case, we write $D = D_1 + D_2$, where $D_1 \in (\ker W)^\perp$, $D_2 \in \ker W$. Since $u = u^*$ satisfies (6.4) and (6.7), we get

$$W D_1 = W D = \sum_{j=1}^{k_1} e_m^{B(k_1-m-j)} b v^\top(j) D = \xi$$

for ξ given by (6.12). Note that the left-hand side of the last equality is a finite linear combination of vectors from M since $e_m^{Bk} b \in M$ $\forall k \in \mathbb{Z}_{-m}^{\infty}$, i.e., $W\mathbb{R}^n \subset M$. Clearly also $\xi \in M$ by the assumptions on x^* and φ. The assumption on $rank\,W$ is equivalent to $\dim(\ker W)^{\perp} = p$, which yields that $W|_{(\ker W)^{\perp}} : (\ker W)^{\perp} \to M$ is bijective. Therefore, $D_1 = W|_{(\ker W)^{\perp}}^{-1} \xi$ and the proof is complete. $\qquad\square$

Of course, Remark 6.3 can be also applied (see Corollary 6.3 below).

Now, we generalize Lemma 6.1 to nonsingular matrices to obtain a necessary condition for $rank\,W = p$.

Lemma 6.2. *If $rank\,\widetilde{W} = p$ for $\widetilde{W}$ of (6.19), then there exist at most $n - p$ linearly independent constant vectors $\mu_1, \mu_2, \ldots, \mu_{n-p} \in \mathbb{R}^n$ such that*

$$\mu_i^{\top} v(j) = 0, \quad \forall i \in \mathbb{Z}_1^{n-p}, \ j \in \mathbb{Z}_1^{k_1}.$$

Proof. Suppose that $\mu_1, \mu_2, \ldots, \mu_{n-p+1} \in \mathbb{R}^n$ are linearly independent constant vectors such that

$$\mu_i^{\top} v(j) = 0, \quad \forall i \in \mathbb{Z}_1^{n-p+1}, \ j \in \mathbb{Z}_1^{k_1}.$$

Then

$$\sum_{j=1}^{k_1} w(j) v^{\top}(j) \mu_i = \widetilde{W} \mu_i = 0, \quad \forall i \in \mathbb{Z}_1^{n-p+1},$$

i.e., $\dim \ker \widetilde{W} \geq n - p + 1$. Equivalently, $\dim(\ker \widetilde{W})^{\perp} \leq p - 1$, which is a contradiction with p linearly independent rows of $\widetilde{W}$ taken as vectors in $(\ker \widetilde{W})^{\perp}$. $\qquad\square$

Remark 6.4. All $n - p$ vectors from the previous lemma do not have to exist. For instance, set $w(1) = w(2) = (1, 2)^{\top}$, $v(1) = (1, 0)^{\top}$, $v(2) = (0, 1)^{\top}$. Then $W = \left(\begin{smallmatrix} 1 & 1 \\ 2 & 2 \end{smallmatrix}\right)$ has rank 1, but only the trivial vector $\mu_1 = (0, 0)^{\top}$ satisfies $\mu_1^{\top} v(j) = 0$ for each $j = 1, 2$.

The following statement shows that the set of control functions for problem (6.5)–(6.7) remains nonempty if conditions (6.8)–(6.9) are replaced with the assumptions of Theorem 6.5 and conditions (6.21)–(6.22), as it constructs a concrete u^* given more generally by (6.30).

Corollary 6.3. *Let M, x^*, φ, b be as in Theorem 6.5 and let conditions (6.21)–(6.22) be valid. A control function $u = u^*$ for the problem (6.5)–(6.7) can be expressed in the form*

$$u^*(k) = b^{\top} e_m^{B^{\top}(k_1 - m - k - 1)} \left(\sum_{j=1}^{k_1} e_m^{B(k_1 - m - j)} bb^{\top} e_m^{B^{\top}(k_1 - m - j)} \right)\Bigg|_M^{-1} \xi$$

for $k \in \mathbb{Z}_0^{k_1 - 1}$.

Proof. Denote

$$\widetilde{G} := \sum_{j=1}^{k_1} e_m^{B(k_1-m-j)} bb^\top e_m^{B^\top(k_1-m-j)}. \tag{6.31}$$

If we prove that $rank\,\widetilde{G} = p$ and $\ker\,\widetilde{G} = M^\perp$, we will be able to apply Theorem 6.6 with $v(k) = e_m^{B(k_1-m-k)} b$.

Clearly, $\widetilde{G}$ has the form of (6.29), so $\widetilde{G}\mathbb{R}^n \subset M$ as discussed in the proof of Theorem 6.6, i.e., $rank\,\widetilde{G} \leq \dim M = p$. In particular, $\widetilde{G}M \subset M$.

Now, take the matrix S of (6.23). From the proof of Theorem 6.5 we know that $S^{-1}BS = \begin{pmatrix} \widetilde{B} & C_1 \\ \Theta_{n-p,p} & C_2 \end{pmatrix}$. Consequently,

$$e_m^{Bk} = S(S^{-1}e_m^{Bk}S)S^{-1} = S \begin{pmatrix} e_m^{\widetilde{B}k} & C(k) \\ \Theta_{n-p,p} & e_m^{C_2k} \end{pmatrix} S^{-1}$$

for some $p \times (n - p)$ matrix function $C(k)$. Analogically,

$$e_m^{Bk} b = S(S^{-1}e_m^{Bk}S)(S^{-1}b)$$

$$= S \begin{pmatrix} e_m^{\widetilde{B}k} & C(k) \\ \Theta_{n-p,p} & e_m^{C_2k} \end{pmatrix} \begin{pmatrix} \widetilde{b} \\ \theta_{n-p} \end{pmatrix} = S \begin{pmatrix} e_m^{\widetilde{B}k}\widetilde{b} \\ \theta_{n-p} \end{pmatrix}$$

and

$$\widetilde{G} = S \sum_{j=1}^{k_1} \begin{pmatrix} e_m^{\widetilde{B}(k_1-m-j)}\widetilde{b} \\ \theta_{n-p} \end{pmatrix} \begin{pmatrix} \widetilde{b}^\top e_m^{\widetilde{B}^\top(k_1-m-j)} & \theta_{n-p}^\top \end{pmatrix} S^\top$$

$$= S \sum_{j=1}^{k_1} \begin{pmatrix} e_m^{\widetilde{B}(k_1-m-j)}\widetilde{b}\widetilde{b}^\top e_m^{\widetilde{B}^\top(k_1-m-j)} & \Theta_{p,n-p} \\ \Theta_{n-p,p} & \Theta_{n-p,n-p} \end{pmatrix} S^\top.$$

Suppose that $\widetilde{G}M \neq M$. Then there exists a vector $\zeta \in M$ such that $\widetilde{G}\zeta = \theta$. Accordingly,

$$0 = \zeta^\top \widetilde{G}\zeta = (S^\top \zeta)^\top \sum_{j=1}^{k_1} \begin{pmatrix} e_m^{\widetilde{B}(k_1-m-j)}\widetilde{b}\widetilde{b}^\top e_m^{\widetilde{B}^\top(k_1-m-j)} & \Theta_{p,n-p} \\ \Theta_{n-p,p} & \Theta_{n-p,n-p} \end{pmatrix} S^\top \zeta$$

$$= \zeta_1^\top \sum_{j=1}^{k_1} e_m^{\widetilde{B}(k_1-m-j)}\widetilde{b}\widetilde{b}^\top e_m^{\widetilde{B}^\top(k_1-m-j)} \zeta_1, \tag{6.32}$$

where $S^\top \zeta = (\zeta_1, \zeta_2)^\top \in \mathbb{R}^p \times \mathbb{R}^{n-p}$. Note that $\zeta_1 \neq \theta_p$ whenever $\zeta \in M$. Clearly, vectors $\widetilde{b}, \widetilde{B}\widetilde{b}, \ldots, \widetilde{B}^{p-1}\widetilde{b}$ are linearly independent if and only if vectors

$$\begin{pmatrix} \widetilde{b} \\ \theta_{n-p} \end{pmatrix} = S^{-1}b, \quad \begin{pmatrix} \widetilde{B}\widetilde{b} \\ \theta_{n-p} \end{pmatrix} = S^{-1}Bb, \quad \ldots, \quad \begin{pmatrix} \widetilde{B}^{p-1}\widetilde{b} \\ \theta_{n-p} \end{pmatrix} = S^{-1}B^{p-1}b$$

are linearly independent. Hence, $rank(\widetilde{b}, \widetilde{B}\widetilde{b}, \ldots, \widetilde{B}^{p-1}\widetilde{b}) = p$, and we can apply [4, Lemma 3.4] (see also Remark 6.2) saying that the matrix

$$H := \sum_{j=1}^{k_1} e_m^{\widetilde{B}(k_1-m-j)} \widetilde{b}\widetilde{b}^\top e_m^{\widetilde{B}^\top (k_1-m-j)}$$

is nonsingular for each $k_1 \geq (p-1)(m-1)+1$. More precisely, in the proof of that lemma it is shown that H is positive definite. So the right-hand side of (6.32) is positive and a contradiction follows. Therefore, $\widetilde{G}M = M$, i.e., $rank\widetilde{G} \geq p$.

In conclusion, $rank\widetilde{G} = p = \dim M$ implying $M = im\widetilde{G} = im\widetilde{G}^\top = (\ker \widetilde{G})^\perp$ since $\widetilde{G}$ is symmetric.

Moreover, for any $F \in \ker \widetilde{G} = M^\perp$ we have

$$F^\top e_m^{Bk} b = b^\top e_m^{B^\top k} F = 0, \quad \forall k \in \mathbb{Z}_{-m}^\infty.$$

Thus by setting $v(k) = e_m^{B(k_1-m-k)}b$, the control function of (6.30) is independent of F, and the statement of the corollary coincides with the statement of Theorem 6.6. $\qquad\square$

6.1.1.3 Numerical example and discussion

Example 6.2. Consider the following boundary value problem for $x = (x_1, x_2, x_3, x_4)^\top \in \mathbb{R}^4$:

$$\begin{aligned}
\Delta x_1(k) &= x_1(k-2) + u(k), \quad k \in \mathbb{Z}_0^3, \\
\Delta x_2(k) &= x_2(k-2) + u(k), \quad k \in \mathbb{Z}_0^3, \\
\Delta x_3(k) &= x_1(k-2) - x_2(k-2) + x_3(k-2), \quad k \in \mathbb{Z}_0^3, \\
\Delta x_4(k) &= x_1(k-2) + x_4(k-2), \quad k \in \mathbb{Z}_0^3, \\
x(-2) &= (0, 0, 0, 0)^\top, \\
x(k) &= (0, 0, 0, 1)^\top, \quad k = -1, 0, \\
x(4) &= (1, 1, 0, -1)^\top.
\end{aligned}$$

In this case $n = 4$, $m = 2$, $k_1 = 4$,

$$\varphi(k) = \begin{cases} (0,0,0,0)^\top, & k = -2, \\ (0,0,0,1)^\top, & k = -1,0, \end{cases} \qquad B = \begin{pmatrix} 1 & 0 & 0 & 0 \\ 0 & 1 & 0 & 0 \\ 1 & -1 & 1 & 0 \\ 1 & 0 & 0 & 1 \end{pmatrix},$$

and $b = (1,1,0,0)^\top$. It is easy to verify that

$$M = [(1,1,0,0)^\top, (0,0,0,1)^\top] = [b, Bb]$$

is a B-invariant linear space of dimension $p = 2$, containing x^*, $\varphi(k) \,\forall k \in \mathbb{Z}^0_{-2}$, and b. Thus conditions (6.21) and (6.22) are immediately verified.

From the definition of the discrete delayed matrix exponential one gets

$$e_m^{B(k_1-m-j)} b = \begin{cases} (2,2,0,1)^\top, & j = 1, \\ (1,1,0,0)^\top, & j = 2,3,4. \end{cases}$$

Matrix $\widetilde{G}$ defined by (6.31) and ξ of (6.12) have the form

$$\widetilde{G} = \begin{pmatrix} 7 & 7 & 0 & 2 \\ 7 & 7 & 0 & 2 \\ 0 & 0 & 0 & 0 \\ 2 & 2 & 0 & 1 \end{pmatrix}, \qquad \xi = \begin{pmatrix} 1 \\ 1 \\ 0 \\ -5 \end{pmatrix},$$

respectively. To find $\widetilde{G}|_M^{-1}\xi$, we solve equation

$$\alpha \widetilde{G} \begin{pmatrix} 1 \\ 1 \\ 0 \\ 0 \end{pmatrix} + \beta \widetilde{G} \begin{pmatrix} 0 \\ 0 \\ 0 \\ 1 \end{pmatrix} = \xi$$

for $\alpha, \beta \in \mathbb{R}$. This is the same as solving the system

$$14\alpha + 2\beta = 1,$$
$$4\alpha + \beta = -5.$$

Hence, $\alpha = \frac{11}{6}$, $\beta = -\frac{37}{3}$, and $\widetilde{G}|_M^{-1}\xi = (\frac{11}{6}, \frac{11}{6}, 0, \frac{-74}{6})^\top$. Now, we apply Corollary 6.3 to derive the control function u^* and, consequently, the solution of Example 6.2. These values are shown in Table 6.4.

TABLE 6.4 Control function and the corresponding solution of Example 6.2 obtained by Corollary 6.3.

k	-2	-1	0	1	2	3	4
$u^*(k)$			-5	$\frac{11}{3}$	$\frac{11}{3}$	$\frac{11}{3}$	
$x(k)$	$\begin{pmatrix} 0 \\ 0 \\ 0 \\ 0 \end{pmatrix}$	$\begin{pmatrix} 0 \\ 0 \\ 0 \\ 1 \end{pmatrix}$	$\begin{pmatrix} 0 \\ 0 \\ 0 \\ 1 \end{pmatrix}$	$\begin{pmatrix} -5 \\ -5 \\ 0 \\ 1 \end{pmatrix}$	$\begin{pmatrix} -\frac{4}{3} \\ -\frac{4}{3} \\ 0 \\ 2 \end{pmatrix}$	$\begin{pmatrix} \frac{7}{3} \\ \frac{7}{3} \\ 0 \\ 3 \end{pmatrix}$	$\begin{pmatrix} 1 \\ 1 \\ 0 \\ -1 \end{pmatrix}$

Now, for instance let us set

$$
v(k) = \begin{cases}
(0, 0, 0, 1)^\top, & k = 1, \\
(1, 0, 1, 0)^\top, & k = 2, \\
(0, 1, 1, 0)^\top, & k = 3, \\
(0, 0, 0, 0)^\top, & k = 4.
\end{cases}
$$

Then the matrix W of (6.29) has the form

$$
W = \begin{pmatrix}
1 & 1 & 2 & 2 \\
1 & 1 & 2 & 2 \\
0 & 0 & 0 & 0 \\
0 & 0 & 0 & 1
\end{pmatrix}
$$

and it is easy to check that

$$
\ker W = [(1, -1, 0, 0)^\top, (-2, 0, 1, 0)^\top],
$$
$$
(\ker W)^\perp = [(1, 1, 2, 2)^\top, (0, 0, 0, 1)^\top].
$$

We find $W|^{-1}_{(\ker W)^\perp} \xi$ using equation

$$
\alpha W \begin{pmatrix} 1 \\ 1 \\ 2 \\ 2 \end{pmatrix} + \beta W \begin{pmatrix} 0 \\ 0 \\ 0 \\ 1 \end{pmatrix} = \xi
$$

for $\alpha, \beta \in \mathbb{R}$. This is equivalent to

$$
10\alpha + 2\beta = 1,
$$
$$
2\alpha + \beta = -5,
$$

resulting in $\alpha = \frac{11}{6}$, $\beta = -\frac{26}{3}$. Hence $W|_{(\ker W)^{\perp}}^{-1}\xi = (\frac{11}{6}, \frac{11}{6}, \frac{11}{3}, -5)^{\top}$. Choosing a general vector $F = (f_1 - 2f_2, -f_1, f_2, 0)^{\top} \in \ker W$, $f_{1,2} \in \mathbb{R}$, and Theorem 6.6, we obtain control function u^* and the corresponding solution of Example 6.2 shown in Table 6.5.

TABLE 6.5 Control function and the corresponding solution of Example 6.2 obtained by Theorem 6.6.

k	-2	-1	0	1	2	3	4
$u^*(k)$			-5	$\frac{11}{2} + f_1 - f_2$	$\frac{11}{2} - f_1 + f_2$	0	
$x(k)$	$\begin{pmatrix} 0 \\ 0 \\ 0 \\ 0 \end{pmatrix}$	$\begin{pmatrix} 0 \\ 0 \\ 0 \\ 1 \end{pmatrix}$	$\begin{pmatrix} 0 \\ 0 \\ 0 \\ 1 \end{pmatrix}$	$\begin{pmatrix} -5 \\ -5 \\ 0 \\ 1 \end{pmatrix}$	$\begin{pmatrix} \frac{1}{2} + f_1 - f_2 \\ \frac{1}{2} + f_1 - f_2 \\ 0 \\ 2 \end{pmatrix}$	$\begin{pmatrix} 6 \\ 6 \\ 0 \\ 3 \end{pmatrix}$	$\begin{pmatrix} 1 \\ 1 \\ 0 \\ -1 \end{pmatrix}$

6.1.1.4 Conclusions

New control functions are derived for linear difference equations with delay, which can be used to construct a nontrivial solution of a boundary value problem with zero boundary conditions. A special case of invariant linear subspace is considered and corresponding control functions are constructed. Results for weakly nonlinear problems are also discussed. The results in this part are motivated from [9].

6.2 Iterative learning control for fixed trial lengths

6.2.1 Iterative learning control for linear systems

In this section, we consider the ILC problem for the following discrete controlled systems with single delay:

$$\begin{cases} x_k(t+1) = Ax_k(t) + A_1 x_k(t-m) + Bu_k(t), \ t \in \mathbb{Z}_0^T, \\ x_k(t) = \varphi(t), \ t \in \mathbb{Z}_{-m}^0, \\ y_k(t) = Cx_k(t) + Du_k(t), \end{cases} \tag{6.33}$$

where $m \geq 1$ is a prefixed integer and $A, A_1 \in \mathbb{R}^{n \times n}$ satisfying $AA_1 = A_1 A$ with $\operatorname{rank}(A) = n$, which implies A^{-1} exists. The index $k = 1, 2, \ldots$ denotes the k-th iteration and T is a prefixed positive integer. The variable $x_k : \mathbb{Z}_{-m}^T \to \mathbb{R}^n$ denotes the state, $u_k : \mathbb{Z}_0^T \to \mathbb{R}^r$ denotes the dominant input, and $y_k : \mathbb{Z}_0^T \to \mathbb{R}^{m_*}$ denote the output. In addition, B, C, and D denote $n \times r$, $m_* \times n$, and $m_* \times r$ constant matrices, respectively.

Recently, Diblík and Khusainov [2] investigated the representation of solutions of the following Cauchy problem of linear discrete systems with single delay:

$$\begin{cases} x(t+1) = Ax(t) + A_1 x(t-m) + f(t), \ t \in \mathbb{Z}_0^\infty, \\ x(t) = \varphi(t), \ t \in \mathbb{Z}_{-m}^0, \end{cases} \tag{6.34}$$

where A, A_1, and m are defined in (6.33), $x : \mathbb{Z}_{-m}^\infty \to \mathbb{R}^n$, $f : \mathbb{Z}_0^\infty \to \mathbb{R}^n$, and $\varphi : \mathbb{Z}_{-m}^0 \to \mathbb{R}^n$.

With the aid of the discrete matrix delayed exponential function, [2] developed the classical idea in ODEs to derive the representation of solutions of linear discrete systems. Here we collect a matrix-formed solution of (6.34), which is useful to investigate ILC problems and asymptotical behavior of solutions.

Lemma 6.3. *(see [2, Theorem 3.5]) The solution $x(t)$, $t \in \mathbb{Z}_{-m}^\infty$, of (6.34) has the form*

$$x(t) = A^t e_m^{B_1 t} A^{-m} \varphi(-m) + \sum_{j=-m+1}^{0} A^{(t-j)} e_m^{B_1(t-m-j)} \left[\varphi(j) - A\varphi(j-1) \right]$$

$$+ \sum_{j=1}^{t} A^{(t-j)} e_m^{B_1(t-m-j)} f(j-1), \tag{6.35}$$

where $B_1 = A^{-1} A_1 A^{-m}$ provided that A^{-1} exists and $e_m^{B_1 \cdot}$ is defined in (6.1).

With the help of (6.1) and (6.35), we apply the representation of solution in (6.35) to study the ILC problem for discrete controlled systems with single delay (6.33).

By (6.35), one can see that the state $x_k(t)$ of (6.33) has the following form:

$$x_k(t) = A^t e_m^{B_1 t} A^{-m} \varphi(-m) + \sum_{j=-m+1}^{0} A^{(t-j)} e_m^{B_1(t-m-j)} \left[\varphi(j) - A\varphi(j-1) \right]$$

$$+ \sum_{j=1}^{t} A^{(t-j)} e_m^{B_1(t-m-j)} B u_k(j-1). \tag{6.36}$$

The main novelty of this part is that we do not turn our original ILC problem for (6.33) to a Roesser model to derive the convergence result, which is very different from the existing literature. In other words, we fully use formulation (6.35) for the solution to system (6.33) via two given learning laws to generate the control input u_k for the output y_k of the system to track the desired reference trajectory as accurately as possible with k tending to infinity uniformly on a finite time interval.

For a constant matrix $Q \in \mathbb{R}^{n \times n}$, we define $\|Q\| = \max_{\|z\|=1} \|Qz\|$ generated by $\|\cdot\|$, where $\|z\|$ denotes the norm for $z \in \mathbb{R}^n$. For a discrete vector function $x : \mathbb{Z}_0^T \to \mathbb{R}^n$, we define the λ-norm $\|x\|_\lambda = \sup_{t \in \mathbb{Z}_0^T} \{\lambda^t \|x(t)\|\}$, $0 < \lambda < 1$.

Lemma 6.4. *For all $t \in \mathbb{Z}$,*

$$\|e_m^{Gt}\| \le e^{\|G\|(t+m)}. \tag{6.37}$$

Proof. For any $t \in \mathbb{Z}_1^{\infty}$, that is, $t \in \mathbb{Z}_{(s-1)(m+1)+1}^{s(m+1)}$, where $s = 1, 2, \ldots$, taking the matrix norm for (6.1), we get

$$
\begin{aligned}
\|e_m^{Gt}\| &\le 1 + \|G\| \frac{t!}{1!(t-1)!} + \|G\|^2 \frac{(t-m)!}{2!(t-m-2)!} + \cdots \\
&\quad + \|G\|^s \frac{(t-(s-1)m)!}{s!(t-(s-1)m-s)!} \\
&\le 1 + \|G\| \frac{t}{1!} + \|G\|^2 \frac{t^2}{2!} + \cdots + \|G\|^s \frac{t^s}{s!} \\
&\le \sum_{n=0}^{\infty} \frac{(\|G\|t)^n}{n!} = e^{\|G\|t}.
\end{aligned}
$$

Next, (6.1) implies also $\|e_m^{Gt}\| = 0$ for $t \in \mathbb{Z}_{-\infty}^{-m-1}$ and $\|e_m^{Gt}\| = 1$ for $t \in \mathbb{Z}_{-m}^{0}$.
    Summarizing, (6.37) holds for all $t \in \mathbb{Z}$. $\square$

Lemma 6.5. *Let $G \in \mathbb{R}^{n \times n}$, whose eigenvalues $\lambda_1, \lambda_2, \cdots, \lambda_n$ are different from each other. The following relation holds:*

$$e_m^{Gt} = P\,diag(e_m^{\lambda_1 t}, e_m^{\lambda_2 t}, \cdots, e_m^{\lambda_n t})P^{-1}, \ t \in \mathbb{Z},$$

where $P := (\alpha_1, \alpha_2, \cdots, \alpha_n) \in \mathbb{R}^{n \times n}$ is invertible, P^{-1} denotes the invertible matrix of P, and $\alpha_i \in \mathbb{R}^n$ denotes the corresponding eigenvector of λ_i, $i = 1, 2, \cdots, n$.

Proof. Note that $\lambda_i \alpha_i = G\alpha_i$. Then,

$$
\begin{aligned}
GP &= (G\alpha_1, G\alpha_2, \cdots, G\alpha_n) \\
&= (\lambda_1 \alpha_1, \lambda_2 \alpha_2, \cdots, \lambda_n \alpha_n) \\
&= (\alpha_1, \alpha_2, \cdots, \alpha_n)\mathrm{diag}(\lambda_1, \lambda_2, \cdots, \lambda_n) \\
&= P\mathrm{diag}(\lambda_1, \lambda_2, \cdots, \lambda_n). \tag{6.38}
\end{aligned}
$$

Since P is invertible, P^{-1} exists. Multiplying by P^{-1} on both sides of (6.38), one obtains

$$G = P\mathrm{diag}(\lambda_1, \lambda_2, \cdots, \lambda_n)P^{-1}.$$

Further, one can derive the following facts:

$$
\begin{aligned}
G^2 &= P\mathrm{diag}(\lambda_1, \lambda_2, \cdots, \lambda_n)P^{-1}P\mathrm{diag}(\lambda_1, \lambda_2, \cdots, \lambda_n)P^{-1} \\
&= P\mathrm{diag}(\lambda_1^2, \lambda_2^2, \cdots, \lambda_n^2)P^{-1},
\end{aligned}
$$

$$\vdots$$

$$G^n = P\,\mathrm{diag}(\lambda_1^n, \lambda_2^n, \cdots, \lambda_n^n)\,P^{-1}.$$

From the above facts, by (6.1), for any $t \in \mathbb{Z}_{(s-1)(m+1)+1}^{s(m+1)}, s = 1, 2, \cdots$, we get

$$e_m^{Gt} = I + G\frac{t!}{1!(t-1)!} + G^2\frac{(t-m)!}{2!(t-m-2)!} + \cdots + G^s\frac{(t-(s-1)m)!}{s!(t-(s-1)m-s)!}$$

$$= P\begin{pmatrix} 1 + \cdots + \lambda_1^s\frac{(t-(s-1)m)!}{s!(t-(s-1)m-s)!} & & \\ & \ddots & \\ & & 1 + \cdots + \lambda_n^s\frac{(t-(s-1)m)!}{s!(t-(s-1)m-s)!} \end{pmatrix}$$

$$\times P^{-1}$$

$$= P\,\mathrm{diag}(e_m^{\lambda_1 t}, e_m^{\lambda_2 t}, \cdots, e_m^{\lambda_n t})\,P^{-1}.$$

The proof is completed. $\square$

6.2.1.1 ILC design and convergence analysis

Let y_d be a desired reference trajectory and let the k-th iteration error be

$$e_k(t) := y_d(t) - y_k(t). \tag{6.39}$$

Denote $\Delta x_k(t) := x_{k+1}(t) - x_k(t)$ and $\Delta u_k(t) := u_{k+1}(t) - u_k(t)$.

For (6.33), we set

$$\Delta u_k(t) = L_1 e_k(t). \tag{6.40}$$

For (6.33) with $D = \Theta$, we set

$$\Delta u_k(t) = L_2 e_k(t+1), \tag{6.41}$$

where L_1 and L_2 are $r \times m^*$ learning gain parameter matrices that need to be determined in (6.43) and (6.50), respectively, below. For example, one can choose $L_1 = \sigma D^{-1}, \sigma \in [0, 1)$ and $L_1 = \sigma(CB)^{-1}, \sigma \in [0, 1)$.

By (6.36), one has

$$\Delta x_k(t) = \sum_{j=1}^{t} A^{(t-j)} e_m^{B_1(t-m-j)} B\,\Delta u_k(j-1). \tag{6.42}$$

Remark 6.5. In practical applications, we can stop our iteration if there exists a $k \in \mathbb{Z}_1^\infty$ such that $\|e_k(t)\| < \epsilon$, where $\epsilon > 0$ is a prefixed parameter that depends on the specific requirements.

Now consider (6.33) associated with (6.40) and (6.41). We are ready to provide the convergence analysis for error $\|e_k\|_\lambda$.

Theorem 6.7. *Consider (6.33) associated with (6.40). For arbitrary initial input* $u_1(t)$, *we have* $\lim_{k\to\infty} \|e_k\|_\lambda = 0$ *on* $\mathbb{Z}_0^T$ *provided that*

$$\rho(I - DL_1) < 1. \tag{6.43}$$

Proof. For (6.33) with $t \in \mathbb{Z}_0^T$, by (6.39), one has

$$e_{k+1}(t) - e_k(t) = y_k(t) - y_{k+1}(t) = -C\Delta x_k(t) - D\Delta u_k(t).$$

According to (6.40), we have

$$e_{k+1}(t) = (I - DL_1)e_k(t) - C\Delta x_k(t). \tag{6.44}$$

Taking the norm $\|\cdot\|$ on $\mathbb{R}^n$ for (6.44), by Lemma 2.7, one gets

$$\|e_{k+1}(t)\| \leq [\rho(I - DL_1) + \epsilon]\|e_k(t)\| + \|C\|\,\|\Delta x_k(t)\|, \tag{6.45}$$

where ϵ is an arbitrary positive number.

Obviously, $\Delta x_k(0)$ becomes an n-dimensional zero vector. According to (6.43) and (6.45), it is easy to get $\lim_{k\to\infty} \|e_k(0)\| = 0$.

When $t \in \mathbb{Z}_1^T$, multiplying both sides of (6.45) by λ^t and then taking the λ-norm, we have

$$\|e_{k+1}\|_\lambda \leq [\rho(I - DL_1) + \epsilon]\|e_k\|_\lambda + \|C\|\,\|\Delta x_k\|_\lambda. \tag{6.46}$$

Now we estimate the value of $\lambda^t\|\Delta x_k(t)\|$. According to (6.37), (6.40), and (6.42), we obtain

$$\lambda^t\|\Delta x_k(t)\| \leq \lambda^t \sum_{j=1}^t \|A^{(t-j)}\|\,\|e_m^{B_1(t-m-j)}\|\,\|B\|\,\|\Delta u_k(j-1)\|$$

$$\leq \lambda^t A_0^{T-1} e^{\|B_1\|T} \|B\|\,\|L_1\| \sum_{j=1}^t \|e_k(j-1)\|$$

$$\leq \lambda^t A_0^{T-1} e^{\|B_1\|T} \|B\|\,\|L_1\| \sum_{j=1}^t \lambda^{-(j-1)}\lambda^{j-1}\|e_k(j-1)\|$$

$$\leq A_0^{T-1} e^{\|B_1\|T} \|B\|\,\|L_1\|\,\|e_k\|_\lambda \sum_{j=1}^t \lambda^{t-(j-1)}$$

$$\leq \lambda T A_0^{T-1} e^{\|B_1\|T} \|B\|\,\|L_1\|\,\|e_k\|_\lambda, \tag{6.47}$$

where $A_c^d := \sup\limits_{i \in \mathbb{Z}_c^d} \|A^i\|$.

Taking the supremum norm for both sides of (6.47), one can obtain

$$\|\Delta x_k\|_\lambda = \sup_{t \in \mathbb{Z}_0^T} \{\lambda^t \|\Delta x_k(t)\|\} \le \lambda T A_0^{T-1} e^{\|B_1\|T} \|B\| \, \|L_1\| \, \|e_k\|_\lambda. \qquad (6.48)$$

Now linking (6.46) and (6.48), we have

$$\|e_{k+1}\|_\lambda \le [\rho(I - DL_1) + \epsilon + \mu_\lambda] \|e_k\|_\lambda, \qquad (6.49)$$

where $\mu_\lambda := \lambda T A_0^{T-1} e^{\|B_1\|T} \|B\| \, \|C\| \, \|L_1\|$.

Finally, due to (6.43), when $\lambda \in \left(0, \dfrac{1-\rho(I-DL_1)-\epsilon}{T A_0^{T-1} e^{\|B_1\|T} \|B\| \, \|C\| \, \|L_1\|}\right) \cap (0,1)$, one can obtain

$$\|e_{k+1}\|_\lambda < \|e_k\|_\lambda,$$

which implies $\lim\limits_{k \to \infty} \|e_k\|_\lambda = 0$. The proof is finished. $\qquad \square$

Theorem 6.8. *Consider (6.33) with $D = \Theta$ and (6.41). For arbitrary initial input $u_1(t)$, we have $\lim\limits_{k \to \infty} \|e_k\|_\lambda = 0$ on $\mathbb{Z}_1^T$ provided that*

$$\rho(I - CBL_2) < 1. \qquad (6.50)$$

Proof. For (6.33) with $D = 0$ and $t \in \mathbb{Z}_1^T$, we can get the relation between the k-th error and the $(k+1)$-th error via (6.39):

$$e_{k+1}(t) - e_k(t) = y_k(t) - y_{k+1}(t) = -C\Delta x_k(t). \qquad (6.51)$$

By (6.51) via (6.42), we obtain

$$e_{k+1}(t) = e_k(t) - C\sum_{j=1}^{t} A^{(t-j)} e_m^{B_1(t-m-j)} B\Delta u_k(j-1)$$

$$= e_k(t) - Ce_m^{B_1(-m)} B\Delta u_k(t-1)$$

$$- C\sum_{j=1}^{t-1} A^{(t-j)} e_m^{B_1(t-m-j)} B\Delta u_k(j-1).$$

Due to (6.1) and (6.41), we have

$$e_{k+1}(t) = (I - CBL_2)e_k(t) - C\sum_{j=1}^{t-1} A^{(t-j)} e_m^{B_1(t-m-j)} BL_2 e_k(j). \qquad (6.52)$$

Taking the norm $\| \cdot \|$ for (6.52), by Lemma 2.7, we get

$$\|e_{k+1}(1)\| \leq [\rho(I - CBL_2) + \epsilon]\, \|e_k(1)\|$$

for $t = 1$. By (6.50), it is easy to obtain $\lim_{k \to \infty} \|e_k(1)\| = 0$.

When $t \in \mathbb{Z}_2^T$, we arrive at

$$\|e_{k+1}(t)\|$$
$$\leq [\rho(I - CBL_2) + \epsilon]\, \|e_k(t)\|$$
$$+ \|C\| \sum_{j=1}^{t-1} \|A^{(t-j)}\|\, \|e_m^{B_1(t-m-j)}\|\, \|B\|\, \|L_2\|\, \|e_k(j)\|$$
$$\leq [\rho(I - CBL_2) + \epsilon]\, \|e_k(t)\| + A_1^{T-1} e^{\|B_1\|T} \|B\|\, \|C\|\, \|L_2\|\, \|e_k\|_\lambda \sum_{j=1}^{t-1} \lambda^{-j}.$$

Then by taking the λ-norm, we obtain

$$\|e_{k+1}\|_\lambda \leq [\rho(I - CBL_2) + \epsilon]\, \|e_k\|_\lambda$$
$$+ A_1^{T-1} e^{\|B_1\|T} \|B\|\, \|C\|\, \|L_2\|\, \|e_k\|_\lambda \sum_{j=1}^{t-1} \lambda^{t-j}$$
$$\leq [\rho(I - CBL_2) + \epsilon + \nu_\lambda]\, \|e_k\|_\lambda, \tag{6.53}$$

where $0 < \lambda < 1$ and $\nu_\lambda := \lambda(T-1)A_1^{T-1} e^{\|B_1\|T} \|B\|\, \|C\|\, \|L_2\|$.

Similar to Theorem 6.7, choosing a λ from the set

$$\lambda \in \left(0, \frac{1 - \rho(I - CBL_2) - \epsilon}{(T-1)A_1^{T-1} e^{\|B_1\|T} \|B\|\, \|C\|\, \|L_2\|}\right) \bigcap (0, 1),$$

by (6.50),

$$\|e_{k+1}\|_\lambda < \|e_k\|_\lambda,$$

which implies $\lim_{k \to \infty} \|e_k\|_\lambda = 0$. Thus, the proof is completed. $\qquad\square$

Remark 6.6. In Theorem 6.8, if $y_d(0)$ is in the range of C, then taking $C\varphi(0) = y_d(0)$, we have $y_k(0) = Cx_k(0) = C\varphi(0) = y_d(0)$, which implies that $e_k(0) = 0$ for any $k = 1, 2, \ldots$. As a result, $\lim_{k \to \infty} \|e_k\|_\lambda = 0$ on $\mathbb{Z}_0^T$.

Remark 6.7. Obviously, the smaller spectral radius $\rho(\cdot)$ in Theorems 6.7 and 6.8, the better the convergence performance. In addition, by analyzing formulas (6.49) and (6.53), when the selected λ is determined, the smaller the value of the variable t, the better the convergence performance. These conclusions will also be verified in the next section.

6.2.1.2 Numerical examples and discussion

Example 6.3. Set $n = 2$, $r = m_* = 1$, $T = 12$. Consider the following discrete control system with single delay:

$$\begin{cases} x_k(t+1) = Ax_k(t) + A_1 x_k(t-5) + Bu_k(t), \ t \in \mathbb{Z}_0^{12}, \\ x_k(t) = \varphi(t) = (2,1)^\top, \ t \in \mathbb{Z}_{-5}^0, \\ y_k(t) = Cx_k(t) + Du_k(t), \end{cases} \tag{6.54}$$

where

$$x_k(t) = \begin{pmatrix} x_{1,k}(t) \\ x_{2,k}(t) \end{pmatrix}, \ A = \begin{pmatrix} 1 & 0 \\ 0 & 1 \end{pmatrix}, \ A_1 = \begin{pmatrix} 0 & 1 \\ -2 & 3 \end{pmatrix},$$

$$B = \begin{pmatrix} 1 \\ 2 \end{pmatrix}, \ C = \begin{pmatrix} 0.2 & 0.3 \end{pmatrix}, \ D = 1.$$

We set the learning law (6.40) as

$$u_{k+1}(t) = u_k(t) + L_1 \cdot e_k(t),$$

where we choose $L_1 := 0.7 D^{-1} = 0.7$ and give the desired discrete reference trajectory as

$$y_d(t) = 5t \cos(0.5t), \ t \in \mathbb{Z}_0^{12}.$$

Obviously, we can see the state $x_k \in \mathbb{R}^2$, input $u_k \in \mathbb{R}$, and output $y_k \in \mathbb{R}$. Moreover, $AA_1 = A_1 A$ is satisfied and by (6.36), the state of (6.54) has the following form:

$$x_k(t) = e_5^{A_1 t} \varphi(-5) + \sum_{j=1}^t e_5^{A_1(t-5-j)} Bu_k(j-1). \tag{6.55}$$

According to Lemma 6.5, one can obtain

$$e_5^{A_1 t} = P\mathrm{diag}(e_5^t, e_5^{2t})P^{-1} = \begin{pmatrix} 2 \cdot e_5^t - e_5^{2t} & e_5^{2t} - e_5^t \\ 2 \cdot e_5^t - 2 \cdot e_5^{2t} & 2 \cdot e_5^{2t} - e_5^t \end{pmatrix},$$

where $P = \begin{pmatrix} 1 & 1 \\ 1 & 2 \end{pmatrix}$, $p^{-1} = \begin{pmatrix} 2 & -1 \\ -1 & 1 \end{pmatrix}$, and

$$
e_5^t = \begin{cases}
0, & \text{if } t \in \mathbb{Z}_{-\infty}^{-6}, \\
1, & \text{if } t \in \mathbb{Z}_{-5}^{0}, \\
1 + \frac{t!}{1!(t-1)!}, & \text{if } t \in \mathbb{Z}_1^6, \\
1 + \frac{t!}{1!(t-1)!} + \frac{(t-5)!}{2!(t-5-2)!}, & \text{if } t \in \mathbb{Z}_7^{12}, \\
\vdots &
\end{cases}
$$

$$
e_5^{2t} = \begin{cases}
0, & \text{if } t \in \mathbb{Z}_{-\infty}^{-6}, \\
1, & \text{if } t \in \mathbb{Z}_{-5}^{0}, \\
1 + 2\frac{t!}{1!(t-1)!}, & \text{if } t \in \mathbb{Z}_1^6, \\
1 + 2\frac{t!}{1!(t-1)!} + 4\frac{(t-5)!}{2!(t-5-2)!}, & \text{if } t \in \mathbb{Z}_7^{12}. \\
\vdots &
\end{cases}
\tag{6.56}
$$

Thus, (6.55) becomes

$$
\begin{aligned}
x_k(t) &= \begin{pmatrix} 2 \cdot e_5^t - e_5^{2t} & e_5^{2t} - e_5^t \\ 2 \cdot e_5^t - 2 \cdot e_5^{2t} & 2 \cdot e_5^{2t} - e_5^t \end{pmatrix} \begin{pmatrix} 2 \\ 1 \end{pmatrix} \\
&\quad + \sum_{j=1}^{t} \begin{pmatrix} 2 \cdot e_5^{t-5-j} - e_5^{2(t-5-j)} & e_5^{2(t-5-j)} - e_5^{t-5-j} \\ 2\ e_5^{t-5-j} & 2\ e_5^{2(t-5-j)} & 2\ e_5^{2(t-5-j)} & e_5^{t-5-j} \end{pmatrix} \\
&\quad \times \begin{pmatrix} 1 \\ 2 \end{pmatrix} u_k(j-1) \\
&= \begin{pmatrix} 3 \cdot e_5^t - e_5^{2t} + \sum_{j=1}^{t} e_5^{2(t-5-j)} u_k(j-1) \\ 3 \cdot e_5^t - 2 \cdot e_5^{2t} + 2 \cdot \sum_{j=1}^{t} e_5^{2(t-5-j)} u_k(j-1) \end{pmatrix}.
\end{aligned}
$$

Furthermore, $\rho(1 - 1 \cdot 0.7) = 0.3 < 1$. All the conditions of Theorem 6.7 are satisfied; thus, $\lim_{k \to \infty} \|e_k\|_\lambda = 0$ uniformly on $\mathbb{Z}_0^{12}$.

We set the first input $u_1(t) = 0$, $t \in \mathbb{Z}_0^{12}$. Define the l^2-norm of error $\|e_k\|_{l^2} = \left(\sum_{t \in \mathbb{Z}_0^T} \|e_k(t)\|^2 \right)^{1/2}$.

The upper image of Fig. 6.1 shows the reference trajectory $y_d(t)$ (red (light gray in print version) plus sign) and the output (blue (dark gray in print version) fold line) of system (6.54). The lower image of Fig. 6.1 shows the l^2-norm of

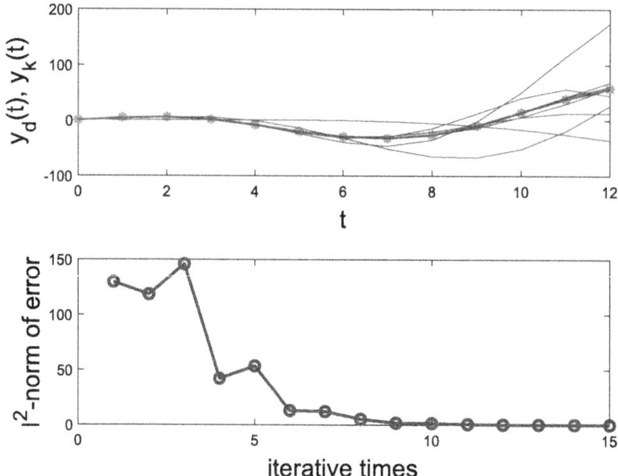

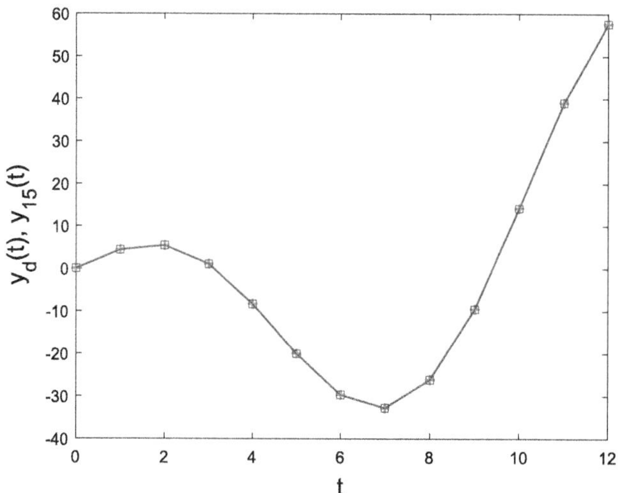

FIGURE 6.1   The tracking performance of system (6.54) and the l^2-norm of the tracking error.

FIGURE 6.2   The reference trajectory $y_d(t)$ and output $y_{20}(t)$ of system (6.54).

the tracking error in each iteration. The reference trajectory (red (light gray in print version) plus sign) and the output $y_{20}(t)$ (blue (dark gray in print version) box) are shown in Fig. 6.2. Obviously, the output $y_k(t)$ can track the reference trajectory $y_d(t)$ effectively. In specific applications, when the accuracy of the error reaches the requirements, we can stop iterating.

Example 6.4. Set $n = r = m_* = 2$, $T = 12$. In this example, we discuss the system of the following form:

$$\begin{cases} x_k(t+1) = Ax_k(t) + A_1 x_k(t-5) + Bu_k(t), \ t \in \mathbb{Z}_0^{12}, \\ x_k(t) = \varphi(t) = (0,0)^\top, \ t \in \mathbb{Z}_{-5}^0, \\ y_k(t) = Cx_k(t), \end{cases} \tag{6.57}$$

where

$$x_k(t) = \begin{pmatrix} x_{1,k}(t) \\ x_{2,k}(t) \end{pmatrix}, \ u_k(t) = \begin{pmatrix} u_{1,k}(t) \\ u_{2,k}(t) \end{pmatrix}, \ y_k(t) = \begin{pmatrix} y_{1,k}(t) \\ y_{2,k}(t) \end{pmatrix},$$

$$A = \begin{pmatrix} 1 & 0 \\ 1 & 1 \end{pmatrix}, \ A_1 = \begin{pmatrix} 1 & 0 \\ 6 & 1 \end{pmatrix}, \ B = \begin{pmatrix} 1 & -0.8 \\ 0 & 1 \end{pmatrix}, \ C = \begin{pmatrix} 1 & 1 \\ 0 & 1 \end{pmatrix}.$$

We set the learning law (6.41) for (6.57) as

$$u_{k+1}(t) = u_k(t) + L_2 e_k(t+1),$$

where we choose

$$L_2 = 0.8(CB)^{-1} = 0.8 \begin{pmatrix} 1 & -0.2 \\ 0 & 1 \end{pmatrix}$$

and give the reference trajectory as

$$y_d(t) = \begin{pmatrix} 3t\cos(0.5t) + 1.5t \\ 0.5\sin(t)(e^{0.25t} - 1) \end{pmatrix}, \ t \in \mathbb{Z}_0^{12},$$

where we mark

$$y_d(t) = \begin{pmatrix} y_{1,d}(t) \\ y_{2,d}(t) \end{pmatrix}, \ e_k(t) = \begin{pmatrix} e_{1,k}(t) \\ e_{2,k}(t) \end{pmatrix}.$$

Clearly, the state $x_k \in \mathbb{R}^2$, input $u_k \in \mathbb{R}^2$, and output $y_k \in \mathbb{R}^2$. Next, $AA_1 = A_1 A$. By (6.36), the state of system (6.57) has the form

$$x_k(t) = A^5 \sum_{j=1}^{t} A^{t-5-j} e_5^{A^{-1}A_1 A^{-5}(t-5-j)} Bu_k(j-1)$$

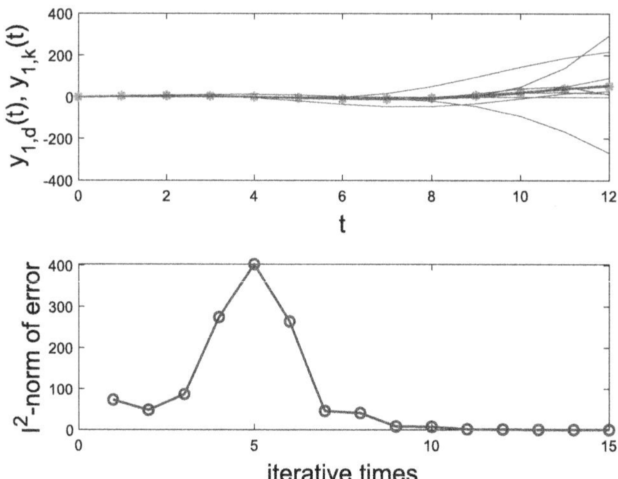

FIGURE 6.3 The tracking performance of system (6.57) and the l^2-norm of the tracking error.

$$
= \sum_{j=1}^{t} \begin{pmatrix} 1 & 0 \\ t-j & 1 \end{pmatrix} \begin{pmatrix} e_5^{t-5-j} & 0 \\ 0 & e_5^{t-5-j} \end{pmatrix} \begin{pmatrix} 1 & -0.8 \\ 0 & 1 \end{pmatrix}
$$
$$
\times \begin{pmatrix} u_{1,k}(j-1) \\ u_{2,k}(j-1) \end{pmatrix}
$$
$$
= \begin{pmatrix} \sum_{j=1}^{t} e_5^{t-5-j} \left[u_{1,k}(j-1) - 0.8 u_{2,k}(j-1) \right] \\ \sum_{j=1}^{t} e_5^{t-5-j} \left[(t-j)u_{1,k}(j-1) + (1-0.8t+0.8j)u_{2,k}(j-1) \right] \end{pmatrix},
$$

where e_5^t is shown in (6.56).

Next, one can see that

$$
\rho(I - CBL_2) = \rho \begin{pmatrix} 0.2 & 0 \\ 0 & 0.2 \end{pmatrix} = 0.2 < 1.
$$

Similar to Theorem 6.7, we also set the first input $u_1(t) = 0$, $t \in \mathbb{Z}_0^{12}$. Thus, Theorem 6.8 guarantees $\lim_{k \to \infty} \|e_k\|_\lambda = 0$ uniformly on $\mathbb{Z}_1^{12}$ for (6.57).

Figs. 6.3 and 6.5 show the tracking performance of system (6.57) and the l^2-norm of the tracking error. Figs. 6.4 and 6.6 show the results of $y_d(t)$ and $y_{15}(t)$. Simulation results show that the iterative output $y_k(t)$ can converge to the desired reference trajectory.

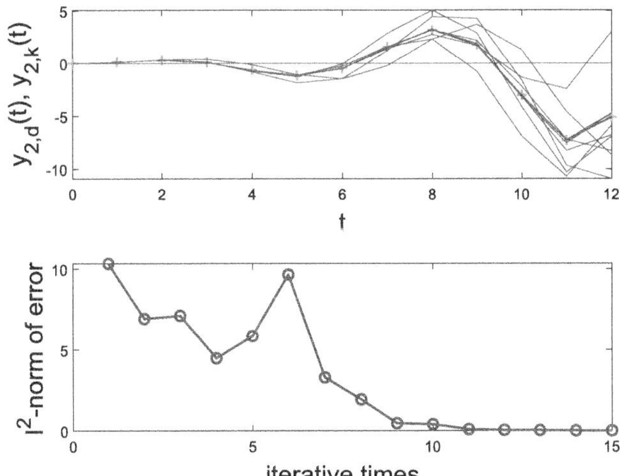

FIGURE 6.4   The reference trajectory $y_{1,d}(t)$ and output $y_{1,15}(t)$ of system (6.57).

FIGURE 6.5   The tracking performance of system (6.57) and the l^2-norm of the tracking error.

6.2.1.3   Conclusion

We adopt a new framework to establish the same convergence theorem for the ILC problem of linear discrete delayed systems by virtue of the representation of solutions via discrete matrix delayed exponential functions. The results in this part are motivated from [116].

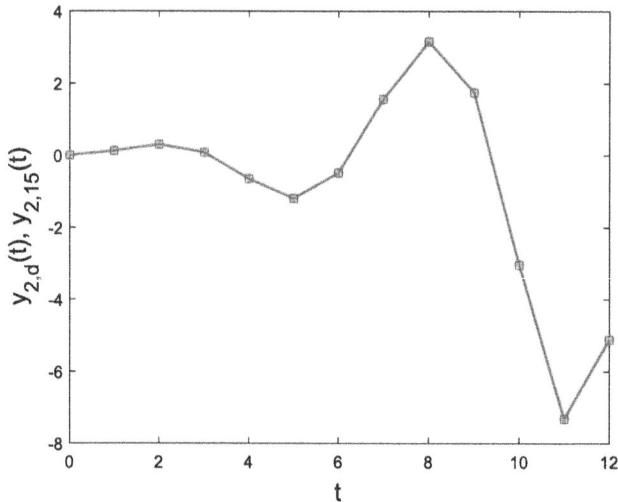

FIGURE 6.6 The reference trajectory $y_{2,d}(t)$ and output $y_{2,15}(t)$ of system (6.57).

6.3 Iterative learning control for varying trial lengths

6.3.1 Iterative learning control for linear discrete delay systems

Let $\mathbf{E}\{X\}$ be the expectation of a stochastic variable X and let $\mathbf{P}[Y]$ be the occurrence probability of event Y. We note that $\mathbb{Z}_0^{T_a}$ means the shortest running interval and $\mathbb{Z}_0^{T_d}$ means the desired running interval. Clearly, $T_d > T_a > 0$.

6.3.1.1 Discrete matrix delayed exponential function

When we consider the nonhomogeneous initial Cauchy problem (6.34), we note that the condition on commutative matrices is acceptable for various matrices, for example, normal matrices and diagonal matrices (see [117, Section 30, Lemma 4] and [118, Corollary 9.8, p. 152]).

The main motivation for $AA_1 = A_1A$ is that the solution formula (6.35) can be found by adopting a similar idea to linear difference systems without delay. In fact, $X(t) := A^t e_m^{B_1 t}$ is an explicit form of the fundamental matrix for $x(t+1) = Ax(t) + A_1x(t-m), t \in \mathbb{Z}_0^\infty$, with $x(t) = A^t, t \in \mathbb{Z}_{-m}^0$. The simple form allows us to generate a sequence of outputs to track the desired reference. We also note that a model of population dynamics with delayed birthrates and delayed logistic terms can be described by delay differential or difference equations with permutable matrices [7,119].

6.3.1.2 Randomly varying trial lengths

A very important difficulty that needs to be dealt with is that the trial lengths are randomly varying for each iteration. Now, we analyze the randomness of trial lengths.

In fact, two situations need to be considered in the iterative process; one is $T_k \geq T_d$ and the other is $T_a \leq T_k < T_d$, where T_d is the desired operation length, T_k is the actual operation length, and T_a is the possible smallest operation length. In the following, we mark the running interval of the k-th iteration as $\mathbb{Z}_0^{T_k}$ ($\mathbb{Z}_0^{T_d}$ and $\mathbb{Z}_0^{T_a}$ can be defined similarly). For the former case, only the data in $\mathbb{Z}_0^{T_d}$ can be used to update the input signal. Therefore, without losing generality, we view this case as $T_k = T_d$. For the latter case, the output does not exist in $\mathbb{Z}_{T_{k+1}}^{T_d}$, i.e., only the date in $\mathbb{Z}_0^{T_k}$ can be used to update the input signal.

In order to characterize this process, we introduce a stochastic variable $\eta_k(t)$, $t \in \mathbb{Z}_0^{T_d}$, satisfying a Bernoulli distribution and taking values 0 and 1 in the k-th iteration. Here $\eta_k(t) = 1$ indicates that the controlled systems can run up to time instant t with a probability of $p(t)$, $0 < p(t) \leq 1$. On the contrary, $\eta_k(t) = 0$ indicates that the controlled systems cannot run up to time instant t with probability $1 - p(t)$. Since the stochastic variable $\eta_k(t)$ obeys the Bernoulli distribution, we obtain the expectation $\mathbf{E}\{\eta_k(t)\} = 0 \cdot (1 - p(t)) + 1 \cdot p(t) = p(t)$.

Hypothesis: Assume the probability $\mathbf{P}[\mathbf{M}_{T_k}] = p_{T_k}$, in which we mark event $\mathbf{M}_{T_k}$ as running terminal at T_k in the k-th iteration.

Obviously, controlled systems can run for the interval $\mathbb{Z}_0^{T_a}$, which implies that $p(t) = 1$ for $t \in \mathbb{Z}_0^{T_a}$. For $t \in \mathbb{Z}_{T_{a+1}}^{T_d}$, we can calculate

$$p(t) = \mathbf{P}[\bigcup_{i=t}^{T_d} \mathbf{M}_i] = \sum_{i=t}^{T_d} \mathbf{P}[\mathbf{M}_i] = \sum_{i=t}^{T_d} p_i.$$

In addition,

$$\mathbf{P}[\bigcup_{i=T_a}^{T_d} \mathbf{M}_i] = \sum_{i=T_a}^{T_d} \mathbf{P}[\mathbf{M}_i] = \sum_{i=T_a}^{T_d} p_i = 1.$$

Thus, we conclude possibility $p(t)$ is

$$p(t) = \begin{cases} 1, \ t \in \mathbb{Z}_0^{T_a}, \\ \sum\limits_{i=t}^{T_d} p_i, \ t \in \mathbb{Z}_{T_{a+1}}^{T_d}. \end{cases}$$

6.3.1.3 ILC design and convergence analysis

In this part, we consider the ILC problem for the following linear discrete delay controlled system:

$$
\begin{cases}
x_k(t+1) = Ax_k(t) + A_1 x_k(t-\sigma) + Bu_k(t), \ t \in \{\eta_k(i) \cdot i, \ i \in \mathbb{Z}_0^{T_d}\}, \\
x_k(t) = \varphi(t), \ t \in \mathbb{Z}_{-\sigma}^0, \\
y_k(t) = Cx_k(t) + Du_k(t),
\end{cases}
$$

$$(6.58)$$

where $A, A_1 \in \mathbb{R}^{n \times n}$ satisfying $AA_1 = A_1 A$ and A^{-1} exist, $B \in \mathbb{R}^{n \times r}$, $C \in \mathbb{R}^{m \times n}$, $D \in \mathbb{R}^{m \times r}$, the delay $\sigma \geq 1$ is a prefixed integer, and $\varphi : \mathbb{Z}_{-\sigma}^0 \to \mathbb{R}^n$ denotes the initial state. The index $k = 1, 2, \dots$ denotes the k-th iteration and the variables $x_k(t) \in \mathbb{R}^n$, $u_k(t) \in \mathbb{R}^r$, and $y_k(t) \in \mathbb{R}^m$ denote the state, input, and output, respectively. In addition, considering the randomness of the trial lengths, $t \in \{\eta_k(i) \cdot i, \ i \in \mathbb{Z}_0^{T_d}\}$ can accurately depict the running interval of the k-th iteration by the stochastic variable $\eta_k(i), i \in Z_0^{T_d}$.

By Lemma 6.3, the state $x_k(t), \ t \in \mathbb{Z}_{-\sigma}^\infty$, of (6.58) can be represented in the form

$$
x_k(t) = A^t e_\sigma^{B_1 t} A^{-\sigma} \varphi(-\sigma) + \sum_{j=-\sigma+1}^{0} A^{(t-j)} e_\sigma^{B_1(t-\sigma-j)} [\varphi(j) - A\varphi(j-1)]
$$

$$
+ \sum_{j=1}^{t} A^{(t-j)} e_\sigma^{B_1(t-\sigma-j)} Bu_k(j-1),
$$

$$(6.59)$$

where $B_1 = A^{-1} A_1 A^{-\sigma}$ and $e_\sigma^{B_1 \cdot}$ is defined similar to (6.1).

Assuming that system (6.58) is stable, controllable, and observable, for arbitrary bounded reference $y_d(t) \in \mathbb{R}^m$, there exists a unique control input $u_d(t) \in \mathbb{R}^r$ such that

$$
\begin{cases}
x_d(t+1) = Ax_d(t) + A_1 x_d(t-m) + Bu_d(t), \ t \in \mathbb{Z}_0^{T_d}, \\
x_d(t) = \varphi(t), \ t \in \mathbb{Z}_{-\sigma}^0, \\
y_d(t) = Cx_d(t) + Du_d(t),
\end{cases}
$$

where $x_d(t) \in \mathbb{R}^n$ is the corresponding referential state.

In ILC, the tracking error of controlled systems is used to update the input, where the tracking error is defined as $e_k(t) := y_d(t) - y_k(t)$. However, the error of interval $\mathbb{Z}_{T_{k+1}}^{T_d}$ does not exist for the k-th iteration in our case where T_k denotes the running terminal of the k-th iteration. Thus, we mark the tracking error for

$t \in \mathbb{Z}_{T_{k+1}}^{T_d}$ as zero, and for $t \in \mathbb{Z}_0^{T_d}$, we define a corrected tracking error as

$$\tilde{e}_k(t) := \eta_k(t)e_k(t) = \begin{cases} e_k(t), \ t \in \mathbb{Z}_0^{T_k}, \\ 0, \ t \in \mathbb{Z}_{T_{k+1}}^{T_d}. \end{cases}$$

Now we can observe the running interval of each iteration as $Z_0^{T_d}$ by using the stochastic variable $\eta_k(t)$. Consider the following ILC updating laws:

$$u_{k+1}(t) = u_k(t) + L_1\tilde{e}_k(t), \ t \in \mathbb{Z}_0^{T_d}, \tag{6.60}$$

and

$$u_{k+1}(t) = u_k(t) + L_2\tilde{e}_k(t+1), \ t \in \mathbb{Z}_0^{T_d}. \tag{6.61}$$

Note $t + 1 = T_d + 1$ when $t = T_d$ in law (6.61), so we consider the desired running interval as $Z_0^{T_d+1}$ to ensure that the error can be used to update input.

Now we are ready to present the first main result in this subsection.

Theorem 6.9. *Consider system (6.58) associated with updating law (6.60). For arbitrary initial input $u_1(\cdot)$, we have*

$$\lim_{k\to\infty} \mathbf{E}\{\|\tilde{e}_k\|_\lambda\} = 0$$

on $\mathbb{Z}_0^{T_d}$ if the Euclidean matrix norm $\|\cdot\|$ related to parameter matrix L_1 satisfies

$$\|I - L_1 D\| < 1. \tag{6.62}$$

Proof. Denote $\delta x_k(t) := x_d(t) - x_k(t)$ as the state error and $\delta u_k(t) := u_d(t) - u_k(t)$ as the input error. We divide the proof into two parts.

Part 1: We show that $\lim_{k\to\infty} \mathbf{E}\{\|\delta u_k\|_\lambda\} = 0$, for $\forall \, t \in \mathbb{Z}_0^{T_d}$.

By updating law (6.60), for $t \in \mathbb{Z}_0^{T_d}$, we have

$$\delta u_{k+1}(t) = \delta u_k(t) - L_1\tilde{e}_k(t) = \delta u_k(t) - \eta_k(t)L_1 e_k(t). \tag{6.63}$$

According to (6.58), one can compute

$$\begin{cases} \delta x_k(t+1) = A\delta x_k(t) + A_1\delta x_k(t-\sigma) + B\delta u_k(t), \ t \in \{\eta_k(i)\cdot i, \ i \in \mathbb{Z}_0^{T_d}\}, \\ \delta x_k(t) = \mathbf{0}, \ t \in \mathbb{Z}_{-\sigma}^0, \\ e_k(t) = C\delta x_k(t) + D\delta u_k(t). \end{cases}$$

$$\tag{6.64}$$

Inserting $e_k(t)$ of (6.64) into (6.63), we get

$$\delta u_{k+1}(t) = [I - \eta_k(t)L_1 D]\delta u_k(t) - \eta_k(t)L_1 C\delta x_k(t). \tag{6.65}$$

Taking the Euclidean norm for (6.65), we have

$$\|\delta u_{k+1}(t)\| \le \|I - \eta_k(t)L_1 D\| \, \|\delta u_k(t)\| + \eta_k(t)\|L_1\| \, \|C\| \, \|\delta x_k(t)\|. \quad (6.66)$$

For $t = 0$, noting that $\eta_k(0) = 1$ and $\delta x_k(0) = \mathbf{0}$, we obtain

$$\|\delta u_{k+1}(0)\| \le \|I - L_1 D\| \, \|\delta u_k(0)\|.$$

By condition (6.62), we get $\|\delta u_{k+1}(0)\| < \|\delta u_k(0)\|$ and

$$\lim_{k \to \infty} \|\delta u_k(0)\| = 0. \quad (6.67)$$

For $t \in \mathbb{Z}_1^{T_d}$, multiplying both sides of (6.66) by $e^{-\lambda t}$, we get

$$e^{-\lambda t}\|\delta u_{k+1}(t)\| \le \|I - \eta_k(t)L_1 D\| \, e^{-\lambda t}\|\delta u_k(t)\|$$
$$+ \eta_k(t)\|L_1\| \, \|C\| \, e^{-\lambda t}\|\delta x_k(t)\|. \quad (6.68)$$

By (6.59), one can derive $\delta x_k(t)$ in (6.64) as follows:

$$\delta x_k(t) = \sum_{j=1}^{t} A^{(t-j)} e_\sigma^{B_1(t-\sigma-j)} B \delta u_k(j-1), \ t \in \{\eta_k(i) \cdot i, \ i \in \mathbb{Z}_0^{T_d}\}. \quad (6.69)$$

Note that we consider the convergence of errors in the sense of expectation. Without loss of generality, we can assume that (6.69) well holds on the full interval $\mathbb{Z}_0^{T_d}$. Again note that $\eta_k(i)$, $i \in \mathbb{Z}_0^{T_d}$, takes the value 0 on the part (not running) interval $\mathbb{Z}_0^{T_d} \setminus \{\eta_k(i) \cdot i, \ i \in \mathbb{Z}_0^{T_d}\}$. Thus, the main contributions to computing the convergence of error in the sense of expectation are the accumulation effects on the running interval $\{\eta_k(i) \cdot i, \ i \in \mathbb{Z}_0^{T_d}\}$.

Now, taking the λ-norm for both sides of (6.69) and using Lemma 6.4, one can obtain

$$\sup_{t \in \mathbb{Z}_1^{T_d}} e^{-\lambda t}\|\delta x_k(t)\| \le \sup_{t \in \mathbb{Z}_1^{T_d}} e^{-\lambda t} \sum_{j=1}^{t} \|A^{(t-j)}\| \, \|e_\sigma^{B_1(t-\sigma-j)}\| \, \|B\| \, \|\delta u_k(j-1)\|$$

$$\le \sup_{t \in \mathbb{Z}_1^{T_d}} e^{-\lambda t} A_0^{T_d-1} e^{\|B_1\|T_d} \|B\| \sum_{j=1}^{t} e^{\lambda(j-1)} e^{-\lambda(j-1)}\|\delta u_k(j-1)\|$$

$$\le \sup_{t \in \mathbb{Z}_1^{T_d}} A_0^{T_d-1} e^{\|B_1\|T_d} \|B\| \, \|\delta u_k\|_\lambda \sum_{j=1}^{t} e^{\lambda(j-1-t)}$$

$$\le e^{\|B_1\|T_d - \lambda} T_d A_0^{T_d-1} \|B\| \, \|\delta u_k\|_\lambda, \quad (6.70)$$

where $A_b^c := \max_{j \in \mathbb{Z}_b^c} \|A^j\|$.

Taking the λ-norm for (6.68) via (6.70), we arrive at

$$\|\delta u_{k+1}\|_{\lambda} \leq \|I - \eta_k(t)L_1 D\| \, \|\delta u_k\|_{\lambda}$$
$$+ \eta_k(t)e^{\|B_1\|T_d - \lambda}T_d A_0^{T_d-1}\|L_1\| \, \|C\| \, \|B\| \, \|\delta u_k\|_{\lambda}. \tag{6.71}$$

Applying the operator $\mathbf{E}\{\cdot\}$ on both sides of (6.71), one can obtain

$$\mathbf{E}\{\|\delta u_{k+1}\|_{\lambda}\}$$
$$\leq \mathbf{E}\{\|I - \eta_k(t)L_1 D\|\}\mathbf{E}\{\|\delta u_k\|_{\lambda}\}$$
$$+ \mathbf{E}\{\eta_k(t)\}e^{\|B_1\|T_d - \lambda}T_d A_0^{T_d-1}\|L_1\| \, \|C\| \, \|B\|\mathbf{E}\{\|\delta u_k\|_{\lambda}\}. \tag{6.72}$$

Noting that $\mathbf{E}\{\eta_k(t)\} = p(t)$ and

$$\mathbf{E}\{\|I - \eta_k(t)L_1 D\|\} = \|I - 0 \cdot L_1 D\| \cdot (1 - p(t)) + \|I - 1 \cdot L_1 D\| \cdot p(t)$$
$$= 1 - p(t) + p(t)\|I - L_1 D\|$$
$$= 1 + p(t)(\|I - L_1 D\| - 1), \tag{6.73}$$

because $0 < p(\cdot) \leq 1$, according to (6.62), we have

$$\vartheta := \max_{t \in \mathbb{Z}_1^{T_d}} (1 + p(t)(\|I - L_1 D\| - 1)) < 1. \tag{6.74}$$

Thus, noting (6.73) and (6.74), (6.72) becomes

$$\mathbf{E}\{\|\delta u_{k+1}\|_{\lambda}\} \leq \tilde{\vartheta}\mathbf{E}\{\|\delta u_k\|_{\lambda}\}, \tag{6.75}$$

where

$$\tilde{\vartheta} := \vartheta + e^{\|B_1\|T_d - \lambda}T_d A_0^{T_d-1}\|L_1\| \, \|C\| \, \|B\|.$$

Next, due to (6.74) we can choose a number

$$\lambda > \max\left\{ \|B_1\|T_d - \ln\left[\frac{1 - \vartheta}{T_d A_0^{T_d-1}\|L_1\| \, \|C\| \, \|B\|}\right], \, 0 \right\}$$

such that $\tilde{\vartheta} < 1$.

Therefore, linking (6.67) and (6.75), we obtain

$$\lim_{k \to \infty} \mathbf{E}\{\|\delta u_k\|_{\lambda}\} = 0, \text{ for } \forall t \in \mathbb{Z}_0^{T_d}. \tag{6.76}$$

Part 2: We prove that $\lim_{k \to \infty} \mathbf{E}\{\|\tilde{e}_k\|_{\lambda}\} = 0$, for $\forall t \in \mathbb{Z}_0^{T_d}$.

Taking the λ-norm for $e_k(\cdot)$ in (6.64) and applying $\mathbf{E}\{\cdot\}$, we get

$$\mathbf{E}\{\|e_k\|_{\lambda}\} \leq \|C\| \, \mathbf{E}\{\|\delta x_k\|_{\lambda}\}$$
$$+ \|D\| \, \mathbf{E}\{\|\delta u_k\|_{\lambda}\}, \, \forall t \in \{\eta_k(i) \cdot i, \, i \in \mathbb{Z}_0^{T_d}\}. \tag{6.77}$$

By (6.70), one can obtain

$$\mathbf{E}\{\|\delta x_k\|_\lambda\} \le e^{\|B_1\|T_d - \lambda} T_d A_0^{T_d-1} \|B\| \, \mathbf{E}\{\|\delta u_k\|_\lambda\}.$$

According to (6.76), for a fixed number λ, we have

$$\lim_{k \to \infty} \mathbf{E}\{\|\delta x_k\|_\lambda\} = 0. \tag{6.78}$$

Combining (6.76) with (6.78), one can obtain $\lim\limits_{k \to \infty} \mathbf{E}\{\|e_k\|_\lambda\} = 0$ via (6.77). Thus, $\lim\limits_{k \to \infty} \mathbf{E}\{\|\tilde{e}_k\|_\lambda\} = 0$ for $\forall\, t \in \mathbb{Z}_0^{T_d}$. The proof is completed. $\qquad\square$

Next we present the second main result in this subsection.

Theorem 6.10. *Consider system* (6.58) *with* $D = \Theta$ *and updating law* (6.61). *For arbitrary initial input* $u_1(\cdot)$,

$$\lim_{k \to \infty} \mathbf{E}\{\|\tilde{e}_k\|_\lambda\} = 0$$

on $\mathbb{Z}_1^{T_d}$ *if the Euclidean matrix norm* $\|\cdot\|$ *related to parameter matrix* L_2 *satisfies*

$$\|I - L_2 C B\| < 1. \tag{6.79}$$

Proof. Considering $D = \Theta$ of system (6.58), we have

$$\begin{cases} \delta x_k(t+1) = A\delta x_k(t) + A_1 \delta x_k(t-\sigma) + B\delta u_k(t), \ t \in \{\eta_k(i) \cdot i, \ i \in \mathbb{Z}_1^{T_d}\}, \\ \delta x_k(t) = \mathbf{0}, \ t \in \mathbb{Z}_{-\sigma}^0, \\ e_k(t) = C\delta x_k(t). \end{cases} \tag{6.80}$$

As mentioned, it is sufficient to consider the error of the accumulation effect on the running interval $\{\eta_k(i) \cdot i, \ i \in \mathbb{Z}_1^{T_d}\}$ when we consider the convergence of error in the sense of expectation.

Without loss of generality, for $t \in \mathbb{Z}_1^{T_d}$, by updating law (6.61) and (6.80), one can obtain

$$\begin{aligned} \delta u_{k+1}(t) &= \delta u_k(t) - L_2 \tilde{e}_k(t+1) \\ &= \delta u_k(t) - \eta_k(t+1)L_2 C\delta x_k(t+1) \\ &= [I - \eta_k(t+1)L_2 C B]\delta u_k(t) \\ &\quad - \eta_k(t+1)L_2 C A\delta x_k(t) - \eta_k(t+1)L_2 C A_1 \delta x_k(t-\sigma). \end{aligned} \tag{6.81}$$

Taking the λ-norm for (6.81), we have

$$\begin{aligned} \|\delta u_{k+1}\|_\lambda &\le \|I - \eta_k(t+1)L_2 C B\| \, \|\delta u_k\|_\lambda + \eta_k(t+1)\|L_2\| \, \|C\| \, \|A\| \, J_1 \\ &\quad + \eta_k(t+1)\|L_2\| \, \|C\| \, \|A_1\| \, J_2, \end{aligned} \tag{6.82}$$

where

$$J_1 := \sup_{t \in \mathbb{Z}_1^{T_d}} e^{-\lambda t} \|\delta x_k(t)\| \text{ and } J_2 := \sup_{t \in \mathbb{Z}_1^{T_d}} e^{-\lambda t} \|\delta x_k(t-\sigma)\|. \tag{6.83}$$

Now, we estimate the values of J_1 and J_2 in (6.83).
Similar to (6.70), we get

$$J_1 \le e^{\|B_1\|T_d-\lambda} T_d A_0^{T_d-1} \|B\| \, \|\delta u_k\|_\lambda, \ \forall \, t \in \mathbb{Z}_1^{T_d}. \tag{6.84}$$

For J_2, when $t \in \mathbb{Z}_1^\sigma$, we get $\|\delta x_k(t-\sigma)\| = 0$ by (6.80). When $t \in \mathbb{Z}_{\sigma+1}^{T_d}$, by (6.59) and Lemma 6.4, one can obtain

$$\sup_{t \in \mathbb{Z}_{\sigma+1}^{T_d}} e^{-\lambda t} \|\delta x_k(t-\sigma)\|$$

$$\le \sup_{t \in \mathbb{Z}_{\sigma+1}^{T_d}} e^{-\lambda t} \sum_{j=1}^{t-\sigma} \|A^{(t-\sigma-j)}\| \, \|e_\sigma^{B_1(t-2\sigma-j)}\| \, \|B\| \, \|\delta u_k(j-1)\|$$

$$\le \sup_{t \in \mathbb{Z}_{\sigma+1}^{T_d}} e^{-\lambda t} A_0^{T_d-\sigma-1} e^{\|B_1\|(T_d-\sigma)} \|B\| \sum_{j=1}^{t-\sigma} e^{\lambda(j-1)} e^{-\lambda(j-1)} \|\delta u_k(j-1)\|$$

$$\le \sup_{t \in \mathbb{Z}_{\sigma+1}^{T_d}} A_0^{T_d-\sigma-1} e^{\|B_1\|(T_d-\sigma)} \|B\| \, \|\delta u_k\|_\lambda \sum_{j=1}^{t-\sigma} e^{\lambda(j-1-t)}$$

$$< e^{\|B_1\|(T_d-\sigma)-(\sigma+1)\lambda}(T_d-\sigma) A_0^{T_d-\sigma-1} \|B\| \, \|\delta u_k\|_\lambda.$$

where $A_b^c := \max_{j \in \mathbb{Z}_b^c} \|A^j\|$.
From the above, one can conclude that

$$J_2 \le e^{\|B_1\|T_d-\lambda} T_d A_0^{T_d-1} \|B\| \, \|\delta u_k\|_\lambda, \ \forall \, t \in \mathbb{Z}_1^{T_d}. \tag{6.85}$$

Substituting (6.84) and (6.85) into (6.82), we have

$$\|\delta u_{k+1}\|_\lambda \le \|I - \eta_k(t+1)L_2 CB\| \, \|\delta u_k\|_\lambda + \eta_k(t+1)\|L_2\| \, \|C\|(\|A\| + \|A_1\|)$$
$$\times e^{\|B_1\|T_d-\lambda} T_d A_0^{T_d-1} \|B\| \, \|\delta u_k\|_\lambda.$$

Then applying the expectation operator $\mathbf{E}\{\cdot\}$ to both sides, we obtain

$$\mathbf{E}\{\|\delta u_{k+1}\|_\lambda\}$$
$$\le \mathbf{E}\{\|I - \eta_k(t+1)L_2 CB\|\} \, \mathbf{E}\{\|\delta u_k\|_\lambda\} + \mathbf{E}\{\eta_k(t+1)\}$$
$$\times e^{\|B_1\|T_d-\lambda} T_d A_0^{T_d-1} \|L_2\| \, \|C\|(\|A\| + \|A_1\|)\|B\| \mathbf{E}\{\|\delta u_k\|_\lambda\}. \tag{6.86}$$

Similar to (6.73), we get

$$\mathbf{E}\{\|I - \eta_k(t+1)L_2CB\|\} = 1 + p(t+1)(\|I - L_2CB\| - 1). \tag{6.87}$$

According to $\mathbf{E}\{\eta_k(t+1)\} = p(t+1)$ and (6.87), (6.86) becomes

$$\begin{aligned}
\mathbf{E}\{\|\delta u_{k+1}\|_\lambda\} \\
\leq [1 + p(t+1)(\|I - L_2CB\| - 1)]\mathbf{E}\{\|\delta u_k\|_\lambda\} + p(t+1) \\
\times e^{\|B_1\|T_d - \lambda} T_d A_0^{T_d-1} \|L_2\| \, \|C\|(\|A\| + \|A_1\|)\|B\|\mathbf{E}\{\|\delta u_k\|_\lambda\}. \tag{6.88}
\end{aligned}$$

Due to $0 < p(t+1) \leq 1$ for $t \in \mathbb{Z}_1^{T_d}$ and (6.79), we get

$$\vartheta := \max_{t \in \mathbb{Z}_1^{T_d}} \{1 + p(t+1)(\|I - L_2CB\| - 1)\} < 1. \tag{6.89}$$

Next, (6.88) becomes

$$\mathbf{E}\{\|\delta u_{k+1}\|_\lambda\} \leq \tilde{\vartheta}\mathbf{E}\{\|\delta u_k\|_\lambda\}, \tag{6.90}$$

where

$$\tilde{\vartheta} := \vartheta + e^{\|B_1\|T_d - \lambda} T_d A_0^{T_d-1} \|L_2\| \, \|C\|(\|A\| + \|A_1\|)\|B\|.$$

Then, according to (6.89) we can choose a number

$$\lambda > \max\left\{ \|B_1\|T_d - \ln\left[\frac{1 - \vartheta}{T_d A_0^{T_d-1} \|L_2\| \, \|C\|(\|A\| + \|A_1\|)\|B\|} \right], 0 \right\}$$

such that $\tilde{\vartheta} < 1$.

Therefore, (6.90) implies

$$\lim_{k \to \infty} \mathbf{E}\{\|\delta u_k\|_\lambda\} = 0, \text{ for } \forall\, t \in \mathbb{Z}_1^{T_d}.$$

Applying $\mathbf{E}\{\cdot\}$ to (6.84), one can obtain

$$\mathbf{E}\{\|\delta x_k\|_\lambda\} \leq e^{\|B_1\|T_d - \lambda} T_d A_0^{T_d-1} \|B\|\mathbf{E}\{\|\delta u_k\|_\lambda\}, \, \forall\, t \in \mathbb{Z}_1^{T_d}.$$

For a fixed λ, we have $\lim_{k \to \infty} \mathbf{E}\{\|\delta x_k\|_\lambda\} = 0$, for $\forall\, t \in \mathbb{Z}_1^{T_d}$.

Taking the λ-norm for $e_k(\cdot)$ in (6.80), we obtain

$$\mathbf{E}\{\|e_k\|_\lambda\} \leq \|C\| \, \mathbf{E}\{\|\delta x_k\|_\lambda\}.$$

Finally, $\lim_{k \to \infty} \mathbf{E}\{\|e_k\|_\lambda\} = 0$. Thus, $\lim_{k \to \infty} \mathbf{E}\{\|\tilde{e}_k\|_\lambda\} = 0$ for $\forall\, t \in \mathbb{Z}_1^{T_d}$. The proof is finished. $\qquad\square$

6.3.1.4 Numerical examples and discussion

Example 6.5. Consider the following discrete delay system:

$$
\begin{cases}
x_k(t+1) = A x_k(t) + A_1 x_k(t-5) + B u_k(t), \ t \in \{\eta_k(i) \cdot i, \ i \in \mathbb{Z}_0^{20}\}, \\
x_k(t) = \varphi(t) = (2,1)^\top, \ t \in \mathbb{Z}_{-5}^0, \\
y_k(t) = C x_k(t) + D u_k(t),
\end{cases}
$$

$$(6.91)$$

where

$$
x_k(t) = \begin{pmatrix} x_{1,k}(t) \\ x_{2,k}(t) \end{pmatrix}, \ A = \begin{pmatrix} 1 & 0 \\ 0 & 1 \end{pmatrix}, \ A_1 = \begin{pmatrix} 0 & 1 \\ -2 & 3 \end{pmatrix},
$$

$$
B = \begin{pmatrix} 1 \\ 2 \end{pmatrix}, \ C = \begin{pmatrix} 0.2 & 0.3 \end{pmatrix}, \ D = 1.
$$

Running terminal T_k of each iteration is randomly selected between 17 and 22. Without loss of generality, set $u_0(t) = 0$. We select the learning law (6.60) as

$$
u_{k+1}(t) = u_k(t) + 0.45\tilde{e}_k(t),
$$

$$(6.92)$$

and the desired reference trajectory is

$$
y_d(t) = 3t \sin(2t), \ t \in \mathbb{Z}_0^{20}.
$$

$$(6.93)$$

Clearly, $AA_1 = A_1 A$ holds. According to (6.59), the state of (6.91) has the following form:

$$
x_k(t) = e_5^{A_1 t}\varphi(-5) + \sum_{j=1}^{t} e_5^{A_1(t-5-j)} B u_k(j-1).
$$

Note that

$$
e_5^{A_1 t} = P\mathrm{diag}(e_5^{2t}, e_5^{t})P^{-1},
$$

where $P = \begin{pmatrix} 1 & 1 \\ 1 & 2 \end{pmatrix}, \ P^{-1} = \begin{pmatrix} 2 & -1 \\ -1 & 1 \end{pmatrix}$. Thus,

$$
e_5^{A_1 t} = \begin{pmatrix} 2 \cdot e_5^{t} - e_5^{2t} & e_5^{2t} - e_5^{t} \\ 2 \cdot e_5^{t} - 2 \cdot e_5^{2t} & 2 \cdot e_5^{2t} - e_5^{t} \end{pmatrix}.
$$

Thanks to Lemma 6.3, one can obtain the following explicit formula of solution:

$$
x_k(t) = \begin{pmatrix} 3 \cdot e_5^t - e_5^{2t} + \sum_{j=1}^{t} e_5^{2(t-5-j)} u_k(j-1) \\[2ex] 3 \cdot e_5^t - 2 \cdot e_5^{2t} + 2 \cdot \sum_{j=1}^{t} e_5^{2(t-5-j)} u_k(j-1) \end{pmatrix}, \tag{6.94}
$$

where e_5^t and e_5^{2t} have the following forms:

$$
e_5^t = \begin{cases} 0, & \text{if } t \in \mathbb{Z}_{-\infty}^{-6}, \\ 1, & \text{if } t \in \mathbb{Z}_{-5}^{0}, \\ 1 + \frac{t!}{(t-1)!}, & \text{if } t \in \mathbb{Z}_{1}^{6}, \\ 1 + \frac{t!}{(t-1)!} + \frac{(t-5)!}{2!(t-7)!}, & \text{if } t \in \mathbb{Z}_{7}^{12}, \\ 1 + \frac{t!}{(t-1)!} + \frac{(t-5)!}{2!(t-7)!} + \frac{(t-10)!}{3!(t-13)!}, & \text{if } t \in \mathbb{Z}_{13}^{18}, \\ 1 + \frac{t!}{(t-1)!} + \frac{(t-5)!}{2!(t-7)!} + \frac{(t-10)!}{3!(t-13)!} + \frac{(t-15)!}{4!(t-19)!}, & \text{if } t \in \mathbb{Z}_{19}^{24}, \\ \vdots \end{cases} \tag{6.95}
$$

and

$$
e_5^{2t} = \begin{cases} 0, & \text{if } t \in \mathbb{Z}_{-\infty}^{-6}, \\ 1, & \text{if } t \in \mathbb{Z}_{-5}^{0}, \\ 1 + 2\frac{t!}{(t-1)!}, & \text{if } t \in \mathbb{Z}_{1}^{6}, \\ 1 + 2\frac{t!}{(t-1)!} + 4\frac{(t-5)!}{2!(t-7)!}, & \text{if } t \in \mathbb{Z}_{7}^{12}, \\ 1 + 2\frac{t!}{(t-1)!} + 4\frac{(t-5)!}{2!(t-7)!} + 8\frac{(t-10)!}{3!(t-13)!}, & \text{if } t \in \mathbb{Z}_{13}^{18}, \\ 1 + 2\frac{t!}{(t-1)!} + 4\frac{(t-5)!}{2!(t-7)!} + 8\frac{(t-10)!}{3!(t-13)!} + 16\frac{(t-15)!}{4!(t-19)!}, & \text{if } t \in \mathbb{Z}_{19}^{24}. \\ \vdots \end{cases}
$$

$$\tag{6.96}$$

With the help of solution (6.94) and discrete matrix delayed exponential functions (6.95) and (6.96), we can present the structure of state x_k very clearly, which also makes the output y_k fully observed on each running time interval associated with the number of time delay. This advantage may avoid overuse of data each time on the whole running time interval. In this case, one can save computation resources and improve the running efficiency in some sense.

Next, noting $L_1 = 0.45$ in learning law (6.92), one can check that $\|I - L_1 D\| = 0.55 < 1$. Now all the conditions of Theorem 6.9 are satisfied; thus, $\lim_{k \to \infty} \mathbf{E}\{\|\tilde{e}_k\|_\lambda\} = 0$ uniformly on $\mathbb{Z}_0^{20}$.

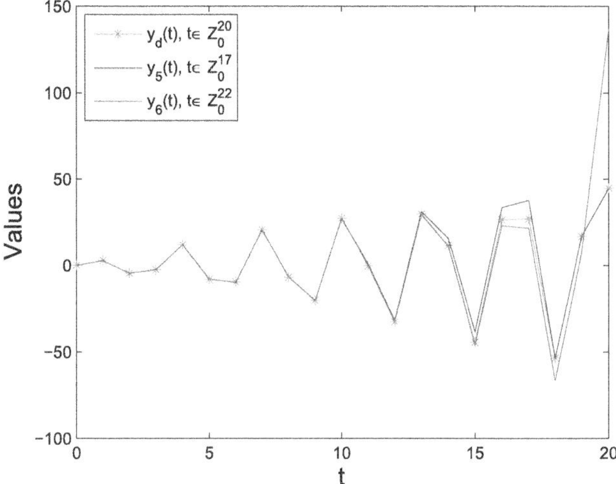

FIGURE 6.7 y_d, y_5, and y_6 of Example 6.5.

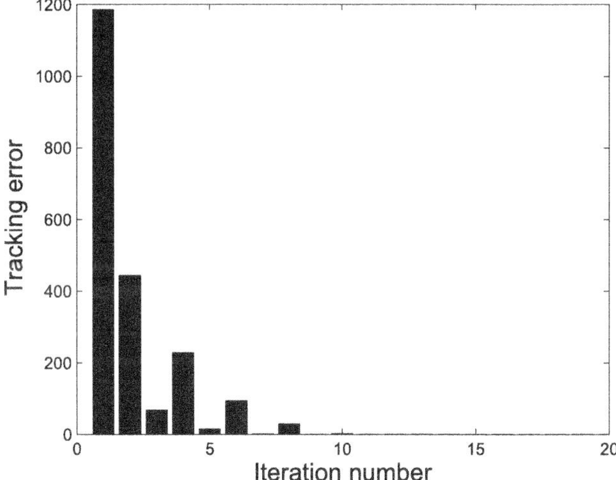

FIGURE 6.8   Tracking error $\|\tilde{e}_k\|_{l^2}$ of Example 6.5.

Fig. 6.7 shows the reference trajectory y_d and output trajectory y_k. When the output does not exist, we let $y_k = y_d$ to make sure $\tilde{e}_k(\cdot) = \eta_k(\cdot)e_k(\cdot)$.

Fig. 6.8 shows the Euclidean distance $\|\tilde{e}_k\|_{l^2}$ of error $\tilde{e}_k$ in each iteration. The error of the 20th iteration is 7.1014×10^{-6} and the running interval of the 20th iteration is $\mathbb{Z}_0^{20}$.

Figs. 6.9 and 6.10 show the reference trajectory y_d, output trajectory y_k, and error Euclidean distance $\|e_k\|_{l^2}$ for the same trial lengths ($\mathbb{Z}_0^{20}$) in each iteration. The error of the 20th iteration is 2.1839×10^{-6}.

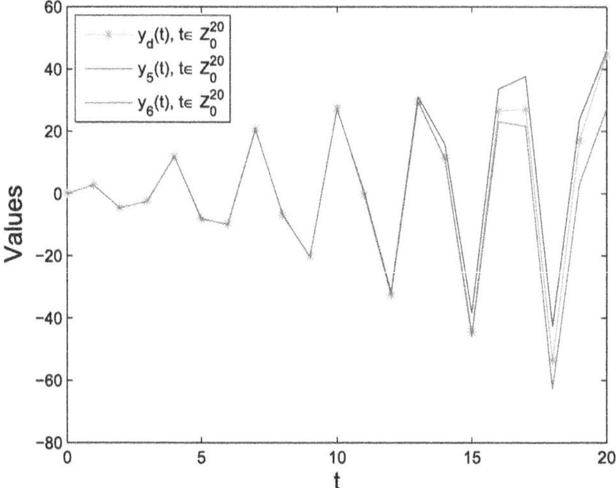

FIGURE 6.9 y_d, y_5, and y_6 of Example 6.5 for the same trial lengths.

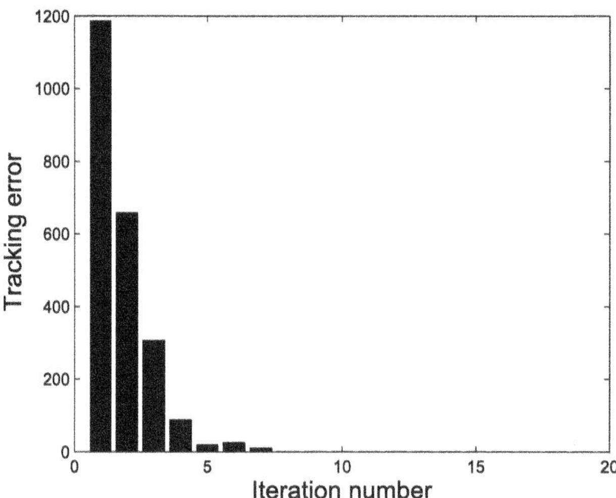

FIGURE 6.10 Tracking error $\|e_k\|_{l^2}$ of Example 6.5 for the same trial lengths.

Example 6.6. Consider

$$
\begin{cases}
x_k(t+1) = x_k(t) + 2x_k(t-5) + 2u_k(t), \ t \in \{\eta_k(i) \cdot i, \ i \in \mathbb{Z}_0^{15}\}, \\
x_k(t) = 0, \ t \in \mathbb{Z}_{-5}^0, \\
y_k(t) = 0.5x_k(t).
\end{cases}
\tag{6.97}
$$

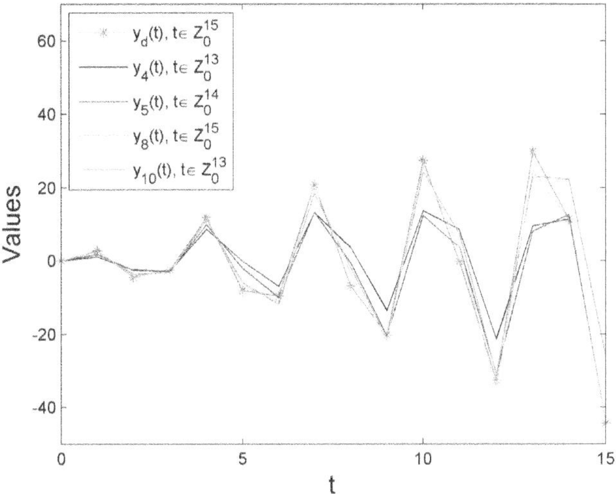

FIGURE 6.11 y_d and y_k ($k = 4, 5, 8, 10$) of Example 6.6.

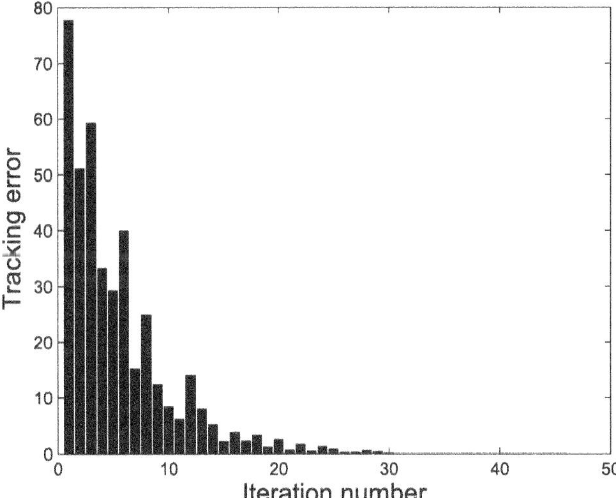

FIGURE 6.12 Tracking error $\|\tilde{e}_k\|_{l^2}$ of Example 6.6.

Set $u_k(0) = 0$ and let T_k randomly take values from $\mathbb{Z}_{13}^{16}$. Select the desired reference trajectory as (6.93) and the learning law (6.61) as

$$u_{k+1}(t) = u_k(t) + 0.4\tilde{e}_k(t+1).$$

We can calculate that $\|E - L_2 CB\| = |1 - 0.4 \times 0.5 \times 2| = 0.6 < 1$. Theorem 6.10 guarantees $\lim_{k \to \infty} \mathbf{E}\{\|\tilde{e}_k\|_\lambda\} = 0$ on $\mathbb{Z}_1^{15}$.

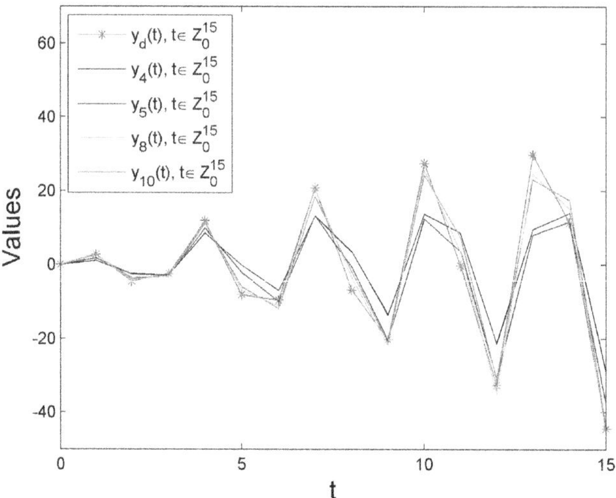

FIGURE 6.13 y_d and y_k ($k = 4, 5, 8, 10$) of Example 6.6 for the same trial lengths.

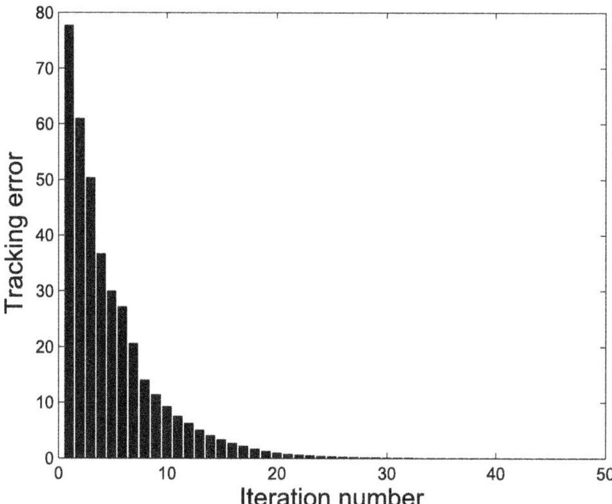

FIGURE 6.14 Tracking error $\|e_k\|_{l^2}$ of Example 6.6 for the same trial lengths.

Fig. 6.11 shows the reference trajectory y_d and output y_k of (6.97). Fig. 6.12 shows the error $\|\tilde{e}_k\|_{l^2}$ for each iteration. The trial length of the 50th iteration is $\mathbb{Z}_0^{13}$ and the error of the 50th iteration is 3.3265×10^{-4}. Figs. 6.13 and 6.14 show the tracking performance of a delay system with the same trial lengths, and the error of the 50th iteration is 6.5889×10^{-4}.

In Figs. 6.11 and 6.13 the lines denote output trajectories of the system; the more iterations, the closer the trajectories are to the desired trajectory.

6.3.1.5 Conclusion

We adopt a new framework to establish the same convergence theorem for ILC problems of linear discrete delayed systems by virtue of the representation of the solution via a discrete matrix delayed exponential function. The results in this part are motivated from [120].

Chapter 7

Stochastic delay systems

7.1 Controllability for first order systems

7.1.1 Null controllability stochastic delay systems

In this part, we study null controllability of the following stochastic systems via delayed perturbation of matrices:

$$\begin{cases} x'(t) = Ax(t) + Bx(t-\tau) + Cu(t) + \int_0^t \Delta(s, x(s))dW(s), \\ t \in J = [0, T], \ \tau \geq 0, \\ x(t) = \psi(t), \ t \in [-\tau, 0], \end{cases} \tag{7.1}$$

where $A, B \in \mathbb{R}^{n \times n}$, by assuming that A and B are permutation matrices and $C \in \mathbb{R}^{n \times m}$, state $x(t) \in \mathbb{R}^n$, control $u(t) \in \mathbb{R}^m$, and initial function $\psi(t) : [-\tau, 0] \longrightarrow \mathbb{R}^n$ and defining the delayed matrix exponential $\exp_\tau(Bt)$ corresponding to delay τ and matrix B and the semilinear function $\Delta \in \mathcal{C}(J \times \mathbb{R}^n, \mathbb{R}^{n \times n})$. The set of all n-dimensional Wiener process $W = (W_1(t), W_2(t), \cdots, W_n(t))^\top$.

Let $(\Omega, \mathfrak{F}, \mathbb{P})$ be a probability space with probability measure $\mathbb{P}$ on Ω and let $\{\mathfrak{F}_t | t \in J\}$ be a filtration generated by $\{W(s) : s \in [0, t]\}$ and $\mathfrak{F} = \mathfrak{F}_T$. Let $L_2(\Omega, \mathfrak{F}_T, \mathbb{R}^n)$ be the Hilbert space of all $\mathfrak{F}_T$-measurable square integrable variables with values in $\mathbb{R}^n$ and let $L_2^{\mathfrak{F}}(J, \mathbb{R}^n)$ be the Hilbert space of all square integrable and $\mathfrak{F}_t$-measurable processes with values in $\mathbb{R}^n$. Let $\mathcal{A} := \mathcal{C}([-\tau, T], L_2(\Omega, \mathfrak{F}_t, \mathbb{P}, \mathbb{R}^n))$ be the Banach space of all square integrable and $\mathfrak{F}_t$-adapted processes $x(t)$ with norm $\|x\|_{\mathcal{C}}^2 = \sup\limits_{t \in [-\tau, T]} \mathbf{E}\|x(t)\|^2 \leq r < \infty$.

We consider the matrix norm $\|A\| = \sup\limits_{\|x\|=1} \|Ax\|$ for the matrix $A : \mathbb{R}^n \to \mathbb{R}^n$ for initial values $\|\psi\|_{\mathcal{C}}^2 = \sup\limits_{s \in [-\tau, 0]} \mathbf{E}\|\psi(s)\|^2$ and $\mathcal{U}_{ad} = L_2^{\mathfrak{F}}(J, \mathbb{R}^m)$ denotes the set of all admissible controls.

Lemma 7.1. *([121]) Let $\Delta : J \times \Omega \to L_2^0$ be a strongly measurable mapping such that*

$$\int_0^T \mathbf{E}\|\Delta(t)\|_{L_2^0}^p dt < \infty.$$

Stability and Controls Analysis for Delay Systems. https://doi.org/10.1016/B978-0-32-399792-8.00013-X
Copyright © 2023 Elsevier Inc. All rights reserved.

Then

$$\mathbf{E}\left\|\int_0^t \Delta(s)dW(s)\right\|^p \le L_\Delta \int_0^t \mathbf{E}\|\Delta(s)\|^p_{L_2^0}ds,$$

for all $t \in J$ and $p \ge 2$, where L_Δ is the constant in p and T (for more details see [122]).

We now introduce the following operators and matrices for further discussion.

The linear bounded operator $L_0^T \in \mathcal{L}(\mathcal{U}_{ad}, L_2(\Omega, \mathfrak{F}_T, \mathbb{R}^n))$ is defined as

$$L_0^T u = \int_0^T e^{A(T-s)} e_\tau^{B_1(T-\tau-s)} Cu(s)ds.$$

Clearly

$$(L_0^T)^\top : L_2(\Omega, \mathfrak{F}_T, \mathbb{R}^n) \longrightarrow \mathcal{U}_{ad}$$

defined by

$$(L_0^T)^\top y(\cdot) = C^\top e_\tau^{B_1^\top (T-\tau-\cdot)} e^{A^\top(T-\cdot)} \mathbf{E}\left\{y\big|\mathfrak{F}.\right\}.$$

The linear controllability operator $\Gamma_\tau[0, T] \in \mathcal{L}(L_2(\Omega, \mathfrak{F}_T, \mathbb{R}^n), L_2(\Omega, \mathfrak{F}_T, \mathbb{R}^n))$ which is associated with the operator L_0^T is defined by

$$\begin{aligned}
\Gamma_\tau[0, T] &= L_0^T (L_0^T)^\top \{\cdot\} \\
&= \int_0^T e^{A(T-s)} e_\tau^{B_1(T-\tau-s)} CC^\top e_\tau^{B_1^\top (T-\tau-s)} e^{A^\top(T-s)} \mathbf{E}\left\{\cdot \Big|\mathfrak{F}_s\right\}ds,
\end{aligned}$$

and we have the following controllability Grammian matrix $W_\tau[0, T] \in \mathcal{L}(\mathbb{R}^n, \mathbb{R}^n)$:

$$W_\tau[0, T] = \int_0^T e^{A(T-s)} e_\tau^{B_1(T-\tau-s)} CC^\top e_\tau^{B_1^\top(T-\tau-s)} e^{A^\top(T-s)}ds. \tag{7.2}$$

Definition 7.1. [123, Definition 2.4] System (7.1) is said to be null controllable on J if for every initial function $\psi \in \mathcal{C}^1([-\tau, 0], \mathbb{R}^n)$, there exists a control $u \in \mathcal{U}_{ad}$ such that system (7.1) has a solution $x \in \mathcal{A}$ that satisfies the initial condition $x(t) = \psi(t)$, $t \in [-\tau, 0]$ and $x(T) = \mathbf{0}$ at finial time T.

Lemma 7.2. *[123, Lemma 4] If a triplet (A, B_1, C) is controllable, i.e., rank $S_n = n$, where*

$$\begin{aligned}
S_n = \Big[& C|AC|A^2C|\cdots|A^{n-1}C|B_1C|AB_1C|A^2B_1C| \\
& \cdots|A^{n-1}B_1C|\cdots|B_1^{n-1}C|\cdots|A^{n-1}B_1^{n-1}C\Big],
\end{aligned}$$

then the matrix function $\omega_\tau(t) = e^{At}e_\tau^{B_1 t}$ consists of elements that are linearly independent on the interval $-\tau \leq t \leq T - \tau$ where $T > (n-1)\tau$, i.e., there exists no nonzero vector $a^\top = (a_1, a_2, \cdots, a_n)$ such that $a^\top \omega_\tau(t) = \mathbf{0}^\top$.

7.1.1.1 Null controllability for linear systems

Consider the following linear stochastic delay system:

$$\begin{cases} x'(t) = Ax(t) + Bx(t-\tau) + Cu(t) + \int_0^t \Delta(s)dW(s), \ t \in J, \ \tau \geq 0, \\ x(t) = \psi(t), \ t \in [-\tau, 0], \end{cases}$$

(7.3)

where the linear stochastic function $\Delta \in C(J, \mathbb{R}^{n \times n})$ is continuous.

By [1, Theorem 1.2], the solution of (7.3) can be expressed in the following form:

$$x(t) = e^{A(t+\tau)}e_\tau^{B_1 t}\psi(-\tau) + \int_{-\tau}^0 e^{A(t-s)}e_\tau^{B_1(t-\tau-s)}[\psi'(s) - A\psi(s)]ds$$

$$+ \int_0^t e^{A(t-s)}e_\tau^{B_1(t-\tau-s)}Cu(s)ds$$

$$+ \int_0^t e^{A(t-s)}e_\tau^{B_1(t-\tau-s)}\left[\int_0^s \Delta(\eta)dW(\eta)\right]ds.$$

(7.4)

In the following Theorem 7.1, we investigate the necessary and sufficient conditions for (7.3) to be null controllable by using controllability matrices.

Theorem 7.1. *System (7.3) is null controllable on J if and only if $W_\tau[0, T]$ is nonsingular.*

Proof. Sufficiency: Since $W_\tau[0, T]$ is nonsingular, its inverse is well defined. For all $\psi(t)$, one can define a control function $u(t)$ as follows:

$$u(t) = C^\top e_\tau^{B_1^\top (T-\tau-t)}e^{A^\top (T-t)}\mathbf{E}\left\{W_\tau^{-1}[0, T]\left(-e^{A(T+\tau)}e_\tau^{B_1 T}\psi(-\tau)\right.\right.$$

$$-\int_{-\tau}^0 e^{A(T-s)}e_\tau^{B_1(T-\tau-s)}[\psi'(s) - A\psi(s)]ds$$

$$\left.\left.-\int_0^T e^{A(T-s)}e_\tau^{B_1(T-\tau-s)}\left[\int_0^s \Delta(\eta)dW(\eta)\right]ds\right)\Big|\mathfrak{F}_t\right\}.$$

(7.5)

Substituting $t = T$ in (7.4) and inserting (7.5) in (7.4), we get

$$x(T) = e^{A(T+\tau)}e_\tau^{B_1 T}\psi(-\tau) + \int_{-\tau}^0 e^{A(T-s)}e_\tau^{B_1(T-\tau-s)}[\psi'(s) - A\psi(s)]ds$$

$$+ \int_0^T e^{A(T-s)}e_\tau^{B_1(T-\tau-s)}CC^\top e_\tau^{B_1^\top (T-\tau-s)}e^{A^\top (T-s)}$$

$$
\times \mathbf{E}\Bigg\{ W_\tau^{-1}[0, T]\Bigg(-e^{A(T+\tau)}e_\tau^{B_1 T}\psi(-\tau)
$$

$$
-\int_{-\tau}^{0} e^{A(T-s)}e_\tau^{B_1(T-\tau-s)}[\psi'(s) - A\psi(s)]ds
$$

$$
-\int_{0}^{T} e^{A(T-s)}e_\tau^{B_1(T-\tau-s)}\left[\int_{0}^{s}\Delta(\eta)dW(\eta)\right]ds\Bigg)\Bigg|\mathfrak{F}_s\Bigg\}ds
$$

$$
+\int_{0}^{T} e^{A(T-s)}e_\tau^{B_1(T-\tau-s)}\left[\int_{0}^{s}\Delta(\eta)dW(\eta)\right]ds.
$$

Hence $x(T) = 0$. Further, the initial condition $x(t) = \psi(t)$, $t \in [-\tau, 0]$, holds by Eqs. (7.3) and (7.4). Thus (7.3) is null controllable on J.

Necessity: We prove the necessary part by contradiction. Assume that (7.3) is null controllable. We must prove that $W_\tau[0, T]$ is nonsingular. On the other hand, if it is singular, then there exists at least one nonzero vector $\tilde{a} \in \mathbb{R}^n$ such that

$$
0 = \tilde{a}^\top W_\tau[0, T]\tilde{a}
$$

$$
= \int_{0}^{T} \tilde{a}^\top e^{A(T-s)}e_\tau^{B_1(T-\tau-s)}CC^\top e_\tau^{B_1^\top(T-\tau-s)}e^{A^\top(T-s)}\tilde{a}ds
$$

$$
= \int_{0}^{T} \left[\tilde{a}^\top e^{A(T-s)}e_\tau^{B_1(T-\tau-s)}C\right]\left[\tilde{a}^\top e^{A(T-s)}e_\tau^{B_1(T-\tau-s)}C\right]^\top ds
$$

$$
= \int_{0}^{T} \left\|\tilde{a}^\top e^{A(T-s)}e_\tau^{B_1(T-\tau-s)}C\right\|^2 ds,
$$

which implies that

$$
\tilde{a}^\top e^{A(T-s)}e_\tau^{B_1(T-\tau-s)}C = \mathbf{0}^\top, \quad \forall s \in J. \tag{7.6}
$$

Since system (7.3) is null controllable, there exist control functions $u_1(t)$ and $u_2(t)$ that drive the initial state to zero at final time T, that is,

$$
e^{A(T+\tau)}e_\tau^{B_1 T}\psi_1(-\tau) + \int_{-\tau}^{0} e^{A(T-s)}e_\tau^{B_1(T-\tau-s)}[\psi_1'(s) - A\psi_1(s)]ds
$$

$$
+\int_{0}^{T} e^{A(T-s)}e_\tau^{B_1(T-\tau-s)}Cu_1(s)ds
$$

$$
+\int_{0}^{T} e^{A(T-s)}e_\tau^{B_1(T-\tau-s)}\left[\int_{0}^{s}\Delta(\eta)dW(\eta)\right]ds = \mathbf{0} \tag{7.7}
$$

and

$$
e^{A(T+\tau)}e_\tau^{B_1 T}\psi_1(-\tau) + \int_{-\tau}^{0} e^{A(T-s)}e_\tau^{B_1(T-\tau-s)}[\psi_1'(s) - A\psi_1(s)]ds
$$

$$+ \int_0^T e^{A(T-s)} e_\tau^{B_1(T-\tau-s)} C u_2(s) ds$$

$$+ \int_0^T e^{A(T-s)} e_\tau^{B_1(T-\tau-s)} \left[\int_0^s \Delta(\eta) dW(\eta) \right] ds = \mathbf{0}. \tag{7.8}$$

From Eqs. (7.7)–(7.8), one can get

$$\int_0^T e^{A(T-s)} e_\tau^{B_1(T-\tau-s)} C \Big[u_2(s) - u_1(s) \Big] ds = \tilde{a}, \tag{7.9}$$

where

$$\tilde{a} = e^{A(T+\tau)} e_\tau^{B_1 T} \left[\Big(\psi_1(-\tau) - \psi_2(-\tau) \Big) \right]$$

$$+ \int_{-\tau}^0 e^{A(T-s)} e_\tau^{B_1(T-\tau-s)} \left[\Big(\psi_1'(s) - \psi_2'(s) \Big) + A \Big(\psi_2(s) - \psi_1(s) \Big) \right] ds.$$

Multiplying by $\tilde{a}^\top$ on both sides of (7.9), one can get

$$\tilde{a}^\top \tilde{a} = \int_0^T \tilde{a}^\top e^{A(T-s)} e_\tau^{B_1(T-\tau-s)} C \Big[u_2(s) - u_1(s) \Big] ds = 0.$$

By Eq. (7.6), we get $\tilde{a}^\top \tilde{a} = 0$, i.e., $\tilde{a} = \mathbf{0}$, which is contracted to $\tilde{a} \neq \mathbf{0}$. Thus, the Grammian matrix $W_\tau[0, T]$ is nonsingular. The proof is completed. $\square$

Also, we prove the null controllability results in the following Theorem 7.2 by using rank correlation of the Cayley–Hamilton (C-H) theorem.

Theorem 7.2. *System (7.3) is null controllable on J if and only if $T \geq (n-1)\tau$ and $\det S_n \neq 0$, where*

$$S_n = \Big[C|AC|A^2 C| \cdots |A^{n-1} C|B_1 C|AB_1 C|A^2 B_1 C|$$

$$\cdots |A^{n-1} B_1 C| \cdots |B_1^{n-1} C| \cdots |A^{n-1} B_1^{n-1} C \Big] = n.$$

Proof. Necessity: Assume that system (7.3) is null controllable, i.e., the solution of system (7.3) satisfies the conditions $x(T) = \mathbf{0}$ and $x(t) = \psi(t)$, $t \in [-\tau, 0]$. As in Eq. (7.4), the solution of (7.3) at time $t = T$ is

$$x(T) = e^A (T+\tau) e_\tau^{B_1 T} \psi(-\tau) + \int_{-\tau}^0 e^{A(T-s)} e_\tau^{B_1(T-\tau-s)} [\psi'(s) - A\psi(s)] ds$$

$$+ \int_0^T e^{A(T-s)} e_\tau^{B_1(T-\tau-s)} C u(s) ds$$

$$+ \int_0^T e^{A(T-s)} e_\tau^{B_1(T-\tau-s)} \left[\int_0^s \Delta(\eta) dW(\eta) \right] ds. \tag{7.10}$$

Denote an arbitrary vector v, which satisfies the following equations:

$$v = e^A(T+\tau)e_\tau^{B_1 T}\psi(-\tau) + \int_{-\tau}^0 e^{A(T-s)}e_\tau^{B_1(T-\tau-s)}[\psi'(s) - A\psi(s)]ds$$

$$+ \int_0^T e^{A(T-s)}e_\tau^{B_1(T-\tau-s)}\left[\int_0^s \Delta(\eta)dW(\eta)\right]ds, \tag{7.11}$$

$$x(T) - v = \int_0^T e^{A(T-s)}e_\tau^{B_1(T-\tau-s)}Cu(s)ds,$$

using the representation of delayed exponential matrix function $e^{A(T-s)} \times e_\tau^{B_1(T-\tau-s)}$ in Definition 2.4. By change of variables, $T - \tau - s = \mu$, we obtain

$$\int_0^T e^{A(T-s)}e_\tau^{B_1(T-\tau-s)}Cu(s)ds$$

$$= \int_{-\tau}^{T-\tau} e^{A(\tau+\mu)}e_\tau^{B_1\mu}Cu((T-\tau-\mu))d\mu$$

$$= \int_{-\tau}^0 \left(I + A\frac{(\tau+\mu)}{1!} + \cdots + A^{k-1}\frac{(\tau+\mu)^{k-1}}{(k-1)!}\right)ICu(T-\tau-\mu)d\mu$$

$$+ \int_0^\tau \left(I + A\frac{(\tau+\mu)}{1!} + \cdots + A^{k-1}\frac{(\tau+\mu)^{k-1}}{(k-1)!}\right)\left(I + B_1\frac{\mu}{1!}\right)$$

$$\times Cu(T-\tau-\mu)d\mu$$

$$+ \int_\tau^{2\tau} \left(I + A\frac{(\tau+\mu)}{1!} + \cdots + A^{k-1}\frac{(\tau+\mu)^{k-1}}{(k-1)!}\right)$$

$$\times \left(I + B_1\frac{\mu}{1!} + B_1^2\frac{(\mu-\tau)^2}{2!}\right)Cu(T-\tau-\mu)d\mu$$

$$+$$

$$\vdots$$

$$+ \int_{(k-2)\tau}^{T-\tau} \left(I + A\frac{(\tau+\mu)}{1!} + \cdots + A^{k-1}\frac{(\tau+\mu)^{k-1}}{(k-1)!}\right)$$

$$\times \left(I + B_1\frac{\mu}{1!} + B_1^2\frac{(\mu-\tau)^2}{2!} + \cdots + B_1^{k-1}\frac{(\mu-(k-2)\tau)^{k-1}}{(k-1)!}\right)$$

$$\times Cu(T-\tau-\mu)d\mu$$

$$= \int_{-\tau}^0 \left(I + A\frac{(\tau+\mu)}{1!} + \cdots + A^{k-1}\frac{(\tau+\mu)^{k-1}}{(k-1)!}\right)Cu(T-\tau-\mu)d\mu$$

$$+ \int_0^\tau \left(I + A\frac{(\tau+\mu)}{1!} + \cdots + A^{k-1}\frac{(\tau+\mu)^{k-1}}{(k-1)!}\right)$$

$$+ B_1 \frac{\mu}{1!} + AB_1 \frac{\mu}{1!} \frac{(\tau + \mu)}{1!} + \cdots + A^{k-1} B_1 \frac{\mu}{1!} \frac{(\tau + \mu)^{k-1}}{(k-1)!} \Bigg)$$

$$\times Cu(T - \tau - \mu)d\mu$$

$$+ \int_{\tau}^{2\tau} \Bigg(I + A \frac{(\tau + \mu)}{1!} + A^2 \frac{(\tau + \mu)^2}{2!} + \cdots + A^{k-1} \frac{(\tau + \mu)^{k-1}}{(k-1)!}$$

$$+ B_1 \frac{\mu}{1!} + AB_1 \frac{\mu}{1!} \frac{(\tau + \mu)}{1!} + \cdots + A^{k-1} B_1 \frac{\mu}{1!} \frac{(\tau + \mu)^{k-1}}{(k-1)!} + B_1^2 \frac{(\mu - \tau)^2}{2!}$$

$$+ AB_1^2 \frac{(\mu - \tau)^2}{2!} \frac{(\tau + \mu)}{1!} + \cdots + A^{k-1} B_1^2 \frac{(\mu - \tau)^2}{2!} \frac{(\tau + \mu)^{k-1}}{(k-1)!} \Bigg)$$

$$\times Cu(T - \tau - \mu)d\mu$$

$$+$$

$$\vdots$$

$$+ \int_{(k-2)\tau}^{T-\tau} \Bigg(I + A \frac{(\tau + \mu)}{1!} + A^2 \frac{(\tau + \mu)^2}{2!} + \cdots + A^{k-1} \frac{(\tau + \mu)^{k-1}}{(k-1)!}$$

$$+ B_1 \frac{\mu}{1!} + AB_1 \frac{\mu}{1!} \frac{(\tau + \mu)}{1!} + \cdots + A^{k-1} B_1 \frac{\mu}{1!} \frac{(\tau + \mu)^{k-1}}{(k-1)!} + B_1^2 \frac{(\mu - \tau)^2}{2!}$$

$$+ AB_1^2 \frac{(\mu - \tau)^2}{2!} \frac{(\tau + \mu)}{1!} + \cdots + A^{k-1} B_1^2 \frac{(\mu - \tau)^2}{2!} \frac{(\tau + \mu)^{k-1}}{(k-1)!}$$

$$+ \cdots$$

$$+ B_1^{k-1} \frac{(\mu - (k-2)\tau)^{k-1}}{(k-1)!} + AB_1^{k-1} \frac{(\mu - (k-2)\tau)^{k-1}}{(k-1)!} \frac{(\tau + \mu)}{1!} + \cdots$$

$$+ A^{k-1} B_1^{k-1} \frac{(\mu - (k-2)\tau)^{k-1}}{(k-1)!} \frac{(\tau + \mu)^{k-1}}{(k-1)!} \Bigg) Cu(T - \tau - \mu)d\mu.$$

Denote

$$\Phi_{11}(T) = \int_{-\tau}^{T-\tau} u(T - \tau - \mu)d\mu,$$

$$\Phi_{12}(T) = \int_{-\tau}^{T-\tau} \frac{(\tau + \mu)}{1!} u(T - \tau - \mu)d\mu,$$

$$\Phi_{13}(T) = \int_{-\tau}^{T-\tau} \frac{(\tau + \mu)^2}{2!} u(T - \tau - \mu)d\mu, \cdots,$$

$$\Phi_{1k}(T) = \int_{-\tau}^{T-\tau} \frac{(\tau + \mu)^{k-1}}{(k-1)!} u(T - \tau - \mu)d\mu,$$

$$\Phi_{21}(T) = \int_{0}^{T-\tau} \frac{\mu}{1!} u(T - \tau - \mu)d\mu,$$

$$\Phi_{22}(T) = \int_0^{T-\tau} \frac{\mu}{1!} \frac{(\tau+\mu)}{1!} u(T-\tau-\mu)d\mu,$$

$$\Phi_{23}(T) = \int_0^{T-\tau} \frac{\mu}{1!} \frac{(\tau+\mu)^2}{2!} u(T-\tau-\mu)d\mu, \cdots,$$

$$\Phi_{2k}(T) = \int_0^{T-\tau} \frac{\mu}{1!} \frac{(\tau+\mu)^{k-1}}{(k-1)!} u(T-\tau-\mu)d\mu,$$

$$\Phi_{31}(T) = \int_\tau^{T-\tau} \frac{(\mu-\tau)^2}{2!} u(T-\tau-\mu)d\mu,$$

$$\Phi_{32}(T) = \int_\tau^{T-\tau} \frac{(\mu-\tau)^2}{2!} \frac{(\tau+\mu)}{1!} u(T-\tau-\mu)d\mu,$$

$$\Phi_{33}(T) = \int_\tau^{T-\tau} \frac{(\mu-\tau)^2}{2!} \frac{(\tau+\mu)^2}{2!} u(T-\tau-\mu)d\mu, \cdots,$$

$$\Phi_{3k}(T) = \int_\tau^{T-\tau} \frac{(\mu-\tau)^2}{2!} \frac{(\tau+\mu)^{k-1}}{(k-1)!} u(T-\tau-\mu)d\mu,$$

$$\cdots$$

$$\Phi_{m1}(T) = \int_{(k-2)\tau}^{T-\tau} \frac{(\mu-(k-2)\tau)^{k-1}}{(k-1)!} u(T-\tau-\mu)d\mu,$$

$$\Phi_{m2}(T) = \int_{(k-2)\tau}^{T-\tau} \frac{(\mu-(k-2)\tau)^{k-1}}{(k-1)!} \frac{(\tau+\mu)}{1!} u(T-\tau-\mu)d\mu,$$

$$\Phi_{m3}(T) = \int_{(k-2)\tau}^{T-\tau} \frac{(\mu-(k-2)\tau)^{k-1}}{(k-1)!} \frac{(\tau+\mu)^2}{2!} u(T-\tau-\mu)d\mu, \cdots,$$

$$\Phi_{mk}(T) = \int_{(k-2)\tau}^{T-\tau} \frac{(\mu-(k-2)\tau)^{k-1}}{(k-1)!} \frac{(\tau+\mu)^{k-1}}{(k-1)!} u(T-\tau-\mu)d\mu.$$

Note that by (7.10) and (7.11), we have

$$C\Phi_{11}(T) + AC\Phi_{12}(T) + \cdots + A^{k-1}C\Phi_{1k}(T) + B_1 C\Phi_{21}(T)$$
$$+ AB_1 C\Phi_{22}(T) + \cdots + A^{k-1}B_1 C\Phi_{1k}(T) + \cdots + B_1^{k-1} C\Phi_{m1}(T)$$
$$+ AB_1^{k-1} C\Phi_{m2}(T) + \cdots + A^{k-1}B_1^{k-1} C\Phi_{mk}(T)$$
$$= x(T) - v. \tag{7.12}$$

Since the system is null controllable, (7.12) has a solution for an arbitrary vector $x(T) - v$. If $k < n$, then the system is overdetermined and not always has a solution. Therefore, for the system to be null controllable, it is necessary that $T > (k-1)\tau \geq (n-1)\tau$. It follows from the C-H theorem that an arbitrary power $A^i B_1^i$, $i \geq n$, of the matrices A, B_1 can be expressed as a linear combination of matrices (see [89]):

$$\left[A | A^2 | \cdots | A^{n-1} | B_1 | AB_1 | A^2 B_1 | \cdots | A^{n-1} B_1 | \cdots | B_1^{n-1} | \cdots | A^{n-1} B_1^{n-1} \right].$$

Therefore, for $k \geq n$ Eq. (7.12) can be replaced by

$$
\begin{aligned}
C\widehat{\Phi}_{11}(T) &+ AC\widehat{\Phi}_{12}(T) + \cdots + A^{n-1}C\widehat{\Phi}_{1n}(T) + B_1 C\widehat{\Phi}_{21}(T) \\
&+ AB_1 C\widehat{\Phi}_{22}(T) + \cdots + A^{n-1}B_1 C\widehat{\Phi}_{1n}(T) + \cdots + B_1^{n-1} C\widehat{\Phi}_{m1}(T) \\
&+ AB_1^{n-1}C\widehat{\Phi}_{m2}(T) + \cdots + A^{n-1}B_1^{n-1}C\widehat{\Phi}_{mn}(T) \\
&= x(T) - v
\end{aligned}
\tag{7.13}
$$

where $\widehat{\Phi}_{ji}(T)$, $j = 1, 2, \cdots, m$, $i = 1, 2, \cdots, n$ are some functions of T. If (7.13) has the solution $\widehat{\Phi}_{ji}(T)$, $j = 1, 2, \cdots, m, i = 1, 2, \cdots, n$, for arbitrary v, then rank

$$
\begin{aligned}
S_n = \Big[&C\,|\,AC\,|\,A^2 C\,|\cdots|\,A^{n-1}C\,|\,B_1 C\,|\,AB_1 C\,|\,A^2 B_1 C\,|\cdots|\,A^{n-1}B_1 C\,| \\
&\cdots|\,B_1^{n-1}C\,|\cdots|\,A^{n-1}B_1^{n-1}C \Big] = n.
\end{aligned}
$$

Sufficiency: We prove our result by contradiction. Assume that system (7.3) is not null controllable and rank $S_n = n$. By Theorem 7.1, $W_\tau[0, T]$ is singular. Namely, there exists at least one nonzero vector $\tilde{a} \in \mathbb{R}^n$ such that

$$
\begin{aligned}
0 &= \tilde{a}^\top W_\tau[0, T]\tilde{a} \\
&= \int_0^T \tilde{a}^\top e^{A(T-s)} e_\tau^{B_1(T-\tau-s)} CC^\top e_\tau^{B_1^\top(T-\tau-s)} e^{A^\top(T-s)} \tilde{a}\, ds \\
&= \int_0^T \Big[\tilde{a}^\top e^{A(T-s)} e_\tau^{B_1(T-\tau-s)} C \Big] \Big[\tilde{a}^\top e^{A(T-s)} e_\tau^{B_1(T-\tau-s)} C \Big]^\top ds,
\end{aligned}
$$

which implies that

$$
\tilde{a}^\top e^{A(T-s)} e_\tau^{B_1(T-\tau-s)} C = \mathbf{0}^\top, \quad s \in J.
$$

Changing variables, $T - \tau - s = \mu$, we have

$$
\tilde{a}^\top e^{A(\tau+\mu)} e_\tau^{B_1 \mu} C = \mathbf{0}^\top, \quad -\tau \leq t \leq T - \tau,
$$

which contradicts the statement of Lemma 7.2 that the matrix function $\omega_\tau(t) = e^{At} e_\tau^{B_1 t}$ consists of elements that are linearly independent on the interval $-\tau \leq t \leq T - \tau$ with $T > (n-1)\tau$. The proof is complete. $\qquad\square$

7.1.1.2 Null controllability for semilinear systems

We derive the sufficient conditions for null controllability of the stochastic semilinear system (7.1). For that we set the following hypotheses.

[H1]. Controllability operator $\Gamma_\tau[0, T]$ has its inverse $\Gamma_\tau^{-1}[0, T] \in \mathcal{U}_{ad}/\ker \Gamma_\tau[0, T]$. Then we take $k_2 = \mathbf{E}\|\Gamma_\tau^{-1}[0, T]\|^2_{\mathcal{L}(L_2(\Omega, \mathfrak{F}_T, \mathbb{R}^n), \mathcal{U}_{ad}/\ker \Gamma)}$. From Wang et al. [87], one can write $k_2 = \sqrt{\|\Gamma_\tau^{-1}[0, T]\|^2}$.

[H2]. $\Delta \in \mathcal{C}(J \times \mathbb{R}^n, \mathbb{R}^n)$, $\exists\, q > 1$ and $M_\Delta(t) \in L_q(J, \mathbb{R}^+)$ such that

$$\|\Delta(t, x) - \Delta(t, y)\|^2 \le M_\Delta(t)\|x - y\|^2, \ x, y \in \mathbb{R}^n.$$

[H3]. Set $4T^2 L_\Delta M_2\left(1 + 3k_1 k_2\right) < 1$.

For brevity, we set $k_1 = \|W_\tau[0, T]\|^2$, $R_\Delta = \sup_{t \in J} \mathbf{E}\|\Delta(t, 0)\|^2$, $L = 2k_1 k_2 T^2 \times L_\Delta M_2 + 2T^2 L_\Delta M_2$, $\vartheta = \|A\|^2 + \|B_1\|^2$, $a = e^{\vartheta(T+\tau)}\mathbf{E}\|\psi(-\tau)\|^2 + \tau \int_{-\tau}^0 e^{\vartheta(T-s)}\mathbf{E}\|\psi'(s) - A\psi(s)\|^2 ds + T^2 L_\Delta R_\Delta \frac{e^{\vartheta T}-1}{\vartheta}$, $M_2 = \left[\frac{e^{\vartheta pT}-1}{\vartheta p}\right]^{\frac{1}{p}} \times \|M_\Delta\|_{L_q(J, \mathbb{R}^+)}$.

The solution of (7.1) can be expressed as

$$x(t) = e^{A(t+\tau)} e_\tau^{B_1 t} \psi(-\tau) + \int_{-\tau}^0 e^{A(t-s)} e_\tau^{B_1(t-\tau-s)}[\psi'(s) - A\psi(s)]ds$$

$$+ \int_0^t e^{A(t-s)} e_\tau^{B_1(t-\tau-s)} Cu(s)ds$$

$$+ \int_0^t e^{A(t-s)} e_\tau^{B_1(t-\tau-s)} \left[\int_0^s \Delta(\eta, x(\eta))dW(\eta)\right]ds,$$

and its control function $u(t)$ is defined as follows:

$$u(t) = C^\top e^{A^\top(T-t)} e_\tau^{B_1^\top(T-\tau-t)} \mathbf{E}\left\{\Gamma_\tau^{-1}[0, T]\left(-e^{A(T+\tau)} e_\tau^{B_1 T} \psi(-\tau)\right.\right.$$

$$- \int_{-\tau}^0 e^{A(T-s)} e_\tau^{B_1(T-\tau-s)}[\psi'(s) - A\psi(s)]ds - \int_0^T e^{A(T-s)} e_\tau^{B_1(T-\tau-s)}$$

$$\left.\left.\times \left[\int_0^s \Delta(\eta, x(\eta))dW(\eta)\right]ds\right)\middle|\mathfrak{F}_t\right\}.$$

We focus on the theoretical results of null controllability of stochastic delay systems via the fixed point technical route. Define the operator

$$\mathcal{T} : \mathcal{A} \to \mathcal{A}$$

by

$$(\mathcal{A}x)(t) = e^{A(t+\tau)} e_\tau^{B_1 t} \psi(-\tau) + \int_{-\tau}^0 e^{A(t-s)} e_\tau^{B_1(t-\tau-s)}[\psi'(s) - A\psi(s)]ds$$

$$+ \int_0^t e^{A(t-s)} e_\tau^{B_1(t-\tau-s)} Cu(s)ds$$

$$+ \int_0^t e^{A(t-s)} e_\tau^{B_1(t-\tau-s)} \left[\int_0^s \Delta(\eta, x(\eta))dW(\eta)\right]ds,$$

which has a fixed point x, which is (7.1). Next, we check that system (7.1) is null controllable on J. Now, we present our theoretical results for semilinear systems by the following Theorem 7.3.

Theorem 7.3. *Suppose that hypotheses [H1]–[H3] hold. Then, (7.1) is null controllable on J provided that*

$$L < 1 \tag{7.14}$$

holds.

Proof. To verify the conditions for the Banach contraction principle, we divide our proof into several steps.

Step 1. First we prove that $\mathcal{T}$ maps $\mathcal{A}$ into itself.

By Hölder's inequality and [H2], one can get

$$\int_0^t e^{\vartheta(t-s)} M_\Delta(s) ds \le \left(\int_0^t e^{\vartheta p(t-s)} ds \right)^{\frac{1}{p}} \left(\int_0^t M_\Delta^q(s) ds \right)^{\frac{1}{q}}$$

$$\le \left[\frac{1}{\vartheta p} (e^{\vartheta p t} - 1) \right]^{\frac{1}{p}} \|M\|_{L_q(J,\mathbb{R}^+)}$$

and

$$\int_0^t e^{\vartheta(t-s)} \mathbf{E}\|\Delta(s,0)\|^2 ds \le M_\Delta \int_0^t e^{\vartheta(t-s)} ds \le M_\Delta \frac{e^{\vartheta t} - 1}{\vartheta}.$$

By using [H1] and [H2], we have

$$\mathbf{E}\|(\mathcal{T}x)(t)\|^2 = \mathbf{E}\left\| e^{A(t+\tau)} e_\tau^{B_1 t} \psi(-\tau) + \int_{-\tau}^0 e^{A(t-s)} e_\tau^{B_1(t-\tau-s)} [\psi'(s) - A\psi(s)] ds \right.$$

$$+ \int_0^t e^{A(t-s)} e_\tau^{B_1(t-\tau-s)} Cu(s) ds$$

$$+ \left. \int_0^t e^{A(t-s)} e_\tau^{B_1(t-\tau-s)} \left[\int_0^s \Delta(\eta, x(\eta)) dW(\eta) \right] ds \right\|^2$$

$$\le 4\mathbf{E}\left\| e^{A(t+\tau)} e_\tau^{B_1 t} \psi(-\tau) \right\|^2$$

$$+ 4\mathbf{E}\left\| \int_{-\tau}^0 e^{A(t-s)} e_\tau^{B_1(t-\tau-s)} [\psi'(s) - A\psi(s)] ds \right\|^2$$

$$+ 4\mathbf{E}\left\| \int_0^t e^{A(t-s)} e_\tau^{B_1(t-\tau-s)} Cu(s) ds \right\|^2$$

$$+ 4\mathbf{E}\left\| \int_0^t e^{A(t-s)} e_\tau^{B_1(t-\tau-s)} \left[\int_0^s \Delta(\eta, x(\eta)) dW(\eta) \right] ds \right\|^2.$$

For better understanding, we estimate the following inequalities:

$$\mathbf{E}\left\|\int_0^t e^{A(t-s)}e_\tau^{B_1(t-\tau-s)}\left[\int_0^s \Delta(\eta,x(\eta))dW(\eta)\right]ds\right\|^2$$

$$\leq \left(\int_0^t ds\right)\left[\int_0^t \left\|e^{A(t-s)}e_\tau^{B_1(t-\tau-s)}\right\|^2 \times L_\Delta \left(\int_0^s \mathbf{E}\left\|\Delta(\eta,x(\eta))\right\|^2 d\eta\right)ds\right]$$

$$\leq T^2 L_\Delta \int_0^t e^{\vartheta(T-s)}\mathbf{E}\left\|\Delta(s,x(s))\right\|^2 ds$$

$$\leq T^2 L_\Delta \left[\int_0^t e^{\vartheta(T-s)}M_\Delta(s)\mathbf{E}\|x(s)\|^2 ds + \int_0^t e^{\vartheta(T-s)}\mathbf{E}\|\Delta(s,0)\|^2 ds\right]$$

$$\leq T^2 L_\Delta r\left[\int_0^t e^{\vartheta(T-s)}M_\Delta(s)ds\right] + T^2 L_\Delta R_\Delta \int_0^t e^{\vartheta(T-s)}ds$$

$$\leq T^2 L_\Delta r\left[\left(\int_0^t e^{\vartheta p(T-s)}ds\right)^{\frac{1}{p}}\left(\int_0^t M_\Delta^q ds\right)^{\frac{1}{q}}\right] + T^2 L_\Delta R_\Delta \frac{\left(e^{\vartheta T}-1\right)}{\vartheta}$$

$$\leq T^2 L_\Delta r\left[\frac{e^{\vartheta pT}-1}{\vartheta p}\right]^{\frac{1}{p}}\|M_\Delta\|_{L_q(J,\mathbb{R}^+)} + T^2 L_\Delta R_\Delta \frac{\left(e^{\vartheta T}-1\right)}{\vartheta}$$

$$\leq T^2 L_\Delta r M_2 + T^2 L_\Delta R_\Delta \frac{\left(e^{\vartheta T}-1\right)}{\vartheta}$$

and

$$\mathbf{E}\left\|\int_0^t e^{A(t-s)}e_\tau^{B_1(t-\tau-s)}Cu(s)ds\right\|^2$$

$$=\mathbf{E}\left\|\int_0^t e^{A(t-s)}e_\tau^{B_1(t-\tau-s)}CC^\top e^{A^\top(T-s)}e_\tau^{B_1^\top(T-\tau-s)}\right.$$

$$\times\left\{\Gamma_\tau^{-1}[0,T]\left(-e^{A(T+\tau)}e_\tau^{B_1 T}\psi(-\tau)\right.\right.$$

$$-\int_{-\tau}^0 e^{A(t-s)}e_\tau^{B_1(t-\tau-s)}[\psi'(s)-A\psi(s)]ds$$

$$\left.\left.-\int_0^T e^{A(t-s)}e_\tau^{B_1(t-\tau-s)}\left[\int_0^s \Delta(\eta,x(\eta))dW(\eta)\right]ds\right)\Big|\mathfrak{F}_s\right\}\right\|^2$$

$$\leq \|W_\tau[0,T]\|^2\|\Gamma_\tau^{-1}[0,T]\|^2 3\left[\left\|e^{A(T+\tau)}e_\tau^{B_1 T}\right\|^2\mathbf{E}\|\psi(-\tau)\|^2\right.$$

$$+\mathbf{E}\left\|\int_{-\tau}^0 e^{A(t-s)}e_\tau^{B_1(t-\tau-s)}[\psi'(s)-A\psi(s)]ds\right\|^2$$

$$+ \mathbf{E} \left\| \int_0^T e^{A(t-s)} e_\tau^{B_1(t-\tau-s)} \left[\int_0^s \Delta(\eta, x(\eta)) dW(\eta) \right] ds \right\|^2$$

$$\leq 3k_1 k_2 \left[e^{\vartheta(T+\tau)} \mathbf{E}\|\psi(-\tau)\|^2 + \tau \int_{-\tau}^0 e^{\vartheta(T-s)} \mathbf{E}\|\psi'(s) - A\psi(s)\|^2 ds \right.$$

$$\left. + T^2 L_\Delta r M_2 + T^2 L_\Delta R_\Delta \frac{\left(e^{\vartheta T} - 1\right)}{\vartheta} \right]$$

$$\leq 3k_1 k_2 \left[a + T^2 L_\Delta M_2 r \right].$$

Therefore,

$$\mathbf{E}\|(\mathcal{T}x)(t)\|^2$$

$$\leq 4e^{\vartheta(t+\tau)} \mathbf{E}\|\psi(-\tau)\|^2 + 4\tau \int_{-\tau}^0 e^{\vartheta(t-s)} \mathbf{E}\|\psi'(s) - A\psi(s)\|^2 ds$$

$$+ 12k_1 k_2 \left[a + T^2 L_\Delta M_2 r \right] + 4T^2 L_\Delta r M_2 + 4T^2 L_\Delta R_\Delta \frac{\left(e^{\vartheta T} - 1\right)}{\vartheta}$$

$$\leq 4 \left[a + 3k_1 k_2 \left(a + T^2 L_\Delta M_2 r \right) + T^2 L_\Delta r M_2 \right]$$

$$\leq 4a \left(1 + 3k_1 k_2 \right) + 4T^2 L_\Delta r M_2 \left(1 + 3k_1 k_2 \right) = r$$

for

$$r = \frac{4a \left(1 + 3k_1 k_2 \right)}{1 - 4T^2 L_\Delta M_2 \left(1 + 3k_1 k_2 \right)} < \infty.$$

By [H3] this implies that $\mathcal{T}$ maps $\mathcal{A}$ into itself.

Step 2. $\mathcal{T}$ is a contraction mapping.

Let $x, y \in \mathcal{A}$. By [H1] and [H2], for each $t \in J$, we have

$$\mathbf{E}\|(\mathcal{T}x)(t) - (\mathcal{T}y)(t)\|^2$$

$$\leq 2\mathbf{E} \left\| \int_0^t e^{A(t-s)} e_\tau^{B_1(t-\tau-s)} CC^\top e^{A^\top(T-s)} e_\tau^{B_1^\top(T-\tau-s)} \right.$$

$$\times \left\{ \Gamma_\tau^{-1}[0, T] \left(-\int_0^T e^{A(t-s)} e_\tau^{B_1(t-\tau-s)} \left[\int_0^s \Delta(\eta, x(\eta)) dW(\eta) \right] ds \right.\right.$$

$$\left.\left. + \int_0^T e^{A(t-s)} e_\tau^{B_1(t-\tau-s)} \left[\int_0^s \Delta(\eta, y(\eta)) dW(\eta) \right] ds \right) \Big| \mathfrak{F}_s \right\} ds$$

$$\left. + \int_0^t e^{A(t-s)} e_\tau^{B_1(t-\tau-s)} \left[\int_0^s \Delta(\eta, x(\eta)) dW(\eta) \right] ds \right.$$

$$\left\| -\int_0^t e^{A(t-s)} e_\tau^{B_1(t-\tau-s)} \left[\int_0^s \Delta(\eta, y(\eta)) dW(\eta) \right] ds \right\|^2$$

$$\leq 2\|W_\tau[0,T]\|^2 \|\Gamma_\tau^{-1}[0,T]\|^2 \mathbf{E} \left\| \int_0^T e^{A(t-s)} e_\tau^{B_1(t-\tau-s)} \right.$$

$$\times \left. \left(\int_0^s [\Delta(\eta, y(\eta)) - \Delta(\eta, x(\eta))] dW(\eta) \right) ds \right\|^2$$

$$+ 2\mathbf{E} \left\| \int_0^t e^{A(t-s)} e_\tau^{B_1(t-\tau-s)} \right.$$

$$\left. \left[\int_0^s [\Delta(\eta, x(\eta)) - \Delta(\eta, y(\eta))] dW(\eta) \right] ds \right\|^2$$

$$\leq 2k_1 k_2 T^2 L_\Delta \int_0^T e^{\vartheta(T-s)} \mathbf{E} \left\| \Delta(s, y(s)) - \Delta(s, x(s)) \right\|^2 ds$$

$$+ 2T^2 L_\Delta \int_0^t e^{\vartheta(t-s)} \mathbf{E} \left\| \Delta(s, x(s)) - \Delta(s, y(s)) \right\|^2 ds$$

$$\leq 2k_1 k_2 T^2 L_\Delta \int_0^T e^{\vartheta(T-s)} M_\Delta(s) ds \left(\sup_{t \in [-\tau, T]} \mathbf{E}\|x(t) - y(t)\|^2 \right)$$

$$+ 2T^2 L_\Delta \int_0^t e^{\vartheta(t-s)} M_\Delta(s) ds \left(\sup_{t \in [-\tau, T]} \mathbf{E}\|x(t) - y(t)\|^2 \right)$$

$$\leq 2k_1 k_2 T^2 L_\Delta \left(\int_0^T e^{\vartheta p(T-s)} ds \right)^{\frac{1}{p}} \left(\int_0^T M_\Delta^q(s) ds \right)^{\frac{1}{q}} \|x - y\|_C^2$$

$$+ 2T^2 L_\Delta \left(\int_0^t e^{\vartheta p(t-s)} ds \right)^{\frac{1}{p}} \left(\int_0^T M_\Delta^q(s) ds \right)^{\frac{1}{q}} \|x - y\|_C^2$$

$$\leq 2k_1 k_2 T^2 L_\Delta \left[\frac{e^{\vartheta pT} - 1}{\vartheta p} \right]^{\frac{1}{p}} \|M_\Delta\|_{L_q(J, \mathbb{R}^+)} \|x - y\|_C^2$$

$$+ 2T^2 L_\Delta \left[\frac{e^{\vartheta pT} - 1}{\vartheta p} \right]^{\frac{1}{p}} \|M_\Delta\|_{L_q(J, \mathbb{R}^+)} \|x - y\|_C^2$$

$$\leq 2k_1 k_2 T^2 L_\Delta M_2 \|x - y\|_C^2 + 2T^2 L_\Delta M_2 \|x - y\|_C^2$$

$$\leq L \|x - y\|_C^2.$$

Hence, from (7.14), we conclude that $\mathcal{T}$ is a contraction mapping on $\mathcal{A}$ and $\mathcal{T}$ has a unique fixed point $x(\cdot) \in \mathcal{A}$, which is a solution of (7.1) satisfying $x(T) = 0$ and initial function $x(t) = \psi(t)$, $t \in [-\tau, 0]$. Thus, system (7.1) is null controllable on J. Hence the proof follows. $\qquad \square$

Remark 7.1. System (7.1) with $u = 0$ and $\Delta = 0$ has been discussed in [1]. Also, relative controllability for system (7.1) with $B = 0$ has been studied in

[38]. Moreover, deterministic semilinear systems with perturbation of matrices are discussed in [87]. However, the authors in [124] investigated the sufficient conditions for semilinear stochastic systems with variable delay only. In this section, we derive sufficient conditions for null controllability of semilinear stochastic systems via delayed perturbation of matrices and verify the obtained theoretical results by numerical simulation.

Remark 7.2. Recently, many interesting works have been done on stochastic calculus based on G-Brownian motion due to its potential applications such as risk measures, super hedging in finance, and so on. Initially, Peng [125] introduced the stochastic differential equations driven by G-Brownian motion, and existence, uniqueness, and stability of the solutions for stochastic differential equations driven by G-Brownian motion have been proved [126]. Moreover, controllability results of fractional order stochastic systems have been well established in finite-dimensional space [127,128] and infinite-dimensional space [127,129]. By employing the ideas and techniques as in those papers, one can study the controllability results for fractional stochastic systems with G-Brownian motion by incorporating $\Delta(t, x(t))dB(t)$ in Eqs. (7.1) instead of $\int_0^t \Delta(s, x(s))dW(s)$. Here $B(\cdot)$ is one dimensional and a quadratic variation process of the G-Brownian motion respectively.

7.1.1.3 Numerical example and discussion

Consider the null controllability of the following semilinear stochastic system with $b = 2$ and $\tau = 0.5$:

$$
\begin{cases}
x_1'(t) = 0.2(x_1(t)) + 0.1(x_1(t - 0.5)) + u_1(t) \\
\qquad + \int_0^t \frac{\exp(s+0.01)}{100} x_1(s)dW_1(s), \ t, s \in [0, 2], \\
x_1(t) = t, \ t \in [-0.5, 0], \\
x_2'(t) = 0.2(x_2(t)) + 0.1(x_2(t - 0.5)) + u_2(t) \\
\qquad + \int_0^t \frac{\exp(s+0.01)}{100} x_2(s)dW_2(s), \ t, s \in [0, 2], \\
x_2(t) = 2t, \ t \in [-0.5, 0].
\end{cases}
\tag{7.15}
$$

This above equation can be rewritten in the form (7.1) with

$x(t) = (x_1(t), x_2(t)) \in \mathbb{R}^2,$

$$
A = \begin{pmatrix} 0.2 & 0 \\ 0 & 0.2 \end{pmatrix}, \ B = \begin{pmatrix} 0.1 & 0 \\ 0 & 0.1 \end{pmatrix}, \ C = \begin{pmatrix} 1 & 0 \\ 0 & 1 \end{pmatrix},
$$

$$
\Delta(t, x(t)) = \begin{pmatrix} \frac{\exp(t+0.01)}{100} x_1(t) \\ \frac{\exp(t+0.01)}{100} x_2(t) \end{pmatrix}, \ \psi(t) = \begin{pmatrix} t & 0 \\ 0 & 2t \end{pmatrix}, \ \psi'(t) = \begin{pmatrix} 1 & 0 \\ 0 & 2 \end{pmatrix}.
$$

Since A and B are diagonal matrices,

$$AB = BA = \begin{pmatrix} 0.02 & 0 \\ 0 & 0.02 \end{pmatrix}.$$

By simple computations, we obtain

$$\exp(A) = \begin{pmatrix} 1.22 & 0 \\ 0 & 1.22 \end{pmatrix}, \quad \exp(-A) = \begin{pmatrix} 0.8187 & 0 \\ 0 & 0.8187 \end{pmatrix},$$

$$B_1 = \begin{pmatrix} 0.09 & 0 \\ 0 & 0.09 \end{pmatrix},$$

$$\exp(B_1) = \begin{pmatrix} 1.09 & 0 \\ 0 & 1.09 \end{pmatrix}, \quad \exp_{0.5}(B_1 \times 2) = \begin{pmatrix} 1.1972 & 0 \\ 0 & 1.1972 \end{pmatrix},$$

$$\exp(A(2+0.5)) = \begin{pmatrix} 1.65 & 0 \\ 0 & 1.65 \end{pmatrix},$$

$$\exp(A(2-s)) = \begin{pmatrix} \exp(0.2(2-s)) & 0 \\ 0 & \exp(0.2(2-s)) \end{pmatrix}.$$

For every $t \in [(n-1)0.5, n0.5)$, $n = \{1, 2, 3, 4\}$, the explicit solution of system (7.1) is

$$x(t) = \exp(A(t+0.5)) \exp_{0.5}(B_1 t) \psi(-0.5)$$
$$+ \int_{-0.5}^{0} \exp(A(t-s)) \exp_{0.5}(B_1(t-0.5-s))[\psi'(s) - A\psi(s)]ds$$
$$+ \int_{0}^{t} \exp(A(t-s)) \exp_{0.5}(B_1(t-0.5-s))Cu(s)ds$$
$$+ \int_{0}^{t} \exp(A(t-s)) \exp_{0.5}(B_1(t-0.5-s)) \left[\int_{0}^{s} \Delta(\eta, x(\eta))dW(\eta) \right] ds,$$

where

$$\exp_{0.5}(B_1 t) = \begin{cases} B_1 \frac{t}{1!}, & t \in [0, 0.5), \\ B_1 \frac{t}{1!} + B_1^2 \frac{(t-0.5)^2}{2!}, & t \in [0.5, 1), \\ B_1 \frac{t}{1!} + B_1^2 \frac{(t-0.5)^2}{2!} + B_1^3 \frac{(t-1)^3}{3!}, & t \in [1, 1.5), \\ B_1 \frac{t}{1!} + B_1^2 \frac{(t-0.5)^2}{2!} + B_1^3 \frac{(t-1)^3}{3!} + B_1^4 \frac{(t-1.5)^4}{4!}, & t \in [1.5, 2), \end{cases}$$

and

$$
\exp_{0.5}(B_1(t - 0.5 - s))
$$

$$
= \begin{cases}
B_1 \frac{(t-0.5)}{1!}, & t \in [0, 0.5), \\[2mm]
B_1 \frac{(t-0.5)}{1!} + B_1^2 \frac{(t-1)^2}{2!}, & t \in [0.5, 1), \\[2mm]
B_1 \frac{(t-0.5)}{1!} + B_1^2 \frac{(t-1)^2}{2!} + B_1^3 \frac{(t-1.5)^3}{3!}, & t \in [1, 1.5), \\[2mm]
B_1 \frac{(t-0.5)}{1!} + B_1^2 \frac{(t-1)^2}{2!} + B_1^3 \frac{(t-1.5)^3}{3!} + B_1^4 \frac{(t-2)^4}{4!}, & t \in [1.5, 2).
\end{cases}
$$

Since $A = A^\top$, $B = B^\top$, and $C = C^\top$, from the above computations, the controllability Grammian matrix of (7.15) by (7.2) has the following explicit form:

$$
W_{0.5}[0, 2] = \int_0^2 \exp(A(2 - s)) \exp_{0.5}(B_1(2 - 0.5 - s))C
$$
$$
\times C^\top \exp_{0.5}(B_1^\top(2 - 0.5 - s)) \exp(A^\top(2 - s))ds
$$
$$
= \int_0^2 \left[\exp(A(2 - s)) \right]^2 \left[\exp_{0.5}(B_1(2 - 0.5 - s)) \right]^2 ds
$$
$$
= W_1 + W_2 + W_3 + W_4.
$$

For every $b - \tau - s \in [(n - 1)\tau, n\tau), n = \{1, 2, 3, 4\}$, we have

$$
W_1 = \int_0^{0.5} \left[\exp(A(2 - s)) \right]^2
$$
$$
\times \left[I + B_1 \frac{(1.5 - s)}{1!} + B_1^2 \frac{(1 - s)^2}{2!} + B_1^3 \frac{(0.5 - s)^3}{3!} \right]^2 ds
$$
$$
= \begin{pmatrix} 1.2556 & 0 \\ 0 & 1.2556 \end{pmatrix},
$$

$$
W_2 = \int_{0.5}^1 \left[\exp(A(2 - s)) \right]^2 \left[I + B_1 \frac{(1.5 - s)}{1!} + B_1^2 \frac{(1 - s)^2}{2!} \right]^2 ds
$$
$$
= \begin{pmatrix} 2.1986 & 0 \\ 0 & 2.1986 \end{pmatrix},
$$

$$
W_3 = \int_1^{1.5} \left[\exp(A(2 - s)) \right]^2 \left[I + B_1 \frac{(1.5 - s)}{1!} \right]^2 ds = \begin{pmatrix} 2.9005 & 0 \\ 0 & 2.9005 \end{pmatrix},
$$

$$
W_4 = \int_{1.5}^2 \left[\exp(A(2 - s)) \right]^2 [I]^2 ds = \begin{pmatrix} 3.0639 & 0 \\ 0 & 3.0639 \end{pmatrix}.
$$

Therefore,

$$W_{0.5}[0, 2] = \begin{pmatrix} 9.4186 & 0 \\ 0 & 9.4186 \end{pmatrix},$$

its inverse $W_{0.5}^{-1}[0, 2] = \begin{pmatrix} 0.1062 & 0 \\ 0 & 0.1062 \end{pmatrix},$

and $k_2 = 0.1062$. Moreover, from Eq. (7.2), one can define a control function $u(t) \in \mathcal{U}_{ad}$ for system (7.15) as

$$u(t) = \exp(A(2 - t)) \exp_{0.5}(B_1(1.5 - t)) W_{0.5}^{-1}[0, 2]\rho$$

$$= \begin{cases} \exp(A(2 - t)) \left[I + B_1 \frac{(1.5-s)}{1!} + B_1^2 \frac{(1-s)^2}{2!} + B_1^3 \frac{(0.5-s)^3}{3!} \right] W_{0.5}^{-1}[0, 2]\rho, \\ \quad t \in [0, 0.5), \\ \exp(A(2 - t)) \left[I + B_1 \frac{(1.5-s)}{1!} + B_1^2 \frac{(1-s)^2}{2!} \right] W_{0.5}^{-1}[0, 2]\rho, \ t \in [0.5, 1), \\ \exp(A(2 - t)) \left[I + B_1 \frac{(1.5-s)}{1!} \right] W_{0.5}^{-1}[0, 2]\rho, \ t \in [1, 1.5), \\ \exp(A(2 - t)) I W_{0.5}^{-1}[0, 2]\rho, \ t \in [1.5, 2), \end{cases}$$

where

$$\rho = -\exp(A(2.5)) \exp_{0.5}(2B_1)\psi(-0.5)$$

$$- \int_{-0.5}^{0} \exp(A(2 - s)) \exp_{0.5}(B_1(1.5 - s))[\psi'(s) - A\psi(s)]ds$$

$$- \int_{0}^{2} \exp(A(2 - s)) \exp_{0.5}(B_1(1.5 - s)) \left[\int_{0}^{s} \Delta(\eta, x(\eta))dW(\eta) \right] ds$$

$$= \begin{pmatrix} 1.64 & 0 \\ 0 & 1.64 \end{pmatrix} \begin{pmatrix} 1.1972 & 0 \\ 0 & 1.1972 \end{pmatrix} \begin{pmatrix} 0.5 & 0 \\ 0 & 1 \end{pmatrix}$$

$$- \int_{-0.5}^{0} \begin{pmatrix} \exp(0.2(2 - s)) & 0 \\ 0 & \exp(0.2(2 - s)) \end{pmatrix}$$

$$\times \left[I + B_1 \frac{(1.5 - s)}{1!} + B_1^2 \frac{(1 - s)^2}{2!} + B_1^3 \frac{(0.5 - s)^3}{3!} \right]$$

$$\times \left[\begin{pmatrix} 1 & 0 \\ 0 & 2 \end{pmatrix} - \begin{pmatrix} 0.2 & 0 \\ 0 & 0.2 \end{pmatrix} \begin{pmatrix} s & 0 \\ 0 & 2s \end{pmatrix} \right] ds$$

$$- \int_{0}^{2} \begin{pmatrix} \exp(0.2(2 - s)) & 0 \\ 0 & \exp(0.2(2 - s)) \end{pmatrix}$$

$$\left[I + B_1 \frac{(1.5 - s)}{1!} + B_1^2 \frac{(1 - s)^2}{2!} + B_1^3 \frac{(0.5 - s)^3}{3!} \right]$$

$$\times \left[\int_0^s \left(\begin{array}{c} \frac{\exp(t+0.01)}{100} \\ \frac{\exp(t+0.01)}{100} \end{array} \right) dW(\eta) \right] ds.$$

Therefore

$$\rho = \left(\begin{array}{cc} 0.9817 & 0 \\ 0 & 1.9634 \end{array} \right) - \left(\begin{array}{cc} 0.9602 & 0 \\ 0 & 1.9204 \end{array} \right) - \left(\begin{array}{c} 0.1930 \\ 0.1930 \end{array} \right)$$

$$= \left(\begin{array}{cc} -0.9387 & 0 \\ 0 & -1.8774 \end{array} \right)$$

and

$$AB_1 = \left(\begin{array}{cc} 0.0180 & 0 \\ 0 & 0.0180 \end{array} \right), \quad \text{rank}\left[C | AB_1 C | A^2 B_1 C \right] = 2.$$

Now, we compute $\|\Delta(t, x(t))\|^2$. We take for any $(x(t), y(t)) \in \mathbb{R}^2$, $t \in [0, 2]$,

$$\|\Delta(t, x) - \Delta(t, y)\|^2 = \frac{\exp(t) + 0.01}{100} (x_1(t) - y_1(t))^2 + (x_2(t) - y_2(t))^2$$

$$\leq \frac{\exp(t) + 0.01}{100} \|x - y\|^2.$$

Hence, $\Delta(t, x(t))$ satisfies [H2], where set $L_\wedge = 0.01$ and $M_\wedge(\cdot) = \frac{\cdot + 0.01}{100} \in L_q([0, 2], \mathbb{R}^+)$. Obviously,

$$\|M_\Delta\| = \frac{1}{100} \left(\frac{(0.1340)^{q+1} - (0.99)^{q+1}}{q + 1} \right) = 1.8108$$

and $R_\Delta = \sup_{t \in [0,2]} \|\Delta(t, 0)\|^2 = 0$. Further,

$$M_2 = \left[\frac{\exp(\vartheta p T) - 1}{\vartheta p} \right]^{\frac{1}{p}} \|M_\Delta\| = 2.9701 \neq 1,$$

when we take $p = q = 2$. Hence we obtain

$$L = 2k_1 k_2 T^2 L_\Delta M_2 + 2T^2 L_\Delta M_2 = 0.3565 < 1,$$

which guarantees that Eq. (7.14) holds. Further, we verify hypothesis [H3] as follows:

$$4T^2 L_\Delta M_2 \left(1 + 3k_1 k_2 \right) = 0.5827 < 1.$$

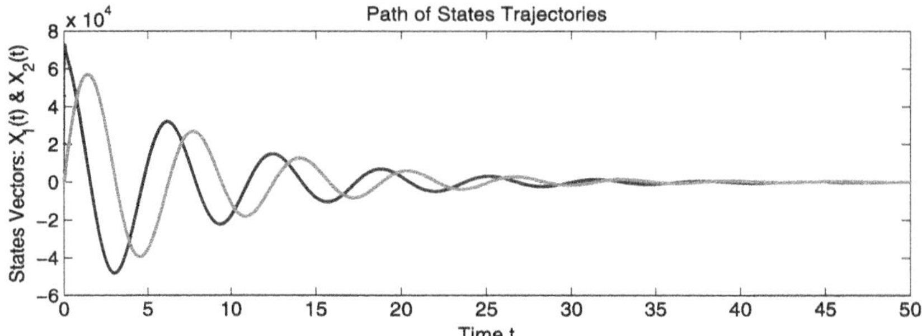

FIGURE 7.1 The trajectories of system (7.3) steer from the initial state.

Thus, all hypotheses of Theorem 7.3 are satisfied. Hence, system (7.15) is null controllable on $[0, 2]$. By using MATLAB®, the controlled trajectories of states and steering control are computed and depicted in Fig. 7.1.

7.1.1.4 Conclusion

This part contributes some meaningful results on null controllability for stochastic linear and semilinear delay systems with perturbation matrices in finite-dimensional settings. Several useful sufficient conditions are established for null controllability of stochastic semilinear system via the fixed point route and a suitable hypothesis on semilinear functions. The results in this part are motivated from [130].

7.2 Controllability for oscillating systems

7.2.1 Controllability of stochastic oscillating delay systems driven by the Rosenblatt distribution

In this section, we consider controllability results for the following second order stochastic nonlinear delay differential systems driven by the Rosenblatt distribution:

$$\begin{cases} \ddot{y}(t) = -A^2 y(t - \tau) + Bu(t) + \Delta(t, y(t))dZ_H(t), \ t \in J := [0, T], \ \tau > 0, \\ y(t) = \psi(t), \\ \dot{y}(t) = \dot{\psi}(t), \ t \in J_1 := [-\tau, 0], \end{cases}$$

$$(7.16)$$

where $A \in \mathbb{R}^{n \times n}$ is a nonsingular matrix, $B \in \mathbb{R}^{n \times m}$, state vector $y(t) \in \mathbb{R}^n$, control function $u : J \longrightarrow \mathbb{R}^m$, $H \in (\frac{1}{2}, 1)$, and $\Delta : J \times \mathbb{R}^n \longrightarrow \mathcal{T}(\mathbb{R}^n)$, where $\mathcal{T}(\mathbb{R}^n)$ denotes the Thorin class which is the smallest class of distributions on $\mathbb{R}^n$ that contains all Gamma distributions and is closed under convolution and weak convergence. The initial function $\psi(t) : J_1 \longrightarrow \mathbb{R}^n$ is an $\mathfrak{F}_0$-measurable $\mathbb{R}^n$-valued stochastic process independent of the Rosenblatt process Z_H with finite second moment.

The Hermite process of order one is the so-called fractional Brownian motion and that of order two is called the Rosenblatt process. Initially, Rosenblatt introduced the following distribution for $t \geq 0$:

$$Z_R(t) = D(R) \int_{R^2} \left(\int_0^t (v - x_1)_+^{\frac{-(1+R)}{2}} (v - x_2)_+^{\frac{-(1+R)}{2}} \, dv \right) dW(x_1) dW(x_2),$$

where $R \in (0, \frac{1}{2})$ and $\{W(x), x \in R\}$ is a standard Brownian motion. The process of $Z_R(1)$ is called the "1 non-Gaussian limiting distribution" (Rosenblatt distribution); for more details see [131]. By employing pathwise Malliavin calculus and the Itô–Wiener multiple integral, the Rosenblatt process was investigated in [121]. In [132] the Rosenblatt process for every $t_1, t_2, \cdots, t_d \geq 0$

$$(Z_R(t_1), Z_R(t_2), \cdots\cdots, Z_R(t_d))\underline{\underline{(d)}}$$

$$\left(\sum_{n=1}^{\infty} \mu_n(t_1)(\zeta_n^2 - 1), \cdots\cdots, \sum_{n=1}^{\infty} \mu_n(t_d)(\zeta_n^2 - 1) \right)$$

was considered, where $\{\zeta_n\}$ are independent and identically distributed $N(0, 1)$ random variables.

Let $\mathcal{A} := C([-\tau, T], L_p(\Omega, \mathfrak{F}_T, \mathcal{P}, \mathbb{R}^n))$ be the Banach space and $\mathfrak{F}_t$-adapted processes $\varphi(t)$ with norm $\| \cdot \|$, where $\|\varphi\|^p = \sup_{t \in J} \left(\mathbf{E}\|\varphi(t)\|^2 \right)^{\frac{p}{2}}$. Also, we introduce the Banach space

$$C^2(J, L_p(\Omega, \mathfrak{F}_T, \mathcal{P}, \mathbb{R}^n))$$
$$= \left\{ y \in C(J, L_p(\Omega, \mathfrak{F}_T, \mathcal{P}, \mathbb{R}^n)) : \ddot{y} \in C(J, L_p(\Omega, \mathfrak{F}_T, \mathcal{P}, \mathbb{R}^n)) \right\}$$

with norm

$$\|y\|_{C^2(J)} = \sup_{t \in J} \left\{ \mathbf{E}\|y(t)\|, \mathbf{E}\|\dot{y}(t)\|, \mathbf{E}\|\ddot{y}(t)\| \right\}.$$

We consider the matrix norm $\|A\| = \sup_{\|y\|=1} \|Ay\|$ for the matrix $A : \mathbb{R}^n \to \mathbb{R}^n$ for initial values $\|\psi\|_C^p = \sup_{s \in J_1} \left(\mathbf{E}\|\psi(s)\|^2 \right)^{\frac{p}{2}}$, $\|\dot{\psi}\|_C^p = \sup_{s \in J_1} \left(\mathbf{E}\|\dot{\psi}(s)\|^2 \right)^{\frac{p}{2}}$, and $\|\ddot{\psi}\|_C^p = \sup_{s \in J_1} \left(\mathbf{E}\|\ddot{\psi}(s)\|^2 \right)^{\frac{p}{2}}$.

Let h be a function such that $\mathbf{E}[h(\zeta_o)] = 0$ and $\mathbf{E}[h(\zeta_o^p)] < \infty$. Suppose the function h has Hermite rank n, i.e., if h allows the Hermite polynomials $h(x) = \sum_{i \geq 0} C_i H_i(x)$, where $C_i = \frac{1}{i!}\mathbf{E}[h(\zeta_0 H_i(\zeta_0))]$, then $n = \min\{i; C_i \neq 0\}$. Since $\mathbf{E}[h(\zeta_o)] = 0$, we have $n \geq 1$. Then by the noncentral limit theorem,

$$\frac{1}{k^H} \sum_{i=1}^{[kt]} h(\zeta_i)$$

converges as $k \longrightarrow \infty$ in the sense of the $\mathbb{R}^n$-distribution to the process

$$Z_H^n(t) = C(H, n) \int_{\mathbb{R}^n} \int_0^t \left(\prod_{i=1}^n (s - y_i)_+^{-(\frac{1}{2} + \frac{1-H}{n})} \right) ds\, dW(y_1), \cdots\cdots, dW(y_n),$$

(7.17)

where $x_+ = \max(x, 0)$ and the above integral is a multiple Wiener–Itô stochastic integral with respect to a Brownian motion $W(y)_{y \in R}$. The constant $C(H, n) > 0$ and it will be taken such that $\mathbf{E}[Z_H^n(1)^2 = 1]$. The process $(Z_H^n(t)_{t \geq 0})$ is called the *Hermite process* and it is H-self-similar in the sense that for any $C > 0$, $(Z_H^n(t))\underline{\underline{(d)}}(C^H Z_H^n(t))$, where $\underline{\underline{(d)}}$ means equivalence of all $\mathbb{R}^n$-distributions and it has stationary increments.

- If $n = 1$, then (7.17) is called fractional Brownian motion with $\frac{1}{2} < H < 1$.
- If $n \geq 2$, then (7.17) is called a non-Gaussian process.
- If $n = 2$, then (7.17) is called a Rosenblatt process.

Consider the linear stochastic control systems driven by the Rosenblatt process of the form

$$\begin{cases} \ddot{y}(t) = -A^2 y(t - \tau) + Bu(t) + \overline{\Delta}(t)dZ_H(t), \ t \in J, \ \tau > 0, \\ y(t) = \psi(t), \\ \dot{y}(t) = \dot{\psi}(t), \ t \in J_1, \end{cases}$$

(7.18)

where $\overline{\Delta} \in C(J, \mathcal{T}(\mathbb{R}^n))$.

Definition 7.2. System (7.16) is said to be controllable if there exists a control function $u^\top : J \longrightarrow \mathbb{R}^m$ such that

$$\ddot{y}(t) = -A^2 y(t - \tau) + Bu^\top(t) + \Delta(t, y(t))dZ_H(t), \ t \in J, \ \tau > 0,$$

has a solution $y = y^\top : [-\tau, T] \longrightarrow \mathbb{R}^n$ satisfying

$$y^\top(t) = \psi(t), \ \dot{y}^\top(t) = \dot{\psi}(t), \ t \in J_1, \ y^\top(T) = y_1,$$

where y_1 is a finite terminal state which belongs to $\mathbb{R}^n$.

From [5, Theorems 1 and 2], the solution of (7.18) can be expressed in the following form:

$$y(t) = (\cos_\tau At)\psi(-\tau) + A^{-1}(\sin_\tau At)\dot{\psi}(-\tau)$$

$$+ A^{-1} \int_{-\tau}^0 \sin_\tau A(t - \tau - s)\ddot{\psi}(s)ds + A^{-1} \int_0^t \sin_\tau A(t - \tau - s)Bu(s)ds$$

$$+ A^{-1} \int_0^t \sin_\tau A(t - \tau - s)\overline{\Delta}(s)dZ_H(s),$$

(7.19)

where $\cos_\tau A : \mathbb{R} \to \mathbb{R}^{n \times n}$ (see (3.3)) and $\sin_\tau A : \mathbb{R} \to \mathbb{R}^{n \times n}$ (see (3.4)).

Let $\mathcal{L}(X, Y)$ denote the space of linear bounded operators from a Banach space X to a Banach space Y. The operator $L_0^T \in \mathcal{L}(L_p^{\mathfrak{F}}(J, \mathbb{R}^m), L_p(\Omega, \mathfrak{F}_T, \mathbb{R}^n))$ is defined by

$$L_0^T u = A^{-1} \int_0^T \sin_\tau A(T - \tau - s) B u(s) ds$$

and its adjoint operator

$$(L_0^T)^\top x = B^\top \sin_\tau A^\top (T - \tau - t) \mathbf{E}\left\{ x \,\middle|\, \mathfrak{F}_t \right\}$$

is mapping from $L_p(\Omega, \mathfrak{F}_T, \mathbb{R}^n)$ to $L_p^{\mathfrak{F}}(J, \mathbb{R}^m)$. We consider the linear controllability operator

$$(\Gamma_\tau)_0^T \{\cdot\} = A^{-1} \int_0^T \sin_\tau A(T - \tau - s) B B^\top \sin_\tau A^\top (T - \tau - s) \mathbf{E}\left\{ \cdot \,\middle|\, \mathfrak{F}_s \right\} ds$$

$$\in \mathcal{L}(L_p(\Omega, \mathfrak{F}_t, \mathbb{R}^n), L_p(\Omega, \mathfrak{F}_t, \mathbb{R}^n))$$

and the delayed controllability Grammian matrix $(W_\tau)_0^T \in \mathcal{L}(\mathbb{R}^n, \mathbb{R}^n)$:

$$(W_\tau)_0^T = A^{-1} \int_0^T \sin_\tau A(T - \tau - s) B B^\top \sin_\tau A^\top (T - \tau - s) ds. \qquad (7.20)$$

Here $\top$ denotes the transpose.

7.2.1.1 Controllability of stochastic linear oscillating delay systems

Theorem 7.4. *The linear system (7.18) is controllable if and only if the delayed controllability Grammian matrix (7.20) is nonsingular.*

Proof. Sufficiency: Since $(W_\tau)_0^T$ is positive definite, i.e., it is nonsingular, and therefore its inverse $\left[(W_\tau)_0^T \right]^{-1}$ is well defined. Define a control function

$$u(t) = B^\top \sin_\tau A^\top (T - \tau - t) \left[(W_\tau)_0^T \right]^{-1} \left[y_1 - (\cos_\tau AT) \psi(-\tau) \right.$$

$$- A^{-1}(\sin_\tau AT) \dot{\psi}(-\tau) - A^{-1} \int_{-\tau}^0 \sin_\tau A(T - \tau - s) \ddot{\psi}(s) ds$$

$$\left. - A^{-1} \int_0^T \sin_\tau A(T - \tau - s) \overline{\Delta}(s) dZ_H(s) \right]. \qquad (7.21)$$

Substituting $t = T$ and inserting (7.21) in (7.19), we get

$$y(T) = (\cos_\tau AT) \psi(-\tau) + A^{-1}(\sin_\tau AT) \dot{\psi}(-\tau)$$

$$
+ A^{-1} \int_{-\tau}^{0} \sin_{\tau} A(T - \tau - s) \ddot{\psi}(s) ds
$$

$$
+ A^{-1} \int_{0}^{T} \sin_{\tau} A(T - \tau - s) B B^{\top} \sin_{\tau} A^{\top}(T - \tau - s) \left[(W_{\tau})_{0}^{T} \right]^{-1}
$$

$$
\times \left[y_1 - (\cos_{\tau} AT)\psi(-\tau) - A^{-1}(\sin_{\tau} AT)\dot{\psi}(-\tau) \right.
$$

$$
- A^{-1} \int_{-\tau}^{0} \sin_{\tau} A(T - \tau - s) \ddot{\psi}(s) ds
$$

$$
\left. - A^{-1} \int_{0}^{T} \sin_{\tau} A(T - \tau - s) \overline{\Delta}(s) d Z_H(s) \right](s) ds
$$

$$
+ A^{-1} \int_{0}^{T} \sin_{\tau} A(T - \tau - s) \overline{\Delta}(s) d Z_H(s).
$$

Now, using the delayed controllability Grammian matrix (7.20) we obtain

$$
y(T) = (\cos_{\tau} AT)\psi(-\tau) + A^{-1}(\sin_{\tau} AT)\dot{\psi}(-\tau)
$$

$$
+ A^{-1} \int_{-\tau}^{0} \sin_{\tau} A(T - \tau - s) \ddot{\psi}(s) ds + \left[(W_{\tau})_{0}^{T} \right] \left[(W_{\tau})_{0}^{T} \right]^{-1}
$$

$$
\times \left[y_1 - (\cos_{\tau} AT)\psi(-\tau) - A^{-1}(\sin_{\tau} AT)\dot{\psi}(-\tau) \right.
$$

$$
- A^{-1} \int_{-\tau}^{0} \sin_{\tau} A(T - \tau - s) \ddot{\psi}(s) ds
$$

$$
\left. - A^{-1} \int_{0}^{T} \sin_{\tau} A(T - \tau - s) \overline{\Delta}(s) d Z_H(s) \right]
$$

$$
+ A^{-1} \int_{0}^{T} \sin_{\tau} A(T - \tau - s) \overline{\Delta}(s) d Z_H(s)
$$

$$
= y_1.
$$

Next, we verify the boundary conditions $y(t) = \psi(t)$ and $\dot{y}(t) = \dot{\psi}(t)$, $t \in J_1$. From (3.4) and (3.3), we have $\sin_{\tau} At = A(t + \tau)$, $\cos_{\tau} At = I$, $t \in J_1$, and we define

$$
\sin_{\tau} A(t - \tau - s) = \begin{cases} \Theta, & s \in (t, 0], \\ A(t - s), & s \in [-\tau, t]. \end{cases}
$$

From the above relation, the solution (7.19) becomes

$$
y(t) = I\psi(-\tau) + A^{-1}A(t + \tau)\dot{\psi}(-\tau) + A^{-1} \int_{-\tau}^{t} \sin_{\tau} A(t - \tau - s) \ddot{\psi}(s) ds.
$$

$$
\tag{7.22}
$$

For simplicity,

$$\int_{-\tau}^{t} \sin_\tau A(t - \tau - s)\ddot{\psi}(s)ds$$

$$= \int_{-\tau}^{t} \sin_\tau A(t - \tau - s)d\dot{\psi}(s)$$

$$= \sin_\tau A(t - \tau - s)\dot{\psi}(s)|_{-\tau}^{t} - \int_{-\tau}^{t} d(\sin_\tau A(t - \tau - s))\dot{\psi}(s)ds$$

$$= -A(t + \tau)\dot{\psi}(-\tau) + A \int_{-\tau}^{t} \dot{\psi}(s)ds$$

$$= -A(t + \tau)\dot{\psi}(-\tau) + A\psi(t) - A\psi(-\tau). \tag{7.23}$$

Substituting (7.23) in (7.22), we get $y(t) = \psi(t)$. Now, $\dot{y}(t) = \dot{\psi}(t)$, $t \in J_1$, holds. Thus, the linear system (7.18) is controllable according to Definition 7.2.

Necessity: On the other hand, if the delayed controllability Grammian matrix (7.20) is singular, then $(W_\tau)_0^T [A^{-1}]^\top$ is also singular, i.e., there exists at least one nonzero state vector x such that

$$0 = x^\top (W_\tau)_0^T [A^{-1}]^\top x$$

$$= \int_0^T x^\top A^{-1} \sin_\tau A(T - \tau - s)BB^\top \sin_\tau A^\top (T - \tau - s)[A^{-1}]^\top x ds$$

$$= \int_0^T \left[x^\top A^{-1} \sin_\tau A(T - \tau - s)B\right]\left[x^\top A^{-1} \sin_\tau A(T - \tau - s)B\right]^\top ds$$

$$= \int_0^T \left\| x^\top A^{-1} \sin_\tau A(T - \tau - s)B \right\|^2 ds,$$

which implies that

$$x^\top A^{-1} \sin_\tau A(T - \tau - \eta)B = 0^\top, \ \forall\, \eta \in J. \tag{7.24}$$

Since the linear system (7.18) is controllable, according to Definition 7.2, there exists a control $u_1(t)$ that steers the initial state to zero at time T, i.e.,

$$y(T) = (\cos_\tau AT)\psi(-\tau) + A^{-1}(\sin_\tau AT)\dot{\psi}(-\tau)$$

$$+ A^{-1} \int_{-\tau}^{0} \sin_\tau A(T - \tau - s)\ddot{\psi}(s)ds$$

$$+ A^{-1} \int_0^T \sin_\tau A(T - \tau - s)Bu_1(s)ds$$

$$+ A^{-1} \int_0^T \sin_\tau A(T - \tau - s)\overline{\Delta}(s)dZ_H(s) = \mathbf{0}, \tag{7.25}$$

where 0 denotes the n-dimensional zero vector. Similarly, there exists a control $u_2(t)$ that steers the initial state to x at time T, i.e.,

$$y(T) = (\cos_\tau AT)\psi(-\tau) + A^{-1}(\sin_\tau AT)\dot\psi(-\tau)$$

$$+ A^{-1} \int_{-\tau}^{0} \sin_\tau A(T - \tau - s)\ddot\psi(s)ds$$

$$+ A^{-1} \int_{0}^{T} \sin_\tau A(T - \tau - s)Bu_2(s)ds$$

$$+ A^{-1} \int_{0}^{T} \sin_\tau A(T - \tau - s)\overline{\Delta}(s)dZ_H(s) = x. \tag{7.26}$$

Then, by (7.25) and (7.26), we have

$$x = A^{-1} \int_{0}^{T} \sin_\tau A(T - \tau - s)B[u_2(s) - u_1(s)]ds.$$

Multiplying by $x^\top$ on both sides of the above equation, we get

$$x^\top x = \int_{0}^{T} x^\top A^{-1} \sin_\tau A(T - \tau - s)B[u_2(s) - u_1(s)]ds.$$

Note (7.24). One can obtain $x^\top x = 0$, i.e., $x = \mathbf{0}$, which is a contradiction. Thus, the delayed controllability Grammian matrix (7.20) is nonsingular. The proof is finished. $\qquad\square$

7.2.1.2 Controllability of stochastic nonlinear oscillating delay systems

Consider the following hypotheses:

[H1]. The linear stochastic delay system (7.18) is controllable on J.

[H2]. Let $p \in (1, \infty)$. The function $\Delta \in C(J \times \mathbb{R}^n, \mathcal{T}(\mathbb{R}^n))$ is continuous, and $\exists M_\Delta \in L_q(J, R^+)$ and $q > 1$ such that

$$\|\Delta(t, y_1) - \Delta(t, y_2)\| \le M_\Delta(t)\|y_1 - y_2\|, \ y_i \in \mathbb{R}^n, \ i = 1, 2.$$

Let $R_\Delta = \sup_{0 \le t \le T} \|\Delta(t, 0)\|^p$.

If hypothesis [H1] holds, then for some $\gamma > 0$ we have $\mathbf{E}\langle(\Gamma_\tau)_0^T x, x\rangle \ge \gamma\mathbf{E}\|x\|^p$, $\forall\, x \in L_p(\Omega, \mathfrak{F}_t, \mathbb{R}^n)$ (see [133, Lemma 2]). Also, $\|[(\Gamma_\tau)_0^T]^{-1}\| \le \frac{1}{\gamma} := M_1$ (see [134]) and we set $M = \max\{\|(W_\tau)_s^T\|^p : s \in J\}$.

The solution of (7.16) can be expressed in the form

$$y(t) = (\cos_\tau At)\psi(-\tau) + A^{-1}(\sin_\tau At)\dot\psi(-\tau)$$

$$+ A^{-1} \int_{0}^{t} \sin_\tau A(t - \tau - s)Bu_y(s)ds$$

$$+ A^{-1} \int_{-\tau}^{0} \sin_\tau A(t - \tau - s)\ddot{\psi}(s)ds$$

$$+ A^{-1} \int_{0}^{t} \sin_\tau A(t - \tau - s)\Delta(s, y(s))dZ_H(s)$$

and its control function $u_y(t)$ is defined as

$$u_y(t) = B^\top \sin_\tau A^\top (T - \tau - t)\mathbf{E}\left\{[(\Gamma_\tau)_0^T]^{-1}\left[y_1 - (\cos_\tau AT)\psi(-\tau)\right.\right.$$

$$- A^{-1}(\sin_\tau AT)\dot{\psi}(-\tau) - A^{-1}\int_{-\tau}^{0} \sin_\tau A(T - \tau - s)\ddot{\psi}(s)ds$$

$$\left.\left. - A^{-1}\int_{0}^{T} \sin_\tau A(T - \tau - s)\Delta(s, y(s))dZ_H(s)\right]\Big|\mathfrak{F}_t\right\}. \tag{7.27}$$

Define the operator $\mathcal{Q} : \mathcal{A} \to \mathcal{A}$ by

$$(\mathcal{Q}y)(t) = (\cos_\tau At)\psi(-\tau) + A^{-1}(\sin_\tau At)\dot{\psi}(-\tau)$$

$$+ A^{-1}\int_{-\tau}^{0} \sin_\tau A(t - \tau - s)\ddot{\psi}(s)ds$$

$$+ A^{-1}\int_{0}^{t} \sin_\tau A(t - \tau - s)Bu_y(s)ds$$

$$+ A^{-1}\int_{0}^{t} \sin_\tau A(t - \tau - s)\Delta(s, y(s))dZ_H(s),$$

and suppose it has a fixed point y. Note $(\mathcal{Q}y)(t) = \psi(t)$, $\frac{d}{dt}(\mathcal{Q}y)(t) = \dot{\psi}(t)$, $t \in J_1$, $(\mathcal{Q}y)(T) = y_1$, from the control u_y, and this means that (7.16) is controllable on J. Define the set

$$\mathbb{B}_r = \left\{ y \in \mathcal{A} : \mathbf{E}\|y\|_C^p = \sup_{t \in [-\tau, T]} \left(\mathbf{E}\|y(t)\|^2\right)^{\frac{p}{2}} \leq r \right\},$$

for each positive number r. Now $\mathbb{B}_r$ is a closed, bounded, and convex set of $\mathcal{A}$ for each r.

Next, we derive sufficient conditions for controllability of the nonlinear system.

Theorem 7.5. *Suppose that hypotheses [H1] and [H2] hold. Then, (7.16) is controllable on J if*

$$M_2[1 + 5^{p-1}MM_1] < 1, \tag{7.28}$$

where

$$M_2 := 5^{p-1}\|A^{-1}\|^p C_H T^{(p-1)H}\left(\frac{e^{\|A\|^p p_1 T} - 1}{2^{p_1} p_1 \|A\|^p}\right)^{\frac{1}{p_1}}$$

$$\times \, \|M_\Delta\|_{L_q(J,R^+)}, \quad p_1 \in (1,\infty).$$

Proof. We verify the conditions of Krasnoselskii's fixed point theorem (see [82]) as follows.

Lemma 7.3. *Under hypotheses [H1]–[H2], there exists an $r > 0$ such that* $\mathcal{Q}(\mathbb{B}_r) \subseteq \mathbb{B}_r$.

Proof. Note

$$
\begin{aligned}
\mathbf{E}&\|(\mathcal{Q}y^r)(t)\|^p \\
&\leq 5^{p-1}\Big[\|\cos_\tau At\|^p \mathbf{E}\|\psi(-\tau)\|^p + \|A^{-1}\|^p \|\sin_\tau At\|^p \mathbf{E}\|\dot\psi(-\tau)\|^p \\
&\quad + \|A^{-1}\|^p \mathbf{E}\left\|\int_{-\tau}^{0} \sin_\tau A(t-\tau-s)\ddot\psi(s)ds\right\|^p \\
&\quad + \|A^{-1}\|^p \mathbf{E}\left\|\int_{0}^{t} \sin_\tau A(t-\tau-s)\Delta(s,y(s))dZ_H(s)\right\|^p \\
&\quad + \mathbf{E}\left\|A^{-1}\int_{0}^{t} \sin_\tau A(t-\tau-s)Bu_y(s)ds\right\|^p\Big] = \sum_{n=1}^{5} I_n.
\end{aligned}
\tag{7.29}
$$

From Lemmas 3.2 and 3.3, we have

$$
\begin{aligned}
I_1 &= 5^{p-1}\|\cos_\tau At\|^p \mathbf{E}\|\psi(-\tau)\|^p \leq 5^{p-1}\cosh(\|A\|^p t)\mathbf{E}\|\psi\|_C^p, \\
I_2 &= 5^{p-1}\|A^{-1}\|^p \|\sin_\tau At\|^p \mathbf{E}\|\dot\psi(-\tau)\|^p \\
&\leq 5^{p-1}\|A^{-1}\|^p \sinh(\|A\|^p(t+\tau))\mathbf{E}\|\dot\psi\|_C^p, \\
I_3 &= 5^{p-1}\|A^{-1}\|^p \mathbf{E}\left\|\int_{-\tau}^{0} \sin_\tau A(t-\tau-s)\ddot\psi(s)ds\right\|^p \\
&\leq 5^{p-1}\|A^{-1}\|^p \tau^{p-1}\mathbf{E}\|\ddot\psi\|_C^p \int_{-\tau}^{0} \|\sin_\tau A(t-\tau-s)\|^p ds \\
&\leq 5^{p-1}\frac{\|A^{-1}\|^p \tau^{p-1}\mathbf{E}\|\ddot\psi\|_C^p}{\|A\|^p}\left[\cosh\left(\|A\|^p(t+\tau)\right) - \cosh\left(\|A\|^p t\right)\right].
\end{aligned}
$$

By employing Lemmas 7.1, 3.2, and 3.3 and hypothesis [H2] we have

$$
\begin{aligned}
I_4 &= 5^{p-1}\|A^{-1}\|^p \mathbf{E}\left\|\int_{0}^{t} \sin_\tau A(t-\tau-s)\Delta(s,y(s))dZ_H(s)\right\|^p \\
&\leq 5^{p-1}\|A^{-1}\|^p C_H T^{(p-1)H}\int_{0}^{t} \mathbf{E}\|\sin_\tau A(t-\tau-s)\Delta(s,y(s))\|^p ds \\
&\leq 5^{p-1}\|A^{-1}\|^p C_H T^{(p-1)H}\int_{0}^{t} \sinh[\|A\|^p(t-s)]\mathbf{E}\|\Delta(s,y(s))\|^p ds \\
&\leq 5^{p-1}\|A^{-1}\|^p C_H T^{(p-1)H}\left[\int_{0}^{t} \sinh[\|A\|^p(t-s)]M_\Delta(s)\mathbf{E}\|y(s)\|^p ds\right.
\end{aligned}
$$

$$+ \int_0^t \sinh[\|A\|^p (t-s)] \mathbf{E} \|\Delta(s,0)\|^p \, ds\Big]$$

$$\leq 5^{p-1} \|A^{-1}\|^p C_H T^{(p-1)H} \Big[r \int_0^t \sinh[\|A\|^p(t-s)] M_\Delta(s) ds$$

$$+ R_\Delta \int_0^t \sinh[\|A\|^p(t-s)] ds \Big].$$

From Hölder's inequality, we have

$$\int_0^t \sinh[\|A\|(t-s)] M_\Delta(s) ds$$

$$\leq \left(\int_0^t (\sinh[\|A\|(t-s)])^{p_1} \, ds \right)^{\frac{1}{p_1}} \left(\int_0^t M_\Delta^q(s) ds \right)^{\frac{1}{q}},$$

where $\frac{1}{p_1} + \frac{1}{q} = 1$ and q is as in [H2]. Thus, we have

$$I_4 \leq 5^{p-1} \|A^{-1}\|^p C_H T^{(p-1)H} \Big[r \left(\int_0^t (\sinh[\|A\|^p(t-s)])^{p_1} ds \right)^{\frac{1}{p_1}}$$

$$\times \left(\int_0^t M_\Delta^q(s) ds \right)^{\frac{1}{q}} + R_\Delta \int_0^t \sinh[\|A\|^p(t-s)] ds \Big]$$

$$\leq 5^{p-1} \|A^{-1}\|^p C_H T^{(p-1)H} \Big[r \left(\int_0^t \frac{\exp(\|A\|^p p_1(t-s))}{2^{p_1}} ds \right)^{\frac{1}{p_1}} \|M_\Delta\|_{L_q(J,R^+)}$$

$$+ \frac{R_\Delta}{\|A\|^p} \cosh(\|A\|^p T) - 1 \Big]$$

$$\leq 5^{p-1} \|A^{-1}\|^p C_H T^{(p-1)H} \Big[r \left(\frac{\exp(\|A\|^p p_1 T) - 1}{2^{p_1} p_1 \|A\|^p} \right)^{\frac{1}{p_1}} \|M_\Delta\|_{L_q(J,R^+)}$$

$$+ \frac{R_\Delta}{\|A\|^p} \cosh(\|A\|^p T) - 1 \Big].$$

From (7.27), Lemmas 7.1, 3.2, and 3.3, [H2], I_1 to I_4, and Hölder's inequality, we obtain

$$I_5 = 5^{p-1} \mathbf{E} \left\| A^{-1} \int_0^t \sin_\tau A(t-\tau-s) Bu_y(s) ds \right\|^p$$

$$\leq 5^{p-1} \|(W_\tau)_0^T\|^p \Big\{ \|[(\Gamma_\tau)_0^T]^{-1}\|^p 5^{p-1} \Big[\mathbf{E} \|y_1\|^p$$

$$+ \|\cos_\tau AT\|^p \mathbf{E}\|\psi(-\tau)\|^p + \|A^{-1}\|^p \|\sin_\tau AT\|^p \mathbf{E}\|\dot\psi(-\tau)\|^p$$

$$+ \|A^{-1}\|^p \mathbf{E} \left\| \int_{-\tau}^0 \sin_\tau A(T-\tau-s)\ddot\psi(s) ds \right\|^p$$

$$+ \|A^{-1}\|^p \mathbf{E}\left\|\int_0^T \sin_\tau A(T-\tau-s)\Delta(s,y(s))dZ_H(s)\right\|^p \Bigg]\Bigg\}$$

$$\leq 5^{2(p-1)} M M_1 \Big[\mathbf{E}\|y_1\|^p + \chi(T) + M_2 r\Big],$$

where

$$\chi(T) := \cosh(\|A\|^p T)\mathbf{E}\|\psi\|_C^p + \|A^{-1}\|^p \sinh(\|A\|^p(T+\tau))\mathbf{E}\|\dot\psi\|_C^p$$
$$+ \frac{\|A^{-1}\|^p \tau^{p-1}\mathbf{E}\|\ddot\psi\|_C^p}{\|A\|^p} \times \Big[\cosh\big(\|A\|^p(T+\tau)\big) - \cosh\big(\|A\|^p T\big)\Big]$$
$$+ \|A^{-1}\|^p C_H T^{(p-1)H} \frac{R_\Delta}{\|A\|^p}\cosh(\|A\|^p T) - 1.$$

From I_1 to I_5, (7.29) becomes

$$\mathbf{E}\|(Qy^r)(t)\|^p$$
$$\leq 5^{p-1}\Bigg\{\cosh(\|A\|^p t)\mathbf{E}\|\psi\|_C^p + \|A^{-1}\|^p \sinh(\|A\|^p(t+\tau))\mathbf{E}\|\dot\psi\|_C^p$$
$$+ \frac{\|A^{-1}\|^p \tau^{p-1}\mathbf{E}\|\ddot\psi\|_C^p}{\|A\|^p}\Big[\cosh(\|A\|^p(t+\tau)) - \cosh(\|A\|^p t)\Big]$$
$$+ 5^{p-1} M M_1\Big(\mathbf{E}\|y_1\|^p + \chi(T) + M_2 r\Big) + \|A^{-1}\|^p C_H$$
$$\times T^{(p-1)H}\Bigg[r\left(\int_0^t \big(\sinh[\|A\|^p(t-s)]\big)^{p_1}ds\right)^{\frac{1}{p_1}}\left(\int_0^t M_\Delta^q(s)ds\right)^{\frac{1}{q}}$$
$$+ R_\Delta \int_0^t \sinh[\|A\|^p(t-s)]ds\Bigg]\Bigg\}$$
$$\leq 5^{p-1}\Bigg\{\cosh(\|A\|^p t)\mathbf{E}\|\psi\|_C^p + \|A^{-1}\|^p \sinh(\|A\|^p(t+\tau))\mathbf{E}\|\dot\psi\|_C^p$$
$$+ \frac{\|A^{-1}\|^p \tau^{p-1}\mathbf{E}\|\ddot\psi\|_C^p}{\|A\|^p}\Big[\cosh(\|A\|^p(t+\tau)) - \cosh(\|A\|^p t)\Big]$$
$$+ 5^{p-1} M M_1\Big(\mathbf{E}\|y_1\|^p + \chi(T) + M_2 r\Big)$$
$$+ \|A^{-1}\|^p C_H T^{(p-1)H}\Bigg[r\left(\frac{\exp(\|A\|^p p_1 t) - 1}{2^{p_1} p_1 \|A\|^p}\right)^{\frac{1}{p_1}}\|M_\Delta\|_{L_q(J,R^+)}$$
$$+ \frac{R_\Delta}{\|A\|^p}\cosh(\|A\|^p t) - 1\Bigg]\Bigg\}$$
$$\leq 5^{p-1}\Bigg\{\chi(T) + 5^{p-1} M M_1 \mathbf{E}\|y_1\|^p$$
$$+ 5^{p-1} M M_1 \chi(T) + 5^{p-1} M M_1 M_2 r + M_2 r\Bigg\}$$

$$\leq 5^{p-1}\left\{\chi(T)\left(1+5^{p-1}MM_1\right)+5^{p-1}MM_1\mathbf{E}\|y_1\|^p\right.$$
$$\left.+M_2r\left(1+5^{p-1}MM_1\right)\right\}.$$

Thus, for some r sufficiently large, from (7.28) we have $(\mathcal{Q}y^r)(t)\in\mathbb{B}_r$, so as a result $\mathcal{Q}(\mathbb{B}_r)\subseteq\mathbb{B}_r$. $\square$

Next, we write the operator $\mathcal{Q}$ as $\mathcal{Q}_1+\mathcal{Q}_2$, where

$$(\mathcal{Q}_1y)(t)=(\cos_\tau At)\psi(-\tau)+A^{-1}(\sin_\tau At)\dot{\psi}(-\tau)$$
$$+A^{-1}\int_{-\tau}^0\sin_\tau A(t-\tau-s)\ddot{\psi}(s)ds$$
$$+A^{-1}\int_0^t\sin_\tau A(t-\tau-s)Bu_y(s)ds$$

and

$$(\mathcal{Q}_2y)(t)=A^{-1}\int_0^t\sin_\tau A(t-\tau-s)\Delta(s,y(s))dZ_H(s). \qquad (7.30)$$

Lemma 7.4. *Assume that hypotheses [H1] and [H2] hold. Then, the operator $\mathcal{Q}_1$ is a contraction mapping.*

Proof. Let $t\in J$. For all $y,z\in\mathbb{B}_r$, we have

$$\mathbf{E}\|(\mathcal{Q}_1y)(t)-(\mathcal{Q}_1z)(t)\|^p$$
$$=\mathbf{E}\left\|A^{-1}\int_0^t\sin_\tau A(t-\tau-s)B[u_y(s)-u_z(s)]ds\right\|^p$$
$$\leq\|(W_\tau)_0^T\|^p\|[(\varGamma_\tau)_0^T]^{-1}\|^p\|A^{-1}\|^p$$
$$\times\mathbf{E}\left\|\int_0^T\sin_\tau A(T-\tau-s)[\Delta(s,z(s))-\Delta(s,y(s))]dZ_H(s)\right\|^p$$
$$\leq\frac{MM_1M_2}{5^{p-1}}\mathbf{E}\|y-z\|_C^p\leq\delta\mathbf{E}\|y-z\|_C^p,$$

where $\delta=\frac{MM_1M_2}{5^{p-1}}$. From (7.28), note that $\delta<1$, which implies that $\mathcal{Q}_1$ is a contraction. $\square$

Lemma 7.5. *The operator $\mathcal{Q}_2:\mathbb{B}_r\to\mathcal{A}$ is continuous and compact.*

Proof. Let $y_n\in\mathbb{B}_r$ with $y_n\longrightarrow y\in\mathbb{B}_r$. For convenience, let $\Delta_n(\cdot)=\Delta(\cdot,y_n(\cdot))$ and $\Delta(\cdot)=\Delta(\cdot,y(\cdot))$, and note

$$\sinh\left[\|A\|^p(\cdot-s)\right]\Delta_n(s)\longrightarrow\sinh\left[\|A\|^p(\cdot-s)\right]\Delta(s),\ s\in J.$$

From [H2], we get

$$\sinh\left[\|A\|^{p}(\cdot - s)\right]\mathbf{E}\|\Delta_n(s) - \Delta(s)\|^{p} \leq 2r\sinh\left[\|A\|^{p}(\cdot - s)\right]M_{\Delta}(s)$$
$$\in L_1(J, R^{+}).$$

Then using (7.30) and Lebesgue's dominated convergence theorem, one can obtain

$$\mathbf{E}\|(\mathcal{Q}_2 y_n)(t) - (\mathcal{Q}_2 y)(t)\|^{p}$$
$$\leq \|A^{-1}\|^{p}C_H T^{(p-1)H}\int_0^t \sinh\left[\|A\|^{p}(t - s)\right]\mathbf{E}\|\Delta_n(s) - \Delta(s)\|^{p}ds \to 0$$

as $n \to \infty$.

Thus, $\mathcal{Q}_2 : \mathbb{B}_r \longrightarrow \mathcal{A}$ is continuous.

Next, we show that $\mathcal{Q}_2(\mathbb{B}_r) \subset \mathcal{A}$ is equicontinuous. For that $y \in \mathbb{B}_r, 0 < t \leq t + h \leq T$, and from (7.30), we obtain

$$(\mathcal{Q}_2 y)(t + h) - (\mathcal{Q}_2 y)(t) = A^{-1}\int_0^{t+h} \sin_\tau A(t + h - \tau - s)\Delta(s)dZ_H(s)$$
$$- A^{-1}\int_0^t \sin_\tau A(t - \tau - s)\Delta(s)dZ_H(s)$$
$$= K_1 + K_2,$$

where

$$K_1 = A^{-1}\int_t^{t+h} \sin_\tau A(t + h - \tau - s)\Delta(s)dZ_H(s)$$

and

$$K_2 = A^{-1}\int_0^t [\sin_\tau A(t + h - \tau - s) - \sin_\tau A(t - \tau - s)]\Delta(s)dZ_H(s).$$

Therefore,

$$\mathbf{E}\|(\mathcal{Q}_2 y)(t + h) - (\mathcal{Q}_2 y)(t)\|^{p} \leq 2^{p-1}\left(\mathbf{E}\|K_1\|^{p} + \mathbf{E}\|K_2\|^{p}\right).$$

Now, we need to check $\mathbf{E}\|K_i\|^{p} \longrightarrow 0$ when $h \longrightarrow 0$, where $i = 1, 2$. For K_1, let q be as in [H2]. With $p_1 = \frac{q}{q-1}$, we have

$$\int_t^{t+h} \sinh\left[\|A\|^{p}(t + h - s)\right]M_{\Delta}(s)ds$$
$$\leq \left[\frac{\exp(\|A\|^{p}hp_1) - 1}{2^{p_1}p_1\|A\|^{p}}\right]^{\frac{1}{p_1}}\|M_{\Delta}\|_{L_q(J, R^{+})}.$$

Using the above inequality, we have

$$\mathbf{E}\|K_1\|^p \le \|A^{-1}\|^p C_H h^{(p-1)H} \int_t^{t+h} \sinh\Big[\|A\|^p(t+h-s)\Big]\mathbf{E}\|\Delta(s)\|^p ds$$

$$\le \|A^{-1}\|^p C_H h^{(p-1)H} \int_t^{t+h} \sinh\Big[\|A\|^p(t+h-s)\Big]$$

$$\times \mathbf{E}\|\Delta(s,y(s)) - \Delta(s,0) + \Delta(s,0)\|^p ds$$

$$\le 2^{p-1}\|A^{-1}\|^p C_H h^{(p-1)H} \int_t^{t+h} \sinh\Big[\|A\|^p(t+h-s)\Big]$$

$$\times \Big(\mathbf{E}\|\Delta(s,y(s)) - \Delta(s,0)\|^p + \mathbf{E}\|\Delta(s,0)\|^p\Big)ds$$

$$\le 2^{p-1}\|A^{-1}\|^p C_H h^{(p-1)H}$$

$$\times \int_t^{t+h} \sinh\Big[\|A\|^p(t+h-s)\Big]M_\Delta(s)\mathbf{E}\|y(s)\|^p ds$$

$$+ 2^{p-1}\|A^{-1}\|^p C_H h^{(p-1)H} \int_t^{t+h} \sinh\Big[\|A\|^p(t+h-s)\Big]R_\Delta(s)ds$$

$$\le 2^{p-1}\|A^{-1}\|^p C_H h^{(p-1)H} r \left[\frac{\exp(\|A\|^p hp_1) - 1}{2^{p_1} p_1 \|A\|^p}\right]^{\frac{1}{p_1}} \|M_\Delta\|_{L_q(J,R^+)}$$

$$+ 2^{p-1}\|A^{-1}\|^p C_H h^{(p-1)H} \frac{R_\Delta}{\|A\|^p}\Big[\cosh(\|A\|^p h) - 1\Big] \to 0 \text{ as } h \to 0.$$

For K_2, let q be as in [H2] and as before $p_1 = \frac{q}{q-1}$. Applying Hölder's inequality we have

$$\mathbf{E}\|K_2\|^p \le \|A^{-1}\|^p C_H T^{(p-1)H} \int_0^t \mathbf{E}\|[\sin_\tau A(t+h-\tau-s)$$

$$- \sin_\tau A(t-\tau-s)]\Delta(s,y(s))\|^p ds$$

$$\le 2^{p-1}\|A^{-1}\|^p C_H T^{(p-1)H}\left[r\left(\int_0^t (\|\sin_\tau A(t+h-\tau-s)\right.\right.$$

$$\left.\left. - \sin_\tau A(t-\tau-s)\|^p)^{p_1} ds\right)^{\frac{1}{p_1}}\right]$$

$$\times \|M_\Delta\|_{L_q(J,R^+)} + 2^{p-1}\|A^{-1}\|^p C_H T^{(p-1)H} R_\Delta$$

$$\times \int_0^t \|\sin_\tau A(t+h-\tau-s) - \sin_\tau A(t-\tau-s)\|^p ds.$$

From (3.4) and (3.3), we know that $\sin_\tau At$ is uniformly continuous for all $t \in J$, and thus, we get $\|\sin_\tau A(t+h-\tau-s) - \sin_\tau A(t-\tau-s)\|^p \longrightarrow 0$ as $h \longrightarrow 0$. Finally, $\mathbf{E}\|(\mathcal{Q}_2 y)(t+h) - (\mathcal{Q}_2 y)(t)\|^p \longrightarrow 0$ as $h \longrightarrow 0$ for all $y \in \mathbb{B}_r$. The other parts are treated similarly. From the Arzela–Ascoli theorem, the operator $\mathcal{Q} : \mathbb{B}_r \longrightarrow \mathcal{A}$ is compact. $\qquad\square$

Therefore, from Krasnoselskii's fixed point theorem (see [82]), $\mathcal{Q}$ has a fixed point y which is the solution of (7.16). Furthermore, $y(T) = y_1$ through the control function $u_y(t)$. Thus, the nonlinear stochastic delay system (7.16) is controllable on J. $\qquad\square$

7.2.1.3 Numerical examples and discussion

Consider the following nonlinear stochastic delay system:

$$\ddot{y}(t) = -A^2 y(t - 0.75) + Bu(t) + \Delta(t, y(t))dZ_H(t), \quad t \in [0, 1.5],$$
$$y(t) = \psi(t), \quad \dot{y}(t) = \dot{\psi}(t), \quad t \in [-0.75, 0], \tag{7.31}$$

where

$$A = \begin{pmatrix} 1 & -1 \\ 0 & 2 \end{pmatrix}, \quad B = \begin{pmatrix} 1 \\ 1 \end{pmatrix}, \quad \Delta(t, y(t)) = \begin{pmatrix} \frac{e^{(t+0.1)}}{10} y_1(t) \\ \frac{e^{(t+0.1)}}{10} y_2(t) \end{pmatrix}.$$

The corresponding linear part is

$$\overline{\Delta}(t) = \begin{pmatrix} \frac{e^{(t+0.1)}}{10} \\ \frac{e^{(t+0.1)}}{10} \end{pmatrix}$$

and

$$\psi(t) = \begin{pmatrix} t \\ 2t \end{pmatrix}, \quad \dot{\psi}(t) = \begin{pmatrix} 1 \\ 2 \end{pmatrix}.$$

Here B is an $n \times m$ matrix with $n = 2$, $m = 1$, control function $u : J \longrightarrow \mathbb{R}^m$, terminal time $T = 1.5$, $\tau = 0.75$. The corresponding delayed controllability Grammian matrix of system (7.31) is

$$[W_{0.75}]_0^{1.5} = A^{-1} \int_0^{1.5} \sin_{0.75} A(0.75 - s) BB^\top \sin_{0.75} A^\top (0.75 - s) ds$$
$$=: W_1 + W_2,$$

where

$$W_1 = A^{-1} \int_0^{0.75} \sin_{0.75} A(0.75 - s) B$$
$$\times B^\top \sin_{0.75} A^\top (0.75 - s) ds, \quad (0.75 - s) \in (0, 0.75),$$
$$W_2 = A^{-1} \int_{0.75}^{1.5} \sin_{0.75} A(0.75 - s) B$$
$$\times B^\top \sin_{0.75} A^\top (0.75 - s) ds, \quad (0.75 - s) \in (-0.75, 0),$$

and

$$
\sin_{0.75}(At) = \begin{cases}
\Theta, & -\infty < t < -0.75, \\
A(t+0.75), & -0.75 \le t < 0, \\
A(t+0.75) - A^3 \frac{t^3}{3!}, & 0 \le t < 0.75, \\
A(t+0.75) - A^3 \frac{t^3}{3!} + A^5 \frac{(t-0.75)^5}{5!}, & 0.75 \le t < 1.5,
\end{cases}
$$

$$
\cos_{0.75}(At) = \begin{cases}
\Theta, & -\infty < t < -0.75, \\
I, & -0.75 \le t < 0, \\
I - A^2 \frac{t^2}{2!}, & 0 \le t < 0.75, \\
I - A^2 \frac{t^2}{2!} + A^4 \frac{(t-0.75)^4}{4!}, & 0.75 \le t < 1.5.
\end{cases}
$$

By simple computations, we obtain the delay Grammian matrix

$$
[W_{0.75}]_0^{1.5} = \begin{pmatrix} 3.2495 & 4.6042 \\ 0.8491 & 2.9060 \end{pmatrix},
$$

where

$$
W_1 = \begin{pmatrix} 0.1131 & 1.8891 \\ 0.0941 & 1.7009 \end{pmatrix}, \quad W_2 = \begin{pmatrix} 3.1364 & 2.7151 \\ 0.7550 & 1.2051 \end{pmatrix}.
$$

Its inverse is

$$
[W_{0.75}^{-1}]_0^{1.5} = \begin{pmatrix} 0.5252 & -0.8320 \\ -0.1534 & 0.5872 \end{pmatrix}.
$$

Thus, the corresponding linear system of (7.31) is controllable on $[0, 1.5]$. More-over, for any final states $y(T) = y_1 = \begin{pmatrix} y_{11} \\ y_{12} \end{pmatrix}$, $\dot{y}(T) = y_1' = \begin{pmatrix} y_{11}' \\ y_{12}' \end{pmatrix}$, and $u_y(t) \in R$,

$$
u_y(t) = B^\top \sin_{0.75} A^\top (0.75 - t)[W_{0.75}^{-1}]_0^{1.5} \chi_1,
$$

where

$$
\chi_1 = y_1 - (\cos_{0.75} A 1.5)\psi(-0.75) - A^{-1}(\sin_{0.75} A 1.5)\dot{\psi}(-0.75)
$$

$$
- A^{-1}\left\{ \int_0^{1.5-0.75} \left[A(1.5-s) - A^3 \frac{(1.5-0.75-s)^3}{6} \right] \overline{\Delta}(s)dZ_H(s) \right.
$$

$$
\left. + \int_{1.5-0.75}^{0.75} [A(1.5-s)]\overline{\Delta}(s)dZ_H(s) + \int_{0.75}^{1.5} [A(1.5-s)]\overline{\Delta}(s)dZ_H(s) \right\}
$$

$$= \begin{pmatrix} y_{11} - 1.5987 \\ y_{12} - 6.4079 \end{pmatrix}.$$

The solution of the corresponding linear systems of (7.31) is

$$y(t) = (\cos_{0.75} At)\psi(-0.75) + A^{-1}(\sin_{0.75} At)\dot{\psi}(-0.75)$$
$$+ A^{-1} \int_0^t \sin_{0.75} A(t - 0.75 - s)BB^\top \sin_{0.75} A^\top(0.75 - s)ds$$
$$\times [W_{0.75}^{-1}]_0^{1.5} \chi_1 + A^{-1} \int_0^t \sin_{0.75} A(t - 0.75)\overline{\Delta}(s)dZ_H(s).$$

Consider the first integral term from the above equation,

$$\int_0^t \sin_{0.75} A(t - 0.75 - s)BB^\top \sin_{0.75} A^\top(0.75 - s)ds.$$

For the interval $t \in (0, 0.75)$, we get $-0.75 < t - 0.75 - s < t - 0.75 < 0$ and $0 < 0.75 - t < 0.75 - s < 0.75$. Then the solution

$$y(t) = \left[I - A^2\frac{t^2}{2}\right]\psi(-0.75) + A^{-1}\left[A(t + 0.75) - A^3\frac{t^3}{6}\right]\dot{\psi}(-0.75)$$
$$+ A^{-1} \int_0^t [A(t - s)]BB^\top\left[A^\top(1.5 - s) - (A^\top)^3\frac{(0.75 - s)^3}{6}\right]ds$$
$$\times [W_{0.75}^{-1}]_0^{1.5} \chi_1 + A^{-1} \int_0^t A(t - s)\overline{\Delta}(s)dZ_H(s). \tag{7.32}$$

Also for the interval $t \in (0.75, 1.5)$, we get $0 < t - 0.75 - s < t - 0.75 < 0.75$ when $0 < s < t - 0.75$ and $-0.75 < t - 0.75 - s < 0$ when $t - 0.75 < s < t$. Furthermore, we obtain $0 < 0.75 - s < 0.75$ when $0 < s < 0.75$ and $-0.75 < 0.75 - t < 0.75 - s < 0$ when $s \in (0.75, t)$. Finally, solution (7.32) can be represented in the following form:

$$y(t) = \left[I - A^2\frac{t^2}{2} + A^4\frac{(t - 0.75)^4}{24}\right]\psi(-0.75) + A^{-1}\left[A(t + 0.75) - A^3\frac{t^3}{6}\right.$$
$$\left. + A^5\frac{(t - 0.75)^5}{120}\right]\dot{\psi}(-0.75) + A^{-1} \int_0^{t-0.75}\left[A(t - s)\right.$$
$$\left. - A^3\frac{(t - 0.75 - s)^3}{6}\right]BB^\top\left[A^\top(1.5 - s) - (A^\top)^3\frac{(0.75 - s)^3}{6}\right]ds$$
$$\times [W_{0.75}^{-1}]_0^{1.5} \chi_1 + A^{-1} \int_{t-0.75}^{0.75}[A(t - s)]BB^\top\left[A^\top(1.5 - s)\right.$$
$$\left. - (A^\top)^3\frac{(0.75 - s)^3}{6}\right]ds[W_{0.75}^{-1}]_0^{1.5} \chi_1$$

$$+ A^{-1} \int_{0.75}^{t} [A(t-s)] B B^{\top} \left[A^{\top}(1.5-s) \right] ds [W_{0.75}^{-1}]_0^{1.5} \chi_1$$

$$+ A^{-1} \int_0^{t-0.75} \left[A(t-s) - A^3 \frac{(t-0.75-s)^3}{6} \right] \overline{\Delta}(s) dZ_H(s)$$

$$+ A^{-1} \int_{t-0.75}^{0.75} [A(t-s)] \overline{\Delta}(s) dZ_H(s)$$

$$+ A^{-1} \int_{0.75}^{t} [A(t-s)] \overline{\Delta}(s) dZ_H(s).$$

Let $p = 2 = p_1$ and $q = 2$. For any $x(t), y(t) \in R^2, t \in [0, 1.5]$,

$$\|\Delta(t, x) - \Delta(t, y)\|^2 = \frac{e^{t+0.1}}{10} \left[(x_1(t) - y_1(t))^2 + (x_2(t) - y_2(t))^2 \right]$$

$$\leq \left| \frac{e^{t+0.1}}{10} \right| \|x - y\|^2.$$

Hence, Δ satisfies [H2], where by set $M_\Delta(\cdot) = \frac{e^{\cdot+0.1}}{10} \in L_q([0, 1.5], R^+)$, $R_\Delta = \sup_{t \in [0,1.5]} \|\Delta(t, 0)\|^2 = 0$. Now

$$\|M_\Delta\|_{L_q([0,1.5],R^+)} = \left(\int_0^{1.5} \left[\frac{e^{s+0.1}}{10} \right]^2 ds \right)^{\frac{1}{2}} = 0.6203$$

and $M_1 = 1.3115$. Furthermore,

$$M_2 := 5^{p-1} \|A^{-1}\|^P C_H T^{(p-1)H} \|M_\Delta\|_{L_q(J,R^+)} \left(\frac{\exp(\|A\|^P p_1 T) - 1}{2^{p_1} p_1 \|A\|^P} \right)^{\frac{1}{p_1}}$$

$$= 0.0812.$$

Hence we obtain

$$M_2 \left[1 + 5^{p-1} M M_1 \right] = 0.7561 < 1,$$

which guarantees that (7.28) holds.

Thus, the hypotheses of Theorem 7.5 are satisfied. Hence, system (7.31) is controllable on [0, 1.5]. By using MATLAB, the controlled trajectories of the states and steering control are computed and depicted in Figs. 7.2, 7.3, and 7.4 in various time intervals and whenever one extends the time interval one can see the different stochastic natures.

7.2.1.4 Conclusions

This part contributes some meaningful results on controllability for second order stochastic nonlinear delay systems driven by the Rosenblatt process with per-

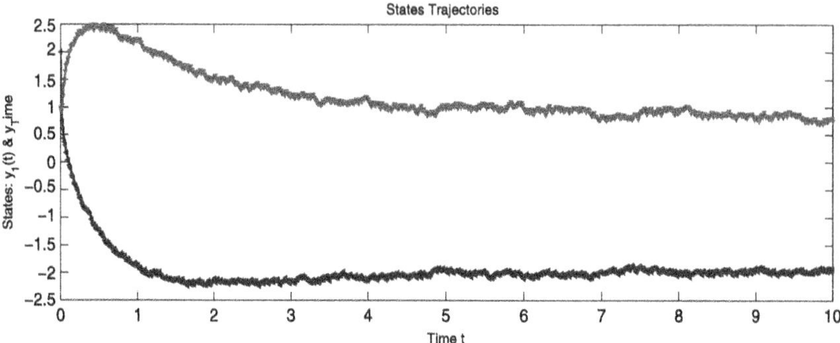

FIGURE 7.2   The trajectory of system (7.31) steers from the initial state.

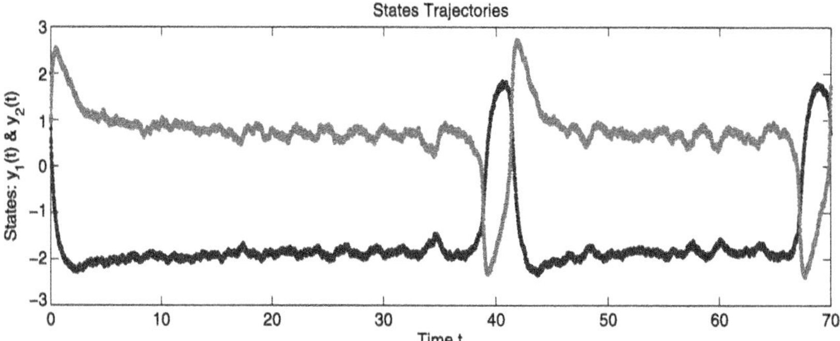

FIGURE 7.3   The trajectory of system (7.31) steers from the initial state.

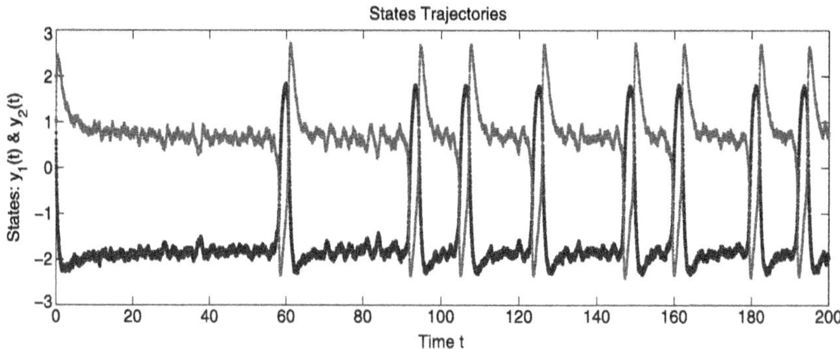

FIGURE 7.4   The trajectory of system (7.31) steers from the initial state.

turbation matrices in finite-dimensional settings. Several sufficient conditions are established for controllability of the stochastic nonlinear system using fixed point theory, delayed Grammian matrices, and appropriate local conditions on the nonlinear term. The results in this part are motivated from [135].

References

[1] D.Ya. Khusainov, G.V. Shuklin, Linear autonomous time-delay system with permutation matrices solving, Stud. Univ. Žilina Math. Ser. 17 (2003) 101–108.

[2] J. Diblík, D.Ya. Khusainov, Representation of solutions of discrete delayed system $x(k+1) = Ax(k) + Bx(k-m) + f(k)$ with commutative matrices, J. Math. Anal. Appl. 318 (2006) 63–76.

[3] J. Diblík, D.Ya. Khusainov, Representation of solutions of linear discrete systems with constant coefficients and pure delay, Adv. Differ. Equ. 2006 (2006) 1–13.

[4] J. Diblík, D.Ya. Khusainov, M. Růžičková, Controllability of linear discrete systems with constant coefficients and pure delay, SIAM J. Control Optim. 47 (2008) 1140–1149.

[5] D.Ya. Khusainov, J. Diblík, M. Růžičková, J. Lukáčová, Representation of a solution of the Cauchy problem for an oscillating system with pure delay, Nonlinear Oscil. 11 (2008) 261–270.

[6] M. Medveď, M. Pospišil, L. Škripková, Stability and the nonexistence of blowing-up solutions of nonlinear delay systems with linear parts defined by permutable matrices, Nonlinear Anal. 74 (2011) 3903–3911.

[7] M. Medveď, M. Pospišil, Sufficient conditions for the asymptotic stability of nonlinear multidelay differential equations with linear parts defined by pairwise permutable matrices, Nonlinear Anal. 75 (2012) 3348–3363.

[8] J. Diblík, M. Fečkan, M. Pospišil, Representation of a solution of the Cauchy problem for an oscillating system with two delays and permutable matrices, Ukr. Math. J. 65 (2013) 58–69.

[9] J. Diblík, M. Fečkan, M. Pospišil, On the new control functions for linear discrete delay systems, SIAM J. Control Optim. 52 (2014) 1745–1760.

[10] J. Diblík, B. Morávková, Discrete matrix delayed exponential for two delays and its property, Adv. Differ. Equ. 2013 (2013) 1–18.

[11] J. Diblík, B. Morávková, Representation of the solutions of linear discrete systems with constant coefficients and two delays, Abstr. Appl. Anal. 2014 (2013) 1–19.

[12] A. Boichuk, J. Diblík, D. Khusainov, M. Růžičková, Fredholm's boundary-value problems for differential systems with a single delay, Nonlinear Anal. 72 (2010) 2251–2258.

[13] M. Pospišil, Representation and stability of solutions of systems of functional differential equations with multiple delays, Electron. J. Qual. Theory Differ. Equ. 54 (2012) 1–30.

[14] M. Li, J. Wang, Finite time stability of fractional delay differential equations, Appl. Math. Lett. 64 (2017) 170–176.

[15] Z. Luo, J. Wang, M. Fečkan, A new method to study ILC problem for time-delay linear systems, Adv. Differ. Equ. 2017 (2017) 1–14.

[16] C. Liang, J. Wang, D. O'Regan, Controllability of nonlinear delay oscillating systems, Electron. J. Qual. Theory Differ. Equ. 2017 (2017) 1–18.

[17] M. Pospíšil, J. Diblík, M. Fečkan, On the controllability of delayed difference equations with multiple control functions, AIP Conf. Proc. 1648 (2015) 58–69.

[18] M. Pospíšil, Representation of solutions of delayed difference equations with linear parts given by pairwise permutable matrices via Z-transform, Appl. Math. Comput. 294 (2017) 180–194.

[19] P. Dorato, Short-Time Stability in Linear Time-Varying Systems, Polytechnic Institute of Brooklyn, 1961.

[20] L. Weiss, E.F. Infante, On the stability of systems defined over a finite-time interval, Proc. Natl. Acad. Sci. USA 54 (1965) 44–48.

[21] L. Weiss, E.F. Infante, Finite time stability under perturbing forces and on product spaces, IEEE Trans. Autom. Control 12 (1967) 54–59.

[22] L. Weiss, Converse theorems for finite-time stability, SIAM J. Appl. Math. 16 (1968) 1319–1324.

[23] F. Amato, M. Ariola, C. Cosentino, Robust finite-time stabilisation of uncertain linear systems, Int. J. Control 84 (2011) 2117–2127.

[24] M.P. Lazarević, A.M. Spasić, Finite-time stability analysis of fractional order time-delay system: Grownwall's approach, Math. Comput. Model. 49 (2009) 475–481.

[25] X. Yang, J. Cao, Finite-time stochastic synchronization of complex networks, Appl. Math. Model. 34 (2010) 3631–3641.

[26] Q. Wang, D.C. Lu, Y.Y. Fang, Stability analysis of impulsive fractional differential systems with delay, Appl. Math. Lett. 40 (2015) 1–6.

[27] Q. Zhu, J. Cao, R. Rakkiyappan, Exponential input-to-state stability of stochastic Cohen-Grossberg neural networks with mixed delays, Nonlinear Dyn. 79 (2015) 1085–1098.

[28] R. Wu, Y. Lu, L. Chen, Finite-time stability of fractional delayed neural networks, Neurocomputing 149 (2015) 700–707.

[29] L. Wang, Y. Shen, Z. Ding, Finite time stabilization of delayed neural networks, Neural Netw. 70 (2015) 74–80.

[30] V.N. Phat, N.H. Muoi, M.V. Bulatov, Robust finite-time stability of linear differential-algebraic delay equations, Linear Algebra Appl. 487 (2015) 146–157.

[31] J. Wang, M. Fečkan, Y. Zhou, Ulam's type stability of impulsive ordinary differential equations, J. Math. Anal. Appl. 395 (2012) 258–264.

[32] J. Wang, Y. Zhou, M. Fečkan, Nonlinear impulsive problems for fractional differential equations and Ulam stability, Comput. Math. Appl. 64 (2012) 3389–3405.

[33] J. Wang, X. Li, M. Fečkan, Y. Zhou, Hermite-Hadamard type inequalities for Riemann-Liouville fractional integrals via two kinds of convexity, Appl. Anal. 92 (2013) 2241–2253.

[34] J. Wang, M. Fečkan, A general class of impulsive evolution equations, Topol. Methods Nonlinear Anal. 46 (2015) 915–934.

[35] J. Wang, M. Fečkan, Y. Zhou, A survey on impulsive fractional differential equations, Fract. Calc. Appl. Anal. 19 (2016) 806–831.

[36] J. Wang, M. Fečkan, Y. Zhou, Center stable manifold for planar fractional damped equations, Appl. Math. Comput. 296 (2017) 257–269.

[37] J. Wang, M. Fečkan, Y. Tian, Stability analysis for a general class of non-instantaneous impulsive differential equations, Mediterr. J. Math. 14 (2017) 1–21.

[38] D.Y. Khusainov, G.V. Shuklin, Relative controllability in systems with pure delay, Int. J. Appl. Math. 2 (2005) 210–221.

[39] M. Pospíšil, J. Diblík, M. Fečkan, On relative controllability of delayed difference equations with multiple control functions, Proc. Int. Conf. Numer. Anal. Appl. Math., AIP Publishing LLC 1648 (2015) 130001.

[40] M. Uchiyama, Formulation of high-speed motion pattern of a mechanical arm by trial, Trans. Soc. Instrum. Control Eng. 14 (1978) 706–712.

[41] T. Seel, T. Schauer, J. Raisch, Monotonic convergence of iterative learning control systems with variable pass length, Int. J. Control 90 (2017) 393–406.

[42] T. Seel, T. Schauer, J. Raisch, Iterative learning control for variable pass length systems, in: Proc. 18th IFAC World Congr., vol. 44, 2011, pp. 4880–4885.

[43] J.X. Xu, J. Xu, On iterative learning for different tracking tasks in the presence of time-varying uncertainties, IEEE Trans. Syst. Man Cybern., Part B, Cybern. 34 (2004) 589–597.

[44] H.S. Ahn, K.L. Moore, Y.Q. Chen, Iterative Learning Control: Robustness and Monotonic Convergence for Interval Systems, Springer Publishing Company, Incorporated, 2007.

[45] Y. Luo, Y.Q. Chen, Fractional order controller for a class of fractional order systems, Automatica 45 (2009) 2446–2450.

[46] Y. Li, Y.Q. Chen, H.S. Ahn, On the PD^α-type iterative learning control for the fractional-order nonlinear systems, in: Proceedings of the American Control Conference, 2011, pp. 4320–4325.

[47] Z.Z. Bien, J.X. Xu (Eds.), Iterative Learning Control: Analysis, Design, Integration and Applications, Springer Science and Business Media, 2012.

[48] Z.G. Li, Y.W. Chang, Y.C. Soh, Analysis and design of impulsive control systems, IEEE Trans. Autom. Control 46 (2001) 894–897.

[49] M.X. Sun, Robust convergence analysis of iterative learning control systems, Control Theory Appl. 15 (1998) 320–326.

[50] H.S. Lee, Z. Bien, Design issues on robustness and convergence of iterative learning controller, Intell. Autom. Soft Comput. 8 (2002) 95–106.

[51] X. Ruan, Z. Bien, Q. Wang, Convergence characteristics of proportional-type iterative learning control in the sense of Lebesgue-p norm, IET Control Theory Appl. 6 (2012) 707–714.

[52] X. Ruan, J. Zhao, Convergence monotonicity and speed comparison of iterative learning control algorithms for nonlinear systems, IMA J. Math. Control Inf. 30 (2013) 473–486.

[53] Y. Li, Y. Chen, H. Ahn, G. Tian, A survey on fractional-order iterative learning control, J. Optim. Theory Appl. 156 (2013) 127–140.

[54] Y. Lan, Y. Zhou, Iterative learning control with initial state learning for fractional order nonlinear systems, Comput. Math. Appl. 64 (2012) 3210–3216.

[55] Y. Lan, Y. Zhou, D-type iterative learning control for fractional order linear time-delay systems, Asian J. Control 15 (2013) 669–677.

[56] D. Hinrichsen, A.J. Pritchard, Mathematical Systems Theory I: Modelling, State Space Analysis, Stability and Robustness, Springer Science & Business Media, 2011.

[57] M. Uchiyama, Formulation of hirh-speed motion pattern of a mechanical arm by trial, Trans. Soc. Instrum. Control Eng. 14 (1978) 706–712.

[58] S. Arimoto, S. Kawamura, Bettering operation of robots by learning, J. Robot. Syst. 1 (1984) 123–140.

[59] D.A. Bristow, M. Tharayil, A.G. Alleyne, A survey of iterative learning control, IEEE Control Syst. Mag. 26 (2006) 96–114.

[60] Q. Zhu, J. Xu, D. Huang, G. Hu, Iterative learning control design for linear discrete-time systems with multiple high-order internal models, Automatica 62 (2015) 65–76.

[61] R. Chi, Y. Liu, Z. Hou, S. Jin, Data-driven terminal iterative learning control with high-order learning law for a class of non-linear discrete-time multiple-input-multiple output systems, IET Control Theory Appl. 9 (2015) 1075–1082.

[62] X. Li, J. Xu, D. Huang, An iterative learning control approach for linear systems with randomly varying trial lengths, IEEE Trans. Autom. Control 59 (2014) 1954–1960.

[63] X. Ruan, Z. Li, Z. Bien, Discrete-frequency convergence of iterative learning control for linear time-invariant systems with higher-order relative degree, Int. J. Autom. Comput. 12 (2015) 281–288.

[64] J. Liu, X. Ruan, Networked iterative learning control design for discrete-time systems with stochastic communication delay in input and output channels, Int. J. Syst. Sci. 48 (2017) 1844–1855.

[65] S.K. Oh, J.M. Lee, Stochastic iterative learning control for discrete linear time-invariant system with batch-varying reference trajectories, J. Process Control 36 (2015) 64–78.

[66] Y.S. Wei, X.D. Li, Iterative learning control for linear discrete-time systems with high relative degree under initial state vibration, IET Control Theory Appl. 10 (2016) 1115–1126.

[67] S. Liu, A. Debbouche, J. Wang, On the iterative learning control for stochastic impulsive differential equations with randomly varying trial lengths, J. Comput. Appl. Math. 312 (2017) 47–57.

[68] S. Liu, A. Debbouche, J. Wang, ILC method for solving approximate controllability of fractional differential equations with noninstantaneous impulses, J. Comput. Appl. Math. 339 (2018) 343–355.

[69] X. Yu, A. Debbouche, J. Wang, On the iterative learning control of fractional impulsive evolution equations in Banach spaces, Math. Methods Appl. Sci. 40 (2015) 6061–6069.

[70] M.P. Lazarević, D. Debeljković, Z. Nenadić, Finite-time stability of delayed systems, IMA J. Math. Control Inf. 17 (2000) 101–109.

[71] S.M. Lozinskii, Error estimate for numerical integration of ordinary differential equations, I., Izv. Vyssh. Uchebn. Zved., Mat. 5 (1958) 52–90.

[72] D. Debeljković, S. Stojanović, A. Jovanović, Further results on finite time and practical stability of linear continuous time delay systems, FME Trans. 41 (2013) 241–249.

[73] G. Dahlquist, Stability and Error Bounds in the Numerical Integration of Ordinary Differential Equations, Almqvist and Wiksell, 1958.

[74] K. Gu, An integral inequality in the stability problem of time-delay systems, in: Decision and Control, Proceedings of the 39th IEEE Conference, vol. 3, IEEE, 2000, pp. 2805–2810.

[75] W.A. Coppel, Stability and Asymptotic Behavior of Differential Equations, DC Heath, Boston, 1965.

[76] D. Debeljković, S. Stojanović, A. Jovanović, Finite-time stability of continuous time delay systems: Lyapunov-like approach with Jensen's and coppel's inequality, Acta Polytech. Hung. 10 (2013) 135–150.

[77] Z. Luo, J. Wang, Finite time stability analysis of systems based on delayed exponential matrix, J. Appl. Math. Comput. 55 (2017) 335–351.

[78] C. Corduneanu, Principles of Differential and Intergral Equations, Allyn and Bacon, MA, USA, 1971.

[79] H.Y. Ye, J.M. Gao, Henry-Gronwall type retarded integral inequalities and their applications to fractional differential equations with delay, Appl. Math. Comput. 218 (2011) 4152–4160.

[80] M.P. Lazarević, Finite-time stability analysis of PD^{α} fractional control of robotic time-delay systems, Mech. Res. Commun. 33 (2006) 269–279.

[81] Z. Luo, W. Wei, J. Wang, Finite time stability of semilinear delay differential equations, Nonlinear Dyn. 89 (2017) 713–722.

[82] M. Krasnoselskii, Topological Methods in the Theory of Nonlinear Integral Equations, Pergamon Press, New York, 1964.

[83] F.M. Kirillova, S.V. Churakova, Relative controllability of linear dynamical systems with delay, Dokl. Akad. Nauk 174 (1968) 1260–1263.

[84] R. Gabasov, F.M. Kirillova, Qualitative Theory of Optimal Processes, M. Nauka, 1971.

[85] M. Pospíšil, Relative controllability of neutral differential equations with a delay, SIAM J. Control Optim. 55 (2017) 835–855.

[86] Y. Kuang, Delay Differential Equations with Applications in Population Dynamics, Academic Press, Boston, 1993.

[87] J. Wang, Z. Luo, M. Fečkan, Relative controllability of semilinear delay differential systems with linear parts defined by permutable matrices, Eur. J. Control 38 (2017) 39–46.

[88] R.A. Horn, C.R. Johnson, Matrix Analysis, Cambridge University Press, 2012.

[89] F.R. Gantmakher, Theory of Matrices, Russian, Nauka, Moscow, 1959.

[90] C. Liang, J. Wang, Stability of delay differential equations via delayed matrix sine and cosine of polynomial degrees, Adv. Differ. Equ. 2017 (2017) 1–17.

[91] J. Diblík, D.Ya. Khusainov, J. Lukáčová, M. Růžičková, Control of oscillating systems with a single delay, Adv. Differ. Equ. 2010 (2010) 1–15.

[92] C. Liang, J. Wang, Analysis of iterative learning control for an oscillating control system with two delays, Trans. Inst. Meas. Control 40 (2018) 1757–1765.

[93] I.A. Bihari, A generalization of a lemma of Bellman and its application to uniqueness problem of differential equation, Acta Math. Acad. Sci. Hung. 7 (1956) 81–94.

[94] J.M. Ortega, W.C. Rheinboldt, Iterative Solution of Nonlinear Equations in Several Variables, Academic Press, 1970.

[95] A.M. Samoilenko, N.A. Perestyuk, Y. Chapovsky, Impulsive Differential Equations, World Scientific, 1995.

[96] D.D. Bainov, S.G. Hristova, Impulsive integral inequalities with a deviation of the argument, Math. Nachr. 171 (1995) 19–27.

[97] Z. You, J. Wang, Stability of impulsive delay differential equations, J. Appl. Math. Comput. 56 (2018) 253–268.

[98] J. Shao, F.W. Meng, Gronwall-Bellman type inequalities and their applications to fractional differential equations, Abstr. Appl. Anal. 2013 (2013) 1056.

[99] J.G. Dong, Stability analysis of switched systems with general nonlinear disturbances, Math. Comput. Model. 58 (2013) 1563–1567.

[100] Q. Feng, F. Meng, B. Zheng, Gronwall-Bellman type nonlinear delay integral inequalities on times scales, J. Math. Anal. Appl. 382 (2011) 772–784.

[101] Z. You, J. Wang, Y. Zhou, M. Fečkan, Representation of solutions and finite time stability for delay differential systems with impulsive effects, Int. J. Nonlinear Sci. Numer. Simul. 20 (2019) 205–221.

[102] W. Wei, X. Xiang, Y. Peng, Nonlinear impulsive integro-differential equation of mixed type and optimal controls, Optimization 55 (2006) 141–156.

[103] Z. You, J. Wang, D. O'Regan, Y. Zhou, Relative controllability of delay differential systems with impulses and linear parts defined by permutable matrices, Math. Methods Appl. Sci. 42 (2019) 954–968.

[104] A.A. Kilbas, H.M. Srivastava, J.J. Trujillo, Theory and Applications of Fractional Differential Equations, Elsevier Science B.V., 2006.

[105] M. Li, J. Wang, Exploring delayed Mittag-Leffler type matrix functions to study finite time stability of fractional delay differential equations, Appl. Math. Comput. 324 (2018) 254–265.

[106] N.I. Mahmudov, Delayed perturbation of Mittag-Leffler functions and their applications to fractional linear delay differential equations, Math. Methods Appl. Sci. 42 (2019) 5489–5497.

[107] J. Wang, M. Fečkan, Y. Zhou, Presentation of solutions of impulsive fractional Langevin equations and existence results, Eur. Phys. J. Spec. Top. 222 (2013) 1857–1874.

[108] S. Liu, J. Wang, W. Wei, Analysis of iterative learning control for a class of fractional differential equations, J. Appl. Math. Comput. 53 (2017) 17–31.

[109] Z. You, M. Fečkan, J. Wang, Relative controllability of fractional delay differential equations via delayed perturbation of Mittag-Leffler functions, J. Comput. Appl. Math. 378 (2020) 112939.

[110] M. Li, J. Wang, Representation of solution of a Riemann-Liouville fractional defferential equation with pure delay, Appl. Math. Lett. 85 (2018) 118–124.

[111] M. Li, J. Wang, Finite time stability and relative controllability of Riemann-Liouville fractional delay differential equations, Math. Methods Appl. Sci. 42 (2019) 6607–6623.

[112] M. Li, A. Debbouche, J. Wang, Relative controllability in fractional differential equations with pure delay, Math. Methods Appl. Sci. 41 (2018) 8906–8914.

[113] F.R. Gantmakher, The Theory of Matrices, American Math. Soc., 2000.

[114] S. Elaydi, An Introduction to Difference Equations, 3rd ed, Springer, New York, 2005.

[115] P.R. Halmos, Finite-Dimensional Vector Spaces, Litton Educational Publishing, New York, 1958.

[116] C. Liang, J. Wang, M. Fečkan, A study on ILC for linear discrete systems with single delay, J. Differ. Equ. Appl. 24 (2018) 358–374.

[117] V.I. Arnold, Geometrical Methods in the Theory of Ordinary Differential Equations, Springer, New York, 1988.

[118] S.N. Chow, Ch. Li, D. Wang, Normal Forms and Bifurcations of Planar Vector Fields, Cambridge Univ. Press, Cambridge, 1994.

[119] Y.C. Zhou, H. Cao, Y.N. Xiao, Difference Equations and Application, Science Press, Beijing, 2014.

[120] C. Liang, J. Wang, D. Shen, Iterative learning control for linear discrete delay systems via discrete matrix delayed exponential function approach, J. Differ. Equ. Appl. 24 (2018) 1756–1776.

[121] C.A. Tudor, Analysis of the Rosenblatt Process, ESAIM Probab. Stat., vol. 12, 2008, pp. 230–257.

[122] N.I. Mahmudov, Approximate controllability of semilinear deterministic and stochastic evolution equations in abstract spaces, SIAM J. Control Optim. 42 (2003) 1604–1622.

[123] R.J. Nirmala, K. Balachandran, J.J. Trujillo, Null controllability of fractional dynamical systems with constrained control, Fract. Calc. Appl. Anal. 20 (2017) 553–565.

[124] N.I. Mahmudov, A. Denker, On controllability of linear stochastic systems, Int. J. Control 73 (2000) 144–151.

[125] S. Peng, Multi-dimensional G-Brownian motion and related stochastic calculus under G-expectation, Stoch. Process. Appl. 118 (2008) 2223–2253.

[126] Y. Ren, X. Jia, R. Sakthivel, The p-th moment stability of solutions to impulsive stochastic differential equations driven by G-Brownian motion, Appl. Anal. 96 (2017) 988–1003.

[127] P. Balasubramaniam, P. Tamilalagan, Approximate controllability of a class of fractional neutral stochastic integro-differential inclusions with infinite delay by using Mainardi's function, Appl. Math. Comput. 256 (2015) 232–246.

[128] T. Sathiyaraj, P. Balasubramaniam, The controllability of fractional damped stochastic integrodifferential systems, Asian J. Control 19 (2017) 1455–1464.

[129] P. Balasubramaniam, V. Vembaraan, T. Senthilkumar, Approximate controllability of impulsive fractional integro-differential systems with nonlocal conditions in Hilbert space, Numer. Funct. Anal. Optim. 35 (2014) 177–197.

[130] T. Sathiyaraj, M. Fečkan, J. Wang, Null controllability results for stochastic delay systems with delayed perturbation of matrices, Chaos Solitons Fractals 138 (2020) 109927.

[131] M. Rosenblatt, Independence and dependence, in: Proc. 4th Berkeley Symp. Math. Statist. Probab., vol. 2, 1961, pp. 431–443.

[132] M. Maejima, C.A. Tudor, On the distribution of the Rosenblatt process, Stat. Probab. Lett. 83 (2013) 1490–1495.

[133] N.I. Mahmudov, S. Zorlu, Controllability of non-linear stochastic systems, Int. J. Control 76 (2003) 95–104.

[134] J. Klamka, Stochastic controllability of linear systems with state delays, Int. J. Appl. Math. Comput. 17 (2007) 5–13.

[135] T. Sathiyaraj, J. Wang, D. O'Regan, Controllability of stochastic nonlinear oscillating delay systems driven by the Rosenblatt distribution, Proc. R. Soc. Edinb., Sect. A, Math. 151 (2021) 217–239.

Index